CRC SERIES IN ANALYSIS FOR ENVIRONMENTAL CONTROL

Editor-in-Chief
James W. Robinson
Professor
College of Chemistry and Physics
Louisiana State University
Baton Rouge, Louisiana

ANALYSIS OF PESTICIDES IN WATER

Volume I: Significance, Principles, Techniques, and Chemistry of Pesticides
Volume II: Chlorine- and Phosphorus-Containing Pesticides
Volume III: Nitrogen-Containing Pesticides

Alfred S. Y. Chau, Senior Editor
B. K. Afghan, Co-Editor
Canada Centre for Inland Waters
Burlington, Ontario
Canada

CHEMICAL ANALYSIS OF INORGANIC CONSTITUENTS OF WATER

Editor
Jon C. Van Loon
Department of Geology and Chemistry
University of Toronto
Canada

ORGANIC ANALYSIS OF WATER POLLUTION: CHEMICAL ANALYSIS

B. K. Afghan, Senior Editor
Alfred S. Y. Chau, Co-Editor
Canada Centre for Inland Waters
Burlington, Ontario
Canada

Analysis of Pesticides in Water

Volume III
Nitrogen-Containing Pesticides

Senior Editor
Alfred S. Y. Chau
Head
Quality Assurance and Methods Section
National Water Research Institute
Canada Centre for Inland Waters
Burlington, Ontario

Co-Editor
B. K. Afghan
Head
Analytical Chemistry Research Section
National Water Research Institute
Canada Centre for Inland Waters
Burlington, Ontario

Editor-in-Chief
CRC Series in Analysis for Environmental Control
James W. Robinson
Department of Chemistry
Louisiana State University
Baton Rouge, Louisiana

CRC Press
Taylor & Francis Group
Boca Raton London New York

CRC Press is an imprint of the
Taylor & Francis Group, an **informa** business

CRC Press
Taylor & Francis Group
6000 Broken Sound Parkway NW, Suite 300
Boca Raton, FL 33487-2742

Reissued 2019 by CRC Press

© 1982 by Taylor & Francis Group, LLC
CRC Press is an imprint of Taylor & Francis Group, an Informa business

No claim to original U.S. Government works

A Library of Congress record exists under LC control number:

Publisher's Note
The publisher has gone to great lengths to ensure the quality of this reprint but points out that some imperfections in the original copies may be apparent.

Disclaimer
The publisher has made every effort to trace copyright holders and welcomes correspondence from those they have been unable to contact.

ISBN 13: 978-0-367-26340-9 (hbk)
ISBN 13: 978-0-367-26342-3 (pbk)
ISBN 13: 978-0-429-29277-4 (ebk)

Visit the Taylor & Francis Web site at http://www.taylorandfrancis.com and the
CRC Press Web site at http://www.crcpress.com

FOREWORD

The assessment of the environmental impact of man's endeavors must be made as rapidly and painlessly as possible. It is not only aesthetically rewarding to retrieve clean lakes, rivers, air, and countrysides; it is possibly vital to man's continued existence. Emotional reaction serves no useful long-range purpose, but in fact eventually boomerangs and disenchants the public. Nevertheless, it is incumbent upon man to safeguard his environment. Clearly, the forces involved are quite beyond our comprehension at this time. When we are operating from a position of ignorance as we are today, it is most important to be sure that the risks we take are minimal, although in many cases a 'fail safe' posture has been adopted which can be extremely expensive. Discussion on the possible short-term and long-term impacts of various pollutants is valuable but probably endless. The truth ultimately emerges from reliable data interpreted with wisdom and understanding. Analytical chemistry provides the invaluable bridge between speculation and firm data. We can generate firm data only with reliable analytical techniques, skilled scientists, and clear minds.

In an effort to provide means of collecting and dispersing this information to all interested parties, we have invited a number of scientists of stature to produce monographs in their field of expertise. The objective of these monographs is to document analytical procedures and techniques that are useful to the environmentalists.

In general, three groups of people will be interested in these monographs: industrial engineers and scientists who are monitoring both liquid and gaseous effluents from, industrial effluents; and environmentalists who are trying to assess pollution levels and amass data on long-term health effects and other effects of pollution.

This book, collected by Mr. Chau and Dr. Afghan, is devoted to the broad and important topic of Pesticides. It examines important facets such as the Significance of the Problem, the Chemistry of Pesticides, and Principles and Techniques. It will provide excellent reference material for producers, users, and testing agencies.

J. W. Robinson
Editor-in-Chief
June 1977

EDITOR-IN-CHIEF

J. W. Robinson is Professor of Chemistry and Chairman of the Analytical Division at Louisiana State University, Baton Rouge, Louisiana. He earned his degrees at The University of Birmingham, England (B.Sc., 1949; Ph.D., 1952; D.Sc., 1977) and obtained American citizenship in 1965.

He worked at both Exxon Research Company and Ethyl Research Corporation for a number of years before returning to Louisiana State University to join its Chemistry Faculty.

Dr. Robinson has written more than 130 publications, as well as two texts: *Undergraduate Instrumental Analysis,* the 3rd edition of which is currently in press, and *Atomic Absorption Spectroscopy,* the 2nd edition of which was published in 1975. He is Editor of two international journals, *Spectroscopy Letters* and *Environmental Science and Engineering.* He is also assistant editor of *Applied Spectroscopy Reviews.* He is a former chairman of the Gordon Research Conference on Analytical Chemistry and of the L.S.U. International Symposium on Analytical Chemistry. He is also director of the Saul Gordon Workshop on Atomic Absorption Spectroscopy.

Dr. Robinson is a Guggenheim Fellow and an Awardee of the Honor Scroll of the American Institute of Chemists.

PREFACE

Recently, there have been an increasing number of xenobiotic materials entering into our environment. Many of them are hazardous to human health and to the ecosystem. Indeed, the problem of environmental protection and pollution control has become one of modern man's preoccupations.

One prerequisite for decision making in environmental protection and pollution control is the ability to identify and measure these xenobiotic materials in our ecosystem. In fact, nearly every phase of environmental protection and pollution control depends upon analytical data. However, it is not sufficient merely to generate data. These data must be reliable and truly represent the situation.

Since the analytical data are used for various stages in the activities of international and national environmental protection and pollution control, the analytical data thus have far-reaching political, scientific, and financial implications and impact. When there is no information on the quality of data, the decisions based on them, at best, are questionable. At worst, if the data are poor, irrational decisions will result. Therefore, an effective quality assurance program is needed to ensure the reliability of data. Suitable analytical methodology is the first consideration in an effective quality assurance program for the generation of reliable data.

Unlike the situation for inorganic pollutants, the nature of organic pollutants is extremely complex and diversified. The number and types of organic pollutants including pesticides are also constantly increasing and changing. Due to the numerous variables (sample matrix, and concentration and types of pollutants), analysis and method development for these materials is a challenge even to the experienced chemist. In fact, for the analysis of many organic pollutants in several environmental substrates, suitable methodology is still lacking.

In these three volumes on pesticides, we have tried to present a detailed survey of the analytical methodology and the essential background information emphasizing the practical aspects derived from evaluation of literature data and the authors' own experience. The pros and cons of the different methods, viewpoints, and approaches are also discussed. Equal amounts of data and discussion from both sides are presented so that there is sufficient information for readers to derive their own conclusions, even though they may not agree with us.

The first volume of this series provides background information on pesticides, while the subsequent two volumes detail analytical methodology on the different classes of pesticides. These volumes were written for the analyst as well as for university students, scientists, and researchers for other disciplines. For the latter groups, an attempt was made, whenever possible, to explain terminology, basic principles, and theories that might be unfamiliar to them.

We wish to acknowledge the pleasant and fruitful relationships with all the contributors. The patience, assistance, and understanding of the publisher are greatly appreciated. I wish to thank my wife Linda for typing and retyping the various versions of my early manuscripts and also for her understanding during the two and one-half years I spent working on these volumes. A special acknowledgment of gratitude is extended to M. Chiba, W. Sans, B. Ripley, M. Forbes, and D. McGregor for their valuable suggestions and critical review of the chapter on the chemical derivatization-gas chromatographic techniques.

A. S. Y. Chau, 1981

THE SENIOR EDITOR

Alfred S. Y. Chau is Head, Quality Assurance Methods Section of the Analytical Methods Division, National Water Research Institute, Canada Centre for Inland Waters.

He obtained his B.Sc. degree from the University of British Columbia in 1961 and the M.Sc. degree from Carleton University in 1966.

From 1965 to 1970, Alfred Chau held the position of pesticide analyst in the Department of Agriculture and later joined the Department of the Environment, first as Head of Organic Laboratories and then as Head, Special Services Section. He has held his current position since 1980.

Alfred Chau is the General Referee and member of the Association of Official Analytical Chemists, a member of the Chemical Institute of Canada, and a task group chairman of the American Society for Testing and Materials. He is included in American Men and Women of Science and Who's Who in Finance and Industry and is the recent recipient of the annual Caledon Award for his contribution to analytical chemistry.

Alfred Chau has engaged in research in a number of areas. In addition to a manual of analysis of pesticides in water, and serving as an associate editor for a book on the analysis of chlorinated hydrocarbons and hydrocarbon, he has published some 90 papers on the analysis of pesticides and other contaminants in water and in sediment. Recently, he has been involved in the development of the first Environmental Standard Reference Materials for organic contaminants such as PCBs in lake sediment.

Furthur, Alfred S. Y. Chau is well known as an accomplished nature artist, being represented in commercial galleries across Canada. His works are in many private and permanent collections including the Dofasco Canadian Art Collection and the Beckett Collection.

THE CO-EDITOR

B. K. Afghan is research scientist in environmental analytical chemistry in the Analytical Chemistry Research Section, Canada Centre for Inland Waters, Burlington, Ontario.

Dr. Afghan received the B.Sc. degree from Sind University, Pakistan, in 1962, and the D.I.C. and Ph.D. degrees in analytical chemistry from the University of London in 1964 and 1969, respectively.

Following positions in research at Dalhousie University (1966 to 1968) and the University of Montreal (1968 to 1969), Dr. Afghan served as a research scientist in analytical methods development in the Department of Energy, Mines, and Resources in Ottawa and has been with the Canadian Centre for Inland Waters since 1972.

Dr. Afghan is a fellow in the Chemical Institute of Canada, member of the editorial board of the *Canadian Journal of Spectroscopy,* and chairman of the task group of the American Society of Testing and Materials. His research has been concerned with modern polarographic and electroanalytical techniques, high speed liquid chromatography, atomic and molecular absorption and fluorescence spectroscopy, trace analysis, and environmental analytical chemistry.

Dr. Afghan has published more than 50 research papers over his research career in the areas of automation, atomic/molecular spectroscopy, luminescence, nutrients, heavy metals, trace organics, and pesticide residues.

CONTRIBUTORS

Professor John W. ApSimon, Ph.D.
Department of Chemistry
Carleton University
Ottawa
Canada

Walter A. Glooschenko, Ph.D.
Research Scientist
Aquatic Ecology Division
National Water Research Institute
Burlington, Ontario
Canada

Raj Grover, Ph.D.
Section Head
Herbicide Behavior in the Environment
Research Station
Agriculture Canada
Regina, Saskatchewan
Canada

Fred K. Kawahara, Ph.D.
Chemist
Environmental Monitoring and Support
  Laboratory
Office of Research and Development
Environmental Protection Agency
Cincinnati, Ohio

Hing-Biu Lee, Ph.D.
Chemist
Analytical Methods Division
National Water Research Institute
Burlington, Ontario
Canada

R. James Maguire, Ph.D.
Research Scientist
National Water Research Institute
Environmental Contaminant Division
Department of Environment
Burlington, Ontario
Canada

Derek Muir, Ph.D.
Research Scientist
Freshwater Institute
Department of Fisheries and Oceans
Winnipeg, Manitoba
Canada

Brian D. Ripley
Chemist, Carbamates and Fungicides
Provincial Pesticide Residue Testing
  Laboratory
Ontario Ministry of Agriculture and
  Food
c/o University of Guelph
Guelph, Ontario
Canada

George J. Sirons
Chemist, Herbicides
Provincial Pesticide Residue Testing
  Laboratory
Guelph, Ontario
Canada

Allan E. Smith, Ph.D.
Research Scientist
Agriculture Canada
Research Station
Regina, Saskatchewan
Canada

W. M. J. Strachan, Ph.D.
National Water Research Institute
Canada Centre for Inland Waters
Burlington, Ontario
Canada

Kazuyuki Yamasaki, Ph.D.
Research Fellow
Department of Chemistry
University of California at Los Angeles
Los Angeles, California

ANALYSIS OF PESTICIDES IN WATER

Volume I
SIGNIFICANCE, PRINCIPLES, TECHNIQUES, AND CHEMISTRY OF PESTICIDES

Environmental Impact and Significance of Pesticides
Basic Principles and Practices in the Analysis of Pesticides
Positive Identification of Pesticide Residues by Chemical Derivatization-
Gas Chromatographic Technique
The Chemistry of Cyclodiene Insecticides

Volume II
CHLORINE- AND PHOSPHORUS-CONTAINING PESTICIDES

Organochlorine Pesticides
Organophosphorus Pesticides
Phenoxyalkyl Acid Herbicides (CPHs)

Volume III
NITROGEN-CONTAINING PESTICIDES

Carbamates
The Substituted Area Herbicides
Triazine Herbicides

TABLE OF CONTENTS

Volume III

Chapter 1
Carbamates. 1
B. D. Ripley and A. S. Y. Chau

Chapter 2
The Substituted Urea Herbicides 183
A. E. Smith and R. Grover

Chapter 3
Triazine Herbicides 213
A. E. Smith, D. C. G. Muir, and R. Grover

Index 241

Chapter 1

CARBAMATE PESTICIDES

Brian D. Ripley and Alfred S. Y. Chau

TABLE OF CONTENTS

I.   Introduction 3

II.   General Chemistry 5
       A.   Nomenclature 6
       B.   Chemical and Physical Properties 16
       C.   Synthesis of Parent Compounds 22
             1.   Aryl *N*-Methylcarbamates 22
             2.   *N,N*-Dimethylcarbamates 24
             3.   *N*-Phenylcarbamates 24
             4.   Oxime *N*-Methylcarbamates 24
             5.   Thiocarbamates 24
             6.   Dithiocarbamates and Ethylenebisdithiocarbamates 25

III.   Mode of Action 26
       A.   Insecticides 26
       B.   Herbicides 27
       C.   Fungicides 28

IV.   Environmental Persistence 29
       A.   Aquatic Environment 29
             1.   Water Quality Objectives 32
       B.   Soil and Sediment 33
       C.   Fish 36
       D.   Plants 36
       E.   Model Ecosystems 38
       F.   Environmental Study 38
       G.   Summary 40

V.   Degradation and Metabolic Processes 40
       A.   Hydrolysis 41
       B.   Oxidation 42
       C.   Conjugation 43
       D.   Typical Metabolic Pathways 44
       E.   Synthesis of Carbamate Pesticide Metabolites 45

VI.   Residue Analysis 49
       A.   Sampling and Sample Preservation 51

VII.   Extraction 66
       A.   *N*-Methylcarbamates 66
             1.   Water 66
                   a.   Typical Extraction Methods for Aqueous Samples 69
                         i.   Separatory Funnel Method 69
                         ii.   XAD-2 Resin Column .69

2.   Soils and Sediment 69
3.   Animal, Bird, and Fish Tissues 71
4.   Plant Tissue 71
B.   Aminophenyl *N*-Methylcarbamates 73
C.   Oxime Carbamates 75
D.   *N*-Phenylcarbamates 75
E.   Thiocarbamate Herbicides 75
F.   Dithiocarbamates and EBDC Fungicides 76
G.   Concentration of Sample Extracts 76

VIII.   Cleanup of Sample Extracts 76
A.   Separation of the Parent Carbamates from Their Phenols 79
1.   Typical Isolation of Carbamate Phenols 80

IX.   Gas Chromatography (GC) 80
A.   Gas Chromatographic Detectors 81
B.   Direct GLC Determination of *N*-Methylcarbamates 82
C.   Direct GLC of Carbamate Phenols 89
D.   Direct GLC of Other Carbamates 90

X.   Derivatization of *N*-Methylcarbamates 94
A.   Derivatives of Intact *N*-Methylcarbamates 95
1.   Typical Acylation of Intact *N*-Methylcarbamates 103
B.   Derivatives of Hydrolysis Products of *N*-Methylcarbamates .103
1.   Derivatization of Carbamate Phenols 104
a.   Typical Methods for Derivatization of Carbamate
Phenols 115
i.   Pentafluorobenzylation 115
ii.   Dinitrophenylation 116
iii. DNT, PFB, or DNP Ethers of Carbamate
Phenols 116
2.   Derivatization of Amine Hydrolysis Products of
N-Methylcarbamates 116
C.   Analytical Hydrolysis of *N*-Methylcarbamates 119
1.   Typical Hydrolysis of *N*-Methylcarbamates to Isolate the
Phenols 121
D.   On-Column Reactions 121

XI.   Derivatization of Other Carbamate Pesticides 123
A.   Derivatization of the Intact Compound 124
B.   Derivatization of Hydrolysis Products 124
C.   Ethylenebisdithiocarbamate (EBDC) Fungicides and Ethylenethiourea
(ETU) 129

XII.   Other Determinative Methods 130
A.   HPLC 130
B.   TLC 136
C.   Enzymatic Techniques 138
D.   Fluorescence 141
E.   Colorimetric Methods 143
1.   Colorimetric Method for Dithiocarbamate and
Ethylenebisdithiocarbamate Fungicides 145

F.   Mass Spectrometry (MS)..146
G.   Other Spectroscopic Techniques 151

XIII. Confirmation of Residues...151

XIV. Appendix 152

References..154

I. INTRODUCTION

The "carbamate" class of pesticides is quickly gaining importance in the field of pest control. Due to the persistence of the organochlorine pesticides and the toxicity of the organophosphorus pesticides and their metabolites, the carbamates offer a viable alternative. Generally the carbamate insecticides demonstrate a high insect toxicity, but have a low toxicity towards warm blooded nontarget species, are more biodegradable and less persistent than the o.c. pesticides, and have relatively less toxic decomposition products. The biological activity of synthetic carbamate pesticides is due to the type of substitution to the basic carbamate moiety that results in these compounds being effective insecticides, herbicides, fungicides, nematicides, miticides, and molluscicides. The insecticidal carbamates are derivatives of carbamic acid and are therefore structurally related. Most of the herbicidal and fungicidal carbamates differ structurally from the carbamate insecticides, being primarily thiocarbamates and dithiocarbamates, respectively.

Although a naturally occurring carbamate ester, the alkaloid physostigmine in calabar seeds, is biologically active,[1] all the carbamate pesticides are synthetic compounds. As early as 1931, the E.I. du Pont de Nemours Company showed that some derivatives of dithiocarbamic acid could control insects,[2] but their superior fungicidal activity resulted in their development as the most widely used class of fungicides; among these compounds are the dithiocarbamates like ferbam, and the ethylenebisdithiocarbamates like zineb and maneb. In the late 1940s and early 1950s, Geigy Chemical Company found that heterocyclic enolic esters of dimethylcarbamic acid possessed insecticidal properties,[3] and this led to the development of the N,N-dimethylcarbamate insecticides such as isolan and dimetilan. Union Carbide Corporation substituted aryl groups for the enols and methylcarbamic acid for the dimethylcarbamic acid and synthesized carbaryl in 1953.[4] Investigators at the University of California, Riverside, examining structure-activity relationships, established the superior insecticidal activity of aryl N-methylcarbamates.[5] It is this group of compounds (e.g., carbaryl, carbofuran) that accounts for most of the production and study among the carbamate pesticides. Biological effects of esters of carbanilic acid (carbamates) were reported by Friesen in 1929,[6] but the herbicidal activity of N-phenylcarbamates (IPC) was not reported until 1945.[7] About 1956, certain dithiocarbamates such as CDEC were shown to be herbicidally active by Monsanto Chemical Company,[8] and subsequently around 1959, Stauffer Chemical Company produced many of the thiocarbamate herbicides (EPTC, butylate).[9] Finally, in 1967, oxime carbamates (aldicarb) were introduced by Union Carbide following their work to synthesize N-methylcarbamates with a spatial resemblance to acetylcholine.[10]

Growth of pesticide production and sales soared in the 1960s, but greater environmental awareness and stricter Federal laws limited this growth in the 1970s.[11] Registrations were restricted or banned for many o.c. pesticides (DDT, aldrin) and use patterns shifted to more specific, less persistent insecticides such as the carbamates. Herbicides and fungicides are inherently less hazardous than insecticides and their increased use continued. Recent evidence indicates that many herbicides and fungicides, or their metabolites or by-products (dioxins, nitrosamines, ethylenethiourea) may be carcinogenic and manufacture is being reduced on many of these compounds.

Many people consider the nonpersistence of carbamates as justification for their use, and since carbamate residues are infrequently found in monitoring studies, is there some substance to this thought? Based on production and use data for carbamates, why are these compounds not found in the environment? Are they in fact environmentally labile, and do they break down to nontoxic end products?

In the past, it was felt that since the carbamates were labile and would not persist in the environment, there was no need to analyze for these compounds. Coupled with this attitude, the general lack of multi-residue methodology precluded their determination. Part of the problem lay with the fact that carbamates do not fit in with the residue methods for the o.c. and o.p. pesticides, and some carbamates tend to break down on-column during GC analysis. Development of chemical derivatization procedures has helped in stabilizing these pesticides for GC analysis and also improved sensitivities.

Residue procedures for o.c. pesticides have been intensively studied, and because of their inherent sensitivity to ECD they are routinely reported at the parts per trillion (10^{-12}) level in water samples. Carbamates are less sensitive, even after derivatization, and are usually reported in the parts per billion (10^{-9}) range. For this reason, many of the monitoring studies do indeed not find carbamate pesticides in environmental substrates, but it cannot be assumed that they are not present. Considering a watershed study, one or two findings of carbamates at the parts per billion level equals many findings at the parts per trillion concentration in terms of actual amount of pesticide present in the waterway.

Furthermore, the intensive use of carbamates and other pesticides in concentrated agricultural areas usually near waterways suggests that continued studies should be done to develop baseline levels in environmental samples. Residues of carbamate pesticides in the environment often appear to be transient, but after point source contaminations (due to spills, misapplication, run-off, etc.) the level of residues can often contribute significantly to the aquatic system and cause direct short-term toxic effects.

Residue methods need to be expanded to allow the monitoring of many carbamates that are routinely used but not often analyzed. Too often the term "carbamate pesticides" has become synonymous with the *N*-methylcarbamate insecticides. Procedures of adequate sensitivity for parent compounds and their metabolites are required for environmental assessment programs. While insecticides are more toxic and more often analyzed, one should not overlook the possibility of contamination and the impact from the less studied compounds like the herbicides and fungicides or many of their metabolites.

This chapter covers some aspects of the general chemistry, environmental persistence, and metabolism of carbamate pesticides, but the major emphasis is concerned with analytical residue methodology. The discussions are devoted primarily to the carbamate insecticides, specifically the *N*-methylcarbamates, but the other carbamate pesticides are considered, albeit in less detail. Although a comprehensive literature review is not intended, most of the major contributions have been considered and several of the more common analytical procedures have been detailed. Unfortunately, many of the methods reported in the literature tend to be specifically designed for one compound although the principle may be expanded to multi-residue applications. Also many of the procedures, such as derivatization, tend to be laboratory specific; a method that works well in one laboratory often does not work in another. For this reason there are few standardized residue methods available and the reader is left with

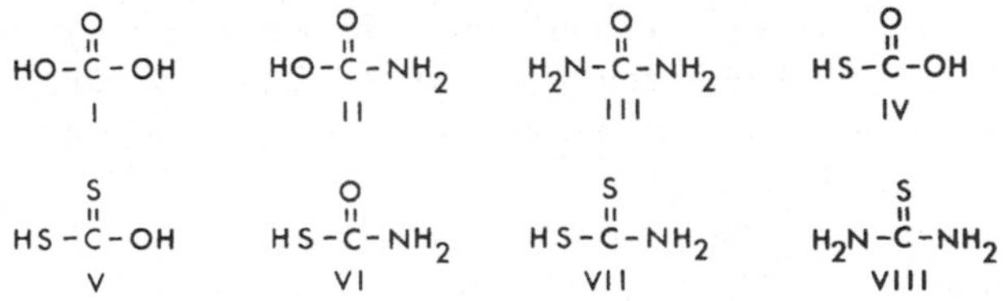

FIGURE 1. Simple carbonic acid derivatives: (I) carbonic acid, (II) carbamic acid, (III) urea, (IV) thiolcarbonic acid, (V) dithiocarbonic acid, (VI) thiolcarbamic acid, (VII) dithiocarbamic acid, and (VIII) thiourea.

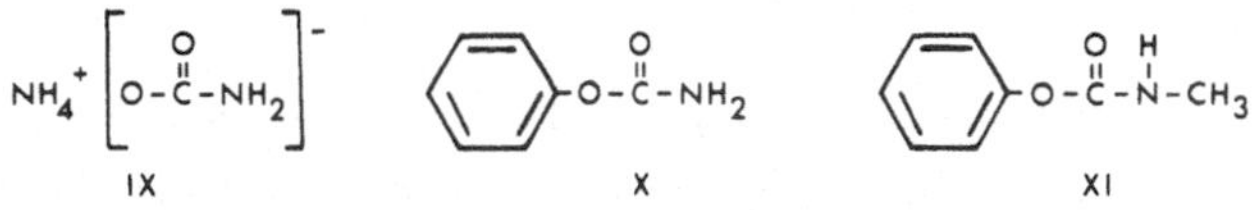

FIGURE 2. Some simple carbamate compounds: (IX) ammonium carbamate, (X) phenyl carbamate, and (XI) phenyl *N*-methylcarbamate.

many procedures to choose from and frequently has difficulty in duplicating previously reported methodology. Acknowledging these problems, several methods are outlined that may be tried, usually with slight modifications, as an initial attempt to analyze carbamate pesticides. Often the procedure may be technique dependent and several attempts should be made before discarding a method; many analysts are at a loss to explain exactly why a procedure does not work for them, but this does not imply it is not an excellent method for another laboratory. Through collaborative studies (although not necessarily using any one method)[12] and in-house quality control[13] each laboratory can optimize its procedure for recovery and sensitivity and be assured that its method is satisfactory. The rewards of successfully applying a method are well worth the effort.

II. GENERAL CHEMISTRY

The fundamental building block of all carbamate pesticides is carbamic acid (II, Figure 1). Carbonic acid (I) is the acid of carbon dioxide from which two amides may be formed: the monoamide, carbamic acid (II), and the diamide, urea (carbamide, III). The sulfur analogs of carbonic acid, thiolcarbonic acid (IV) and dithiocarbonic acid (V), can also form acid amides: thiolcarbamic acid (VI) and dithiocarbamic acid (VII), respectively. The sulfur analog of urea is thiourea (VIII).

None of the above acids exist in the free state since they quickly decompose to simpler molecules; carbamic acid, for example, decomposes to carbon dioxide and ammonia.

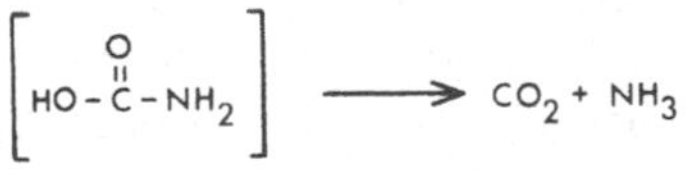

The salts and esters of these acids are stable and it is these compounds that are known as "carbamate" pesticides. An example of a salt of carbamic acid, known as a carbaminate, is ammonium carbamate (IX, Figure 2), which is used with aluminum phosphide as a commercial insecticide and rodenticide. The esters of carbamic acid or its analogs (carbamates) form the basis of almost all the carbamate pesticides. Generally, in plant protection the derivatives of carbamic acid are insecticides whereas those of thiolcarbamic and dithiocarbamic acid are used as herbicides and fungicides, respectively. Urea derivatives are also used as herbicides, but are not considered carbamates and are discussed in another chapter.

The product of the interaction of ammonia and chlorocarbonate is an ester of carbamic acid, generally known as a urethane:[14]

$$Cl-\overset{\overset{O}{\|}}{C}-O-C_2H_5 + NH_3 \longrightarrow NH_2-\overset{\overset{O}{\|}}{C}-O-C_2H_5$$

When a chlorocarbonic ester (acid chloride) reacts with a primary or secondary amine, instead of ammonia, an alkyl urethane results:

$$Cl-\overset{\overset{O}{\|}}{C}-O-C_2H_5 + R-NH_2 \longrightarrow R-\overset{\overset{H}{|}}{N}-\overset{\overset{O}{\|}}{C}-O-C_2H_5$$

Aryl esters of carbamic acid may also be prepared with the simplest one being phenyl carbamate (X). Substituting one of the amide hydrogens in (X) with a methyl group produces phenyl N-methylcarbamate (XI), which has a mild toxicity as an insecticide. This compound can be regarded as the basic structure of all the N-methylcarbamates because most of the commercial carbamate insecticides possess this basic structure and differ only in ring substituents. If the phenyl group in XI is replaced by other aromatic moieties, other aryl N-methylcarbamate insecticides are formed (Table 1).

Other chemical combinations with the basic carbamic acid derivatives (Figure 1) also produce insecticidal compounds. Instead of phenyl carbamate, enolic esters may be prepared and dimethyl substitution of the nitrogen atom results in the N,N-dimethyl-carbamate insecticides such as pirimicarb. Changing the phenolic or enolic substituent of N-methylcarbamates to an aliphatic hydroxy group such as an oxime results in the oxime carbamate insecticides (aldicarb). A phenyl or substituted phenyl group on the nitrogen forms N-phenylcarbamate; aliphatic or aromatic esters of N-phenylcarbamate produce compounds such as IPC that have herbicidal properties. The above relationships to carbamic acid are summarized in Figure 3.

The sulfur analogs of carbamic acid show analogous reactions (Figure 4, Table 1). For example, carbon disulfide can react with ammonia or with primary or secondary amines. Derivatives of thiolcarbamic acid (VI) form herbicidal compounds (EPTC), whereas those of dithiocarbamic acid form herbicides (CDEC) or fungicides (ferbam). Derivatives of ethylenebisdithiocarbamic acid are fungicides (maneb).

Additional details on the chemistry of carbamic acid derivatives may be found in several organic chemistry textbooks and in a review by Melnikov[15] on the chemistry of pesticides.

A. Nomenclature

The simplest characterization of the carbamate pesticides is by their N-substitution or the sulfur analog (Figures 3 and 4, Table 1). These classifications cover most of the commercially available carbamate pesticides, although Schlagbauer and Schlagbauer[16] divided the carbamates into several other subgroups that covered many of the additional experimental carbamate pesticides. Table 1 lists the compounds discussed in this survey which are named by their accepted common name.[11,17-19] Often many of these compounds are better known by other common or trade names and these are included in the Table as well as being cross-referenced in the Appendix. Accepted common names are spelled in lower case letters whereas the trade names are proper nouns (e.g., the active ingredient in Sevin® is carbaryl).

Table 1 also contains the structural formula of the compound and the most common

Table 1
COMMON, TRADE, AND CHEMICAL NAMES AND STRUCTURES OF CARBAMATE PESTICIDES

Common name	Trade or other names	Chemical name and structure
N-methylcarbamates		
bufencarb	Bux®, metalkamate	3:1 Mixture of 3-(1-methylbutyl)phenyl *N*-methylcarbamate and 3-(1-ethylpropyl)phenyl *N*-methylcarbamate
carbanolate	Banol, chlorxylam	2-Chloro-4,5-dimethylphenyl *N*-methylcarbamate
carbaryl	Sevin®	1-Naphthyl *N*-methylcarbamate

Table 1 (continued)
COMMON, TRADE, AND CHEMICAL NAMES AND STRUCTURES OF CARBAMATE PESTICIDES

Common name	Trade or other names	Chemical name and structure
carbofuran	Furadan®, NIA-10242	2,3-Dihydro-2,2-dimethylbenzofuran-7-yl *N*-methylcarbamate
Landrin®	No common name	and 4:1 Mixture of 3,4,5-trimethylphenyl *N*-methylcarbamate and 2,3,5-trimethylphenyl *N*-methylcarbamate
methiocarb	Mesurol®, metmercaptu-ron, mercaptodime-thur	4-Methylthio-3,5-xylyl *N*-methylcarbamate

Mobam

No common
name

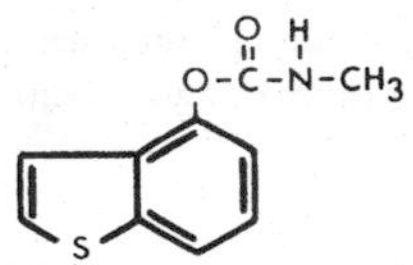

4-Benzothienyl *N*-methylcarbamate

propoxur

Baygon®

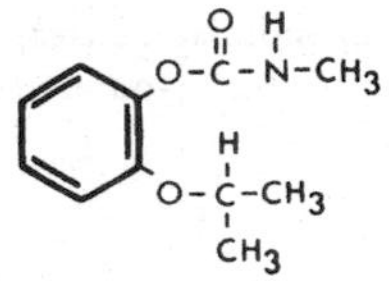

2-Isopropoxyphenyl *N*-methylcarbamate

Aminophenyl *N*-methylcarbamates
aminocarb

Matacil®

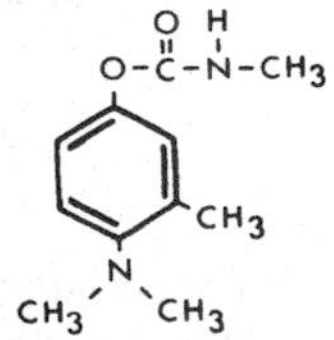

4-Dimethylamino-*m*-tolyl *N*-methylcarbamate

mexacarbate

Zectran®

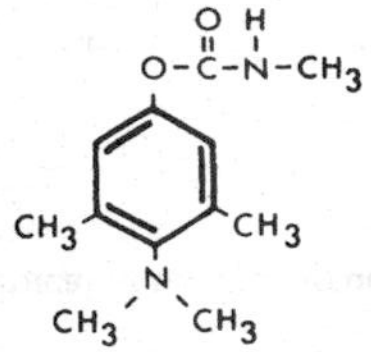

4-Dimethylamino-3,5-xylyl *N*-methylcarbamate

Table 1 (continued)

COMMON, TRADE, AND CHEMICAL NAMES AND STRUCTURES OF CARBAMATE PESTICIDES

Common name	Trade or other names	Chemical name and structure
Oxime *N*-methylcarbamates		
aldicarb	Temik®, UC-21149	$\begin{array}{c} CH_3 \quad\quad O\ H \\ \| \quad\quad \| \| \\ CH_3-S-C-CH=N-O-C-N-CH_3 \\ \| \\ CH_3 \end{array}$
		2-Methyl-2-(methylthio)propionaldehyde *O*-methylcarbamoyloxime
methomyl	Lannate	$\begin{array}{c} CH_3 \quad\quad O\ H \\ \| \quad\quad \| \| \\ CH_3-S-C=N-O-C-N-CH_3 \end{array}$
		1-(Methylthio)acetaldehyde *O*-methylcarbamoyloxime
oxamyl	Vydate, DPX-1410	$\begin{array}{c} CH_3 \quad O \quad\quad O\ H \\ \ \ \setminus \ \| \quad\quad \| \| \\ \quad N-C-C=N-O-C-N-CH_3 \\ \ / \quad\quad \| \\ CH_3 \quad\quad S \\ \quad\quad\quad \| \\ \quad\quad\quad CH_3 \end{array}$
		2-Dimethylamino-1-(methylthio)glyoxal *O*-methylcarbamoylmonoxime
thiofanox	Dacamox	$\begin{array}{c} CH_3 \quad\quad\quad O\ H \\ \| \quad\quad\quad \| \| \\ CH_3-C-\ \ C=N-O-C-N-CH_3 \\ \| \ \ CH_2 \\ CH_3 \ \ \| \\ \quad S-CH_3 \end{array}$
		3,3-Dimethyl-1-(methylthio)-2-butanone *O*-methylcarbamoyloxime

N,N-dimethylcarbamates

dimetilan

Snip

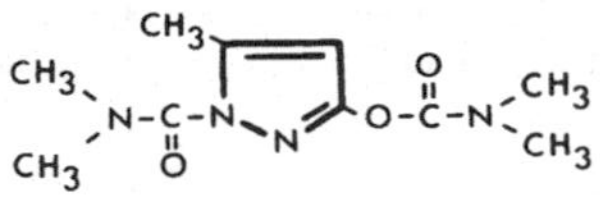

1-Dimethylcarbamoyl-5-methylpyrazoyl-3-yl
N,N-dimethylcarbamate

pirimicarb

Pirimor

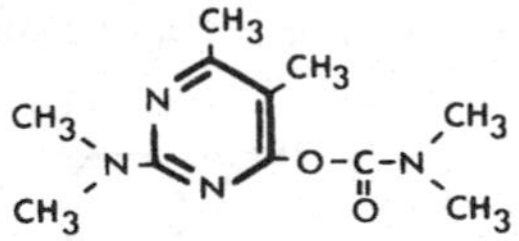

2-Dimethylamino-5,6-dimethylpirimidin-4-yl
N,N-dimethylcarbamate

N-phenylcarbamates

CIPC

chlorpropham,
chloro-IPC

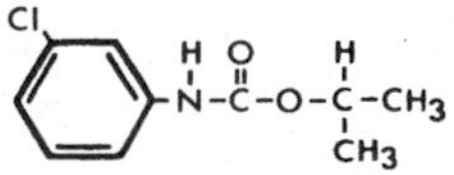

Isopropyl 3-chlorophenylcarbamate
(isopropyl *m*-chlorocarbanilate)

IPC

propham

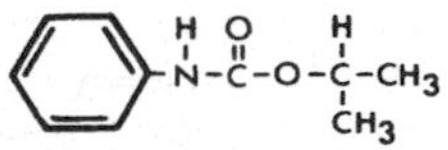

1-Methylethyl phenylcarbamate
(isopropyl carbanilate)

Table 1 (continued)
COMMON, TRADE, AND CHEMICAL NAMES AND STRUCTURES OF CARBAMATE PESTICIDES

Common name	Trade or other names	Chemical name and structure
swep		Methyl 3,4-dichlorophenylcarbamate (methyl 3,4-dichlorocarbanilate)
Thiocarbamates		
butylate	Sutan	S-ethyl di-isobutylthiocarbamate
cycloate	Ro-Neet	S-ethyl cyclohexylethylthiocarbamate
diallate	Avadex	S-2,3-dichloroallyl di-isopropylthiocarbamate (exists as (E) and (Z) isomers)

EPTC	Eptam	$\begin{array}{c}C_3H_7 \\ \diagdown \\ N-\overset{\displaystyle O}{\overset{\|}{C}}-S-C_2H_5 \\ \diagup \\ C_3H_7\end{array}$ *S*-ethyl dipropylthiocarbamate
molinate	Ordram	$\begin{array}{c}CH_2-CH_2-CH_2 \\ \big| \qquad\qquad N-\overset{\displaystyle O}{\overset{\|}{C}}-S-C_2H_5 \\ CH_2-CH_2-CH_2\end{array}$ *S*-ethyl *N,N*-hexamethylenethiocarbamate
pebulate	Tillam	$\begin{array}{c}C_4H_9 \\ \diagdown \\ N-\overset{\displaystyle O}{\overset{\|}{C}}-S-C_3H_7 \\ \diagup \\ C_2H_5\end{array}$ *S*-propyl butylethylthiocarbamate
triallate	Avadex BW	$\begin{array}{c}H \\ \big| \\ (CH_3)_2-C \\ \diagdown \\ N-\overset{\displaystyle O}{\overset{\|}{C}}-S-CH_2-C=\overset{\textstyle Cl}{C}\diagdown Cl \\ \diagup \qquad\qquad\qquad\quad \big| \\ (CH_3)_2-C \qquad\qquad\qquad Cl \\ \big| \\ H\end{array}$ *S*-2,3,3-trichloroallyl di-isopropylthiocarbamate
vernolate	Vernam	$\begin{array}{c}C_3H_7 \\ \diagdown \\ N-\overset{\displaystyle O}{\overset{\|}{C}}-S-C_3H_7 \\ \diagup \\ C_3H_7\end{array}$ *S*-propyl dipropylthiocarbamate

Table 1 (continued)
COMMON, TRADE, AND CHEMICAL NAMES AND STRUCTURES OF CARBAMATE PESTICIDES

Common name	Trade or other names	Chemical name and structure
Dithiocarbamates		
ferbam	Fermate	Ferric dimethyldithiocarbamate
Ziram	Milbam, Zerlate	Zinc dimethyldithiocarbamate
thiram		Tetramethylthiuram disulfide (TMTD)

Ethylenebisdithiocarbamates

mancozeb   Dithane M-45

$$\left[\begin{array}{c} \underset{H}{CH_2-N}-\underset{\underset{S}{\|}}{C}-S-Mn \\ - \\ \underset{H}{CH_2-N}-\underset{\underset{S}{\|}}{C}-S- \end{array} \right]_x Zn_y$$

Co-ordination product of maneb containing 16—20% Mn and 2—2.5% Zn (polymer)

maneb   Dithane M-22, Manzate

$$\left[\begin{array}{c} \underset{H}{CH_2-N}-\underset{\underset{S}{\|}}{C}-S-Mn \\ - \\ \underset{H}{CH_2-N}-\underset{\underset{S}{\|}}{C}-S- \end{array} \right]_x$$

Polymeric manganese ethylenebisdithiocarbamate

nabam   Dithane D-14, Parzate

$$\begin{array}{c} \underset{H}{CH_2-N}-\underset{\underset{S}{\|}}{C}-S-Na \\ - \\ \underset{H}{CH_2-N}-\underset{\underset{S}{\|}}{C}-S-Na \end{array}$$

Disodium ethylenebisdithiocarbamate

zineb   Dithane Z-78, Parzate, Zineb

$$\left[\begin{array}{c} \underset{H}{CH_2-N}-\underset{\underset{S}{\|}}{C}-S-Zn \\ - \\ \underset{H}{CH_2-N}-\underset{\underset{S}{\|}}{C}-S- \end{array} \right]_x$$

Polymeric zinc ethylenebisdithiocarbamate

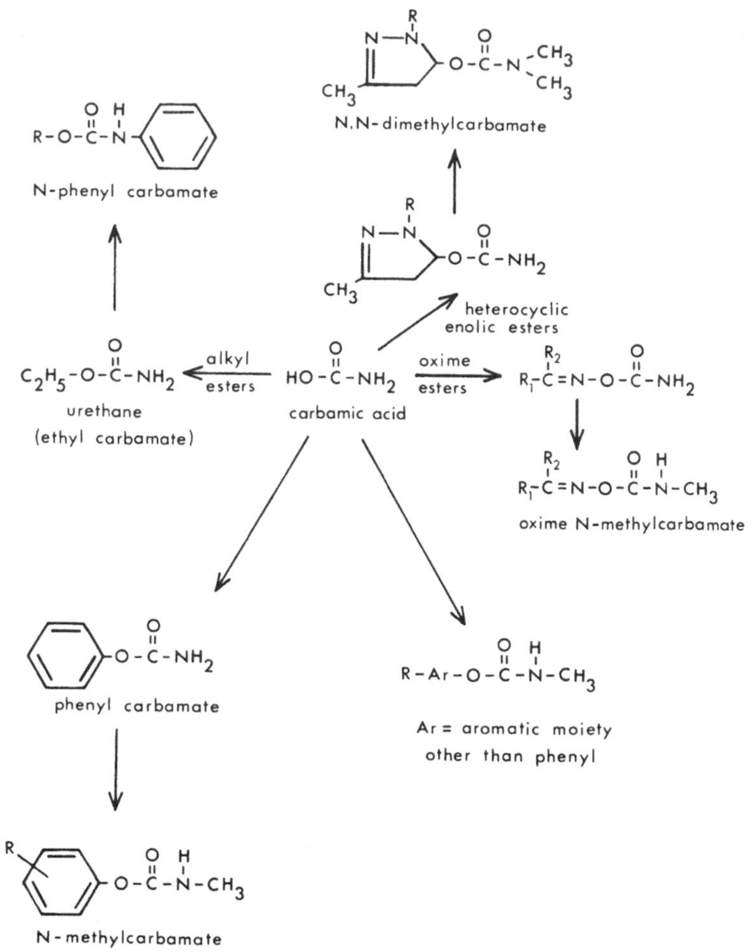

FIGURE 3.   Pesticidal derivatives of carbamic acid.

chemical name. Chemical names are written with the aliphatic, oxime, enolic, or aromatic ester nomenclature first, followed by the *N*- or *S*-carbamate substitution: for example, phenyl *N*-methylcarbamate; 1-naphthyl *N*-methylcarbamate; 2-dimethylamino-5,6-dimethylpyrimidin-4-yl *N,N*-dimethylcarbamate; and ethyl *N,N*-di-*n*-propylthiocarbamate. While the last compound (EPTC) may be more correctly written as thiolcarbamate, the term thiocarbamate is in more general use. Nomenclature for the carbamate pesticides is fairly straightforward and examination of the chemical names and formulae will allow the reader to become familiar with the naming of these compounds.

B. Chemical and Physical Properties

Some physical properties of the carbamate pesticides are shown in Table 2. Solubilities are included, as analysts often require these data to determine appropriate solvents for preparation of standard solutions and for extracting and partitioning of the pesticides. Examination of the physical data reveals the similarity in properties between compounds in the same class yet the dissimilarity between the different classes of carbamate pesticides. These data are available from manufacturers' technical data sheets as well as in various reviews[1,20] and compendia.[18,19,21,22] Additional compilations have been made on such parameters as p-values,[23] solubilities in water,[24] and the chemical and physical properties of herbicides.[25]

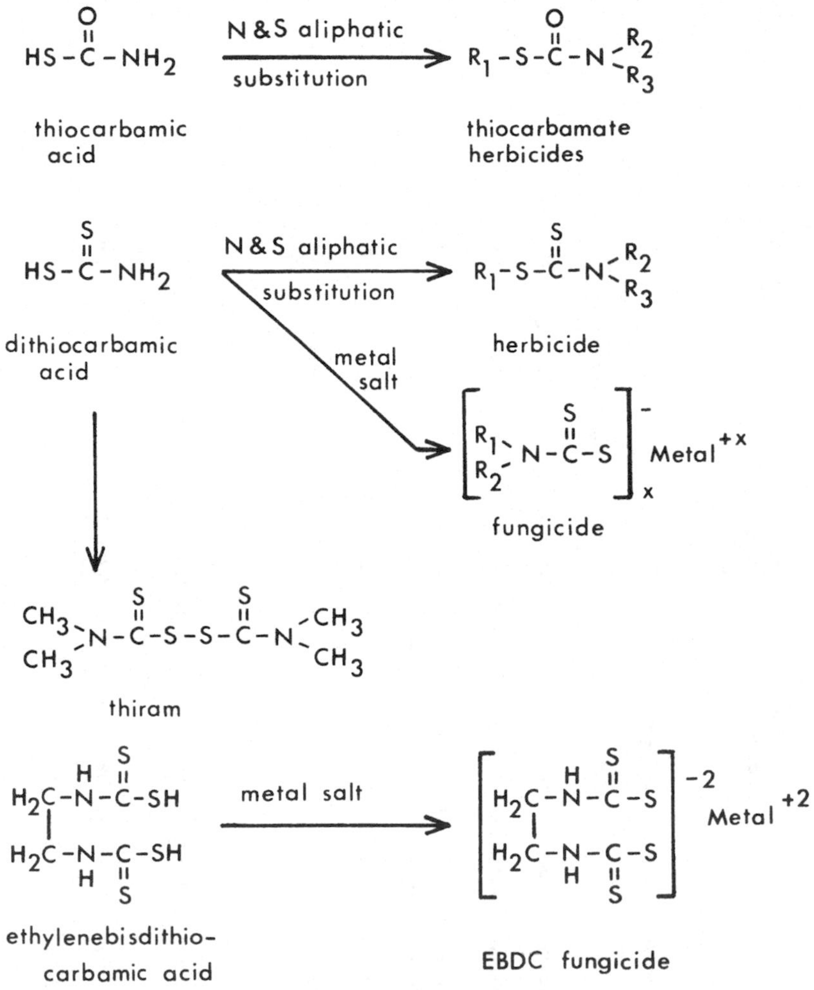

FIGURE 4. Pesticidal derivatives of the sulfur analogs of carbamic acid.

A measure of the toxicity of a compound may be obtained from its LD_{50} value. LD_{50} is defined as the lethal dose of compound required to kill 50% of the test population and is usually expressed in milligrams of chemical per kilogram of body weight. The higher the LD_{50} value, the lower the toxicity. LD_{50} values vary depending on the test animal, its population, and the test conditions such as method of exposure (oral, dermal). N-methylcarbamates are generally moderately toxic whereas the oxime carbamates are highly toxic and the carbamate herbicides and fungicides tend to be less toxic than the insecticides.

Generally, the carbamate insecticides in a pure state are almost odorless, white, crystalline solids that exhibit good shelf stability due to their high melting point and low vapor pressure. However, impure materials (technical compounds) may be colored, semisolid, and impart certain odors (methylisocyanate, sulfur). The nature and extent of these impurities affects the shelf life.

Most of the carbamate insecticides are stable in organic solution although some volatilization and photodecomposition can cause losses of these compounds. Mexacarbate and aminocarb, for example, are partially decomposed when stored for several days as dilute solutions in acetone, carbon tetrachloride, chloroform, or methylene chloride,

Table 2

CHEMICAL AND PHYSICAL PROPERTIES OF SOME CARBAMATE PESTICIDES

Common name[a]	Main[b] use	LD$_{50}$ (rats)	Appearance	Odor	mp bp(mmHg) (°C)	vp (mmHg) (°C)	Solubility[c]	Water solubility (ppm) (°C)
N-methylcarbamates								
bufencarb	I	87	Yellow-amber solid	Faint sweet	49—50	3 × 10^{-5}(30)	Al,xyl	<50
carbanolate	I	30—55	White crystals		130—133		Most organics	
carbaryl	I	540	White crystals	Odorless	142	<0.005(26)	Most polar organics	40(30)
carbofuran	I	8—14	White crystals	Odorless (sl phenol)	150—152	2 × 10^{-5}(33)	Ace, bz, DMF, DMSO, acn	700(25)
Landrin®	I	208	Buff crystals	Mild ester	105—114	5 × 10^{-5}(23)	Ace, al, most organics	60(23)
methiocarb	I	135	White crystals	Mild odor	117—118	negl (rt)	Ace, al, most organics	ins
Mobam	I		White crystals	None	128	1 × 10^{-8}(25)		<0.1%(25)
propoxur	I	128	White crystals	Faint	84—87 (91)	1 × 10^{-2}(120)	Most organics	0.1—0.2% (25)
Aminophenyl N-methylcarbamates								
aminocarb	I	<51(30)	White crystals		93—94	negl (rt)	Polar organics, ar sl	
mexacarbate	I	24	White crystals	Odorless	85	<0.1(139)	Most organics	
Oxime N-methylcarbamates								
aldicarb	I	0.9	White crystals	(Sl sulfur) odorless	98—100	(1 × 10^{-4}(25) <0.05(20)	W, most organics	6000
methomyl	I	17—24	White crystals	Sl sulfur	78—79	5 × 10^{-5}(25)	W, al, ace, tol	5.9g/100ml
oxamyl	I,N	5.4	White crystals	Sl sulfur	100—102 108—110	2.3 × 10^{-4}(25)	W, al, ace, tol	28g/100ml
thiofanox	I	8.5	White crystals	Pungent	57—58	1.7 × 10^{-4}(25)	W, most organics	5200(22)

			Appearance	Odor	Melting point	Vapor pressure	Solubility	
N,N-dimethylcarbamates								
dimetilan	I	<50	Colorless solid		68—71	$1 \times 10^{-4}(20)$	W, chl, DMF, al, ace, xyl, organics	
pirimicarb	I	147	Colorless solid	Odorless	90.5	$3 \times 10^{-5}(30)$	Most organics	0.27g/100ml
N-phenylcarbamates								
CIPC	H	5000—7000	Honey-colored solid	Odorless	41.4	$10^{-5}(25)$	Most organics	89(25)
IPC	H	4500—9000	White-tan crystals		87—87.6		Most organics	250
Swep	H	552	White solid		112—114		Ace, DIBK, DMF	
Thiocarbamates								
butylate	H	5366	Clear, amber liquid	Aromatic	71(10) 138(21)	$1.3 \times 10^{-3}(25)$	Xyl, MIBK	45
cycloate	H	3160	Clear liquid	Aromatic	145(10)	$2 \times 10^{-3}(25)$	Most organics	100(22)
diallate	H	395	Amber, oily liquid	Aromatic	25—30 150(9)	$1.5 \times 10^{-4}(25)$	Most organics, al	14(25)
EPTC	H	1367	Clear, yellow liquid	Amine, aromatic	127(20)	$24 \times 10^{-3}(25)$ 0.1(24)	Bz, al, xyl, tol	365(20)
molinate	H	564	Clear, yellow liquid	Aromatic	137(10)	$8.75 \times 10^{-3}(25)$	Ace, bz, al, xyl	900(21)
pebulate	H	1120	Clear, yellow liquid	Amine, aromatic	142.5(20)	$6.8 \times 10^{-2}(30)$	Ace, bz, al, tol	60(20)
triallate	H	1675—2165	Oily, amber liquid	Aromatic	29—30 148—149(9)	$1.2 \times 10^{-4}(25)$	Most organics	4(25)
vernolate	H	1800	Clear liquid	Aromatic	140(20)	$5.41 \times 10^{-2}(24)$	Most organics	107(21)
Dithiocarbamates								
ferbam	F	17000	Black powder		d > 180	negl (rt)	Chl, pyr, acn	130
ziram	F	1400	White powder	Odorless	240	negl (rt)	Chl, CS$_2$	65(25)
thiram	F	780	Colorless crystals		155—156	negl (rt)	Ace, chl	30

Table 2 (continued)
CHEMICAL AND PHYSICAL PROPERTIES OF SOME CARBAMATE PESTICIDES

Common name[a]	Main[b] use	LD$_{50}$ (rats)	Appearance	Odor	mp bp(mmHg) (°C)	vp (mmHg) (°C)	Solubility[c]	Water solubility (ppm) (°C)
Ethylenebisdithiocarbamates								
manocozeb	F	>8000	Yellow-grey crystals		d		Na$_4$EDTA	
maneb	F	6750	Yellow crystals		d		Na$_4$EDTA	
nabam	F	395	Colorless crystals (·6H$_2$O)		d		W	20%
zineb	F	5200	Light-colored powder		d		Na$_4$EDTA	10

[a] See Table 1.

[b] I, insecticide; H, herbicide; F, fungicide; N, nematicide.

[c] W, water; al, alcohol; ace, acetone; acn, acetonitrile; bz, benzene; chl, chloroform; pyr, pyridine; tol, toluene; xyl, xylene; d, decomposes; rt, room temperature; sl, slightly; ins, insoluble; ar, aromatic.

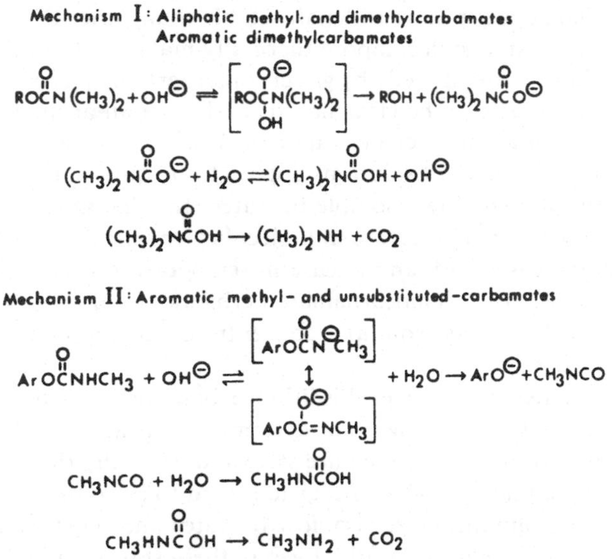

Mechanism I: Aliphatic methyl- and dimethylcarbamates
Aromatic dimethylcarbamates

$$ROCN(CH_3)_2 + OH^\ominus \rightleftharpoons \left[ROCN(CH_3)_2 \atop OH \right] \rightarrow ROH + (CH_3)_2 NCO^\ominus$$

$$(CH_3)_2 NCO^\ominus + H_2O \rightleftharpoons (CH_3)_2 NCOH + OH^\ominus$$

$$(CH_3)_2 NCOH \rightarrow (CH_3)_2 NH + CO_2$$

Mechanism II: Aromatic methyl- and unsubstituted-carbamates

$$ArOCNHCH_3 + OH^\ominus \rightleftharpoons \left[\begin{matrix} ArOCN CH_3 \\ \updownarrow \\ ArOC = NCH_3 \end{matrix} \right] + H_2O \rightarrow ArO^\ominus + CH_3NCO$$

$$CH_3NCO + H_2O \rightarrow CH_3HNCOH$$

$$CH_3HNCOH \rightarrow CH_3NH_2 + CO_2$$

FIGURE 5. Mechanism for alkaline hydrolysis of carbamate chemicals. (From Kuhr, R. J. and Dorough, H. W., *Carbamate Insecticides: Chemistry, Biochemistry and Toxicology*, CRC Press, Boca Raton, Fla., 1976. With permission.)

but they are stable when stored for comparable periods in acetonitrile, benzene, 95% ethanol, *n*-hexane, toluene, and certain other organic solvents.[26] In aqueous solutions these compounds decompose slowly but the rate is enhanced by elevated temperatures such as in boiling water[27] and under alkaline conditions. As such they are particularly prone to hydrolysis[5,28,29] (Figure 5). In certain cases, such as with carbaryl, traces of water and alkali are enough to effect this hydrolysis.

Carbamate insecticides hydrolyze under alkaline conditions; the hydrolysis rate is dependent on the structure of the compound and the hydrolysis conditions (see Sections IV.A and X.C). Generally, mild alkali at room temperature (rt) is sufficient to cause hydrolysis. As expected, the rate of hydrolysis depends on the electron withdrawing effect of substituents on the aromatic ring and also the substitution to the carbamate nitrogen. In general, the stronger the electron withdrawing (Cl, NO_2) substituent in phenyl *N*-methylcarbamates, the faster the hydrolysis under alkaline conditions correlating with Hammett sigma values.[1] With respect to *N*-substitution, the monosubstituted compounds hydrolyze faster than those disubstituted compounds with the same or similar substituents. In fact, several orders of magnitude (10^3 to 10^7) difference in hydrolysis rates have been reported for these compounds.[1,5,28,29] *N*-phenylcarbamates are more stable than the *N*-methylcarbamates.[5,30-32]

In acid solution the hydrolysis rate is generally extremely slow. In fact, small amounts of acid are often added to water samples that are maintained at low temperatures (such as 4°C) to retard hydrolysis or degradation prior to analysis.

Carbamate insecticides tend to break down at elevated temperatures. *N*-methylcarbamates are particularly prone to this problem as they decompose to methylisocyanate and the phenol:[27,33,34]

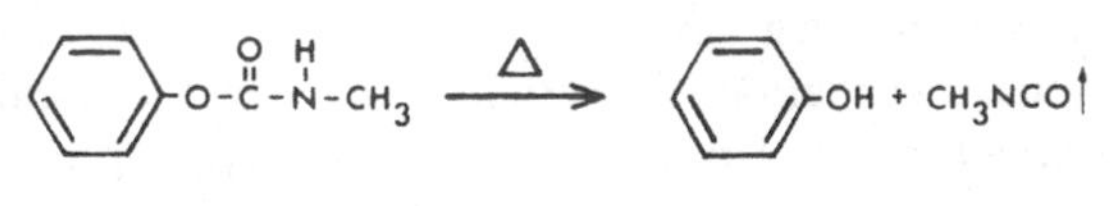

$$\text{C}_6\text{H}_5\text{-O-C-N-CH}_3 \xrightarrow{\Delta} \text{C}_6\text{H}_5\text{-OH} + CH_3NCO\uparrow \qquad (1)$$

N,N-dimethylcarbamates are more thermally stable than either the *N*-methyl or oxime carbamates. This tendency to decompose thermally makes it difficult to analyze many of the carbamate insecticides directly by gas chromatography (see Section IX).

The thiocarbamates tend to be clear liquids and have an aromatic odor. Although these compounds have an appreciable vapor pressure and volatilization is a problem under field conditions, standard solutions appear to be quite stable. Alkyl carbamates and thiocarbamates are sparingly soluble in water. In general, the solubility of these compounds decreases with an increase in molecular weight; however, the solubility of alkyl carbamates increases with an increase in temperature whereas with the thiocarbamates the reverse is true.[35] Thiocarbamate herbicides tend to be stable towards hydrolysis in dilute acid and base solution; concentrated acid at elevated temperature is required for hydrolysis.[36-38]

The dithiocarbamates and ethylenebisdithiocarbamates tend to decompose at elevated temperatures making melting point determinations and GC difficult. It is difficult to obtain these compounds in the pure state and generally the analyst must suffice with "formulation standards". These fungicides tend to be somewhat unstable to heat, light, and moisture. Solubilities are very low in water and most organic solvents; appropriate solvents are given in the Table. Even in these solvents the compounds hydrolyze readily,[39,40] necessitating preparation of standard solutions daily. All these compounds are unstable in dilute acid:

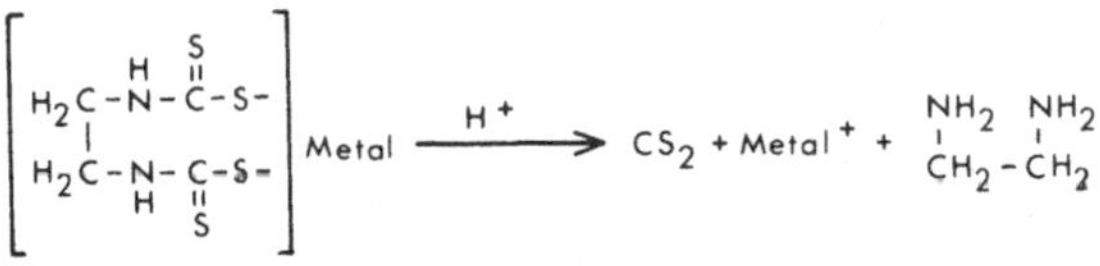

C. Synthesis of Parent Compounds

The preparation of commercial carbamate pesticides has been reviewed previously[1,15,16] and is only briefly discussed here. Details of the major commercial synthetic processes are available.[18,19] The parent carbamate pesticides are generally produced in quantitative yields from low cost starting materials through one- or two-step reactions. The analytical chemist seldom needs to synthesize parent compounds since "analytical standards" are usually available from manufacturers; however, these reactions are instructive as carbamate reaction sequences and may be required by those laboratories wishing to prepare some radiolabeled compounds.

1. Aryl N-Methylcarbamates

Simple aryl esters of *N*-methylcarbamic acid may be synthesized through esterification of the phenol or substituted phenol with methylisocyanate in the presence of a base such as triethylamine. This reaction can be visualized as the reverse process of thermal decomposition or hydrolysis of these carbamates as discussed previously.

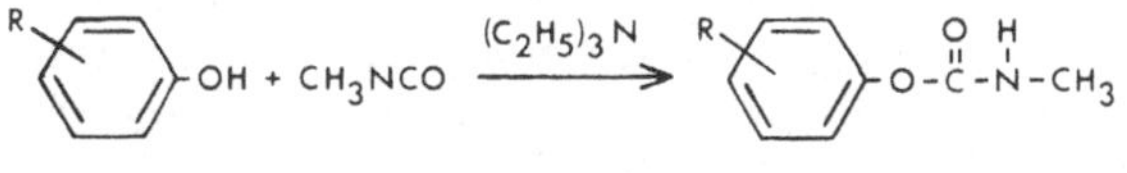

This reaction is usually fast and occurs at low temperatures to produce near quantitative yields and fairly clean products. Many of the experimental aryl *N*-methylcarbamates can easily be formed using this reaction.[41-44]

Two of the most widely used carbamate insecticides, carbaryl and carbofuran, are produced in a similar manner using 1-naphthol[4] and the appropriate heterocyclic phenol (2,3-dihydro-2,2-dimethyl-7-benzofuranol),[45] respectively. The synthesis of car-

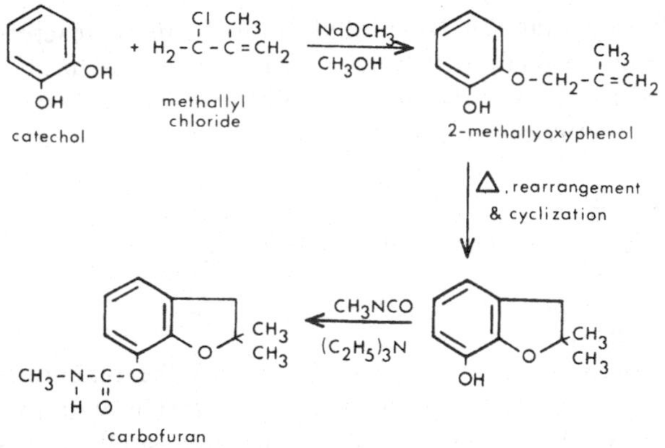

FIGURE 6. Synthesis of carbofuran.[46]

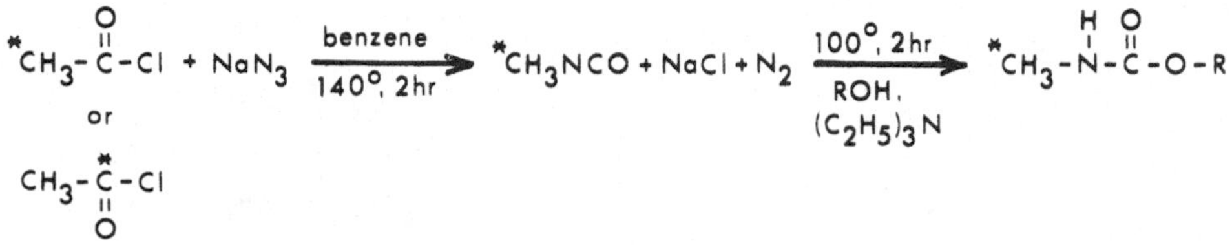

FIGURE 7. Synthesis of methyl or carbonyl labeled N-methylcarbamates. Adapted with permission from Krishna, J. G., Dorough, H. W., and Casida, J. E., *J. Agric. Food Chem.*, 10, 462, 1962. Copyright 1962, American Chemical Society.

bofuran[45,46] provides an interesting study for the organic chemist (Figure 6). The starting material is 2-methallyloxyphenol which is prepared by refluxing catechol and methallyl chloride in the presence of sodium methoxide and methanol, or in acetone with potassium carbonate and potassium iodide. The 2-methallyloxyphenol is heated, and thermally rearranged and cyclized to produce 2,3-dihydro-2,2-dimethyl-7-benzofuranol. This product is then reacted with methylisocyanate and a trace of triethylamine with 2,2-dimethyl-2,3-dihydrobenzofuranyl-7-N-methylcarbamate (carbofuran) being precipitated out of the reaction mixture.

Since this synthetic reaction (Equation 2) is reversible depending on temperature, it provides a convenient means of preparing methyl or carbonyl-labeled compounds[27] (Figure 7). For example, carbaryl can be mixed with acetyl-1-^{14}C or acetyl-2-^{14}C chloride and sodium azide in a sealed reaction tube. When the temperature is raised, acetyl azide is formed followed by its decomposition to methylisocyanate; carbaryl will also be thermally decomposed to 1-naphthol. On cooling, the labeled and nonlabeled methylisocyanate recombine with the 1-naphthol to yield the labeled N-methylcarbamate. Recovery of several ^{14}C carbamates formed in this manner was routinely 40 to 70%. Ring-labeled carbaryl has been synthesized through the reaction of the labeled 1-naphthol with methylisocyanate (Figure 7).[47-49]

Two other less important reactions may also be employed to synthesize aryl N-methylcarbamates. A two-step process whereby the appropriate phenol is reacted with phosgene to form the chloroformate followed by reaction with methylamine is also commercially feasible for some of the carbamates. This reaction is shown below in an alternative preparation of carbaryl.[50]

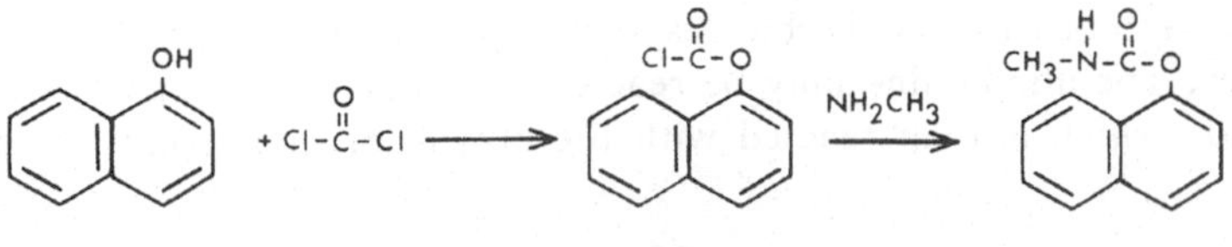

Knaak et al.[49] showed that the 1-naphthyl chloroformate reacted with methyl-^{14}C-amine-HCl to produce *N*-methyl-^{14}C carbaryl; this procedure has been used to prepare several radiolabeled *N*-methylcarbamates.[46,51] In the third synthetic scheme, the phenol may be reacted with methylcarbamoyl chloride in the presence of a weak base.[1,15]

2. N,N-Dimethylcarbamates

N,N-Dimethylcarbamates are synthesized through the reaction of a dimethylcarbamic acid derivative, usually dimethylcarbamoyl chloride, with the appropriate enolized reaction component. The starting materials are usually refluxed in an organic solvent containing a weak inorganic base which acts as an acid acceptor. In the synthesis of pirimicarb, the enolized reactant is 2-dimethylamino-5,6-dimethyl-4-pyrimidone:[18]

3. N-Phenylcarbamates

The *N*-phenylcarbamate herbicides may be prepared by reacting an aniline with an alkyl chloroformate in the presence of sodium hydroxide as shown for IPC:[52]

Alternatively, these herbicides may be prepared by condensation of the alkyl alcohol with the appropriate phenyl isocyanate as shown for CIPC:[53]

4. Oxime N-Methylcarbamates

The oxime *N*-methylcarbamates are synthesized through the reaction of methylisocyanate with the appropriate oxime. Often this esterification proceeds rapidly without a catalyst. Typically, aldicarb may be produced as follows:[10,15,54]

5. Thiocarbamates

Details on the synthesis of 265 thiocarbamates have been given by Tilles.[55] Most of the commercial thiocarbamate herbicides may be prepared using one of the two following processes. The alkylamine may be reacted with phosgene to produce alkylcarbamoyl chloride which is then reacted with the alkylthiol (mercaptan) to produce the

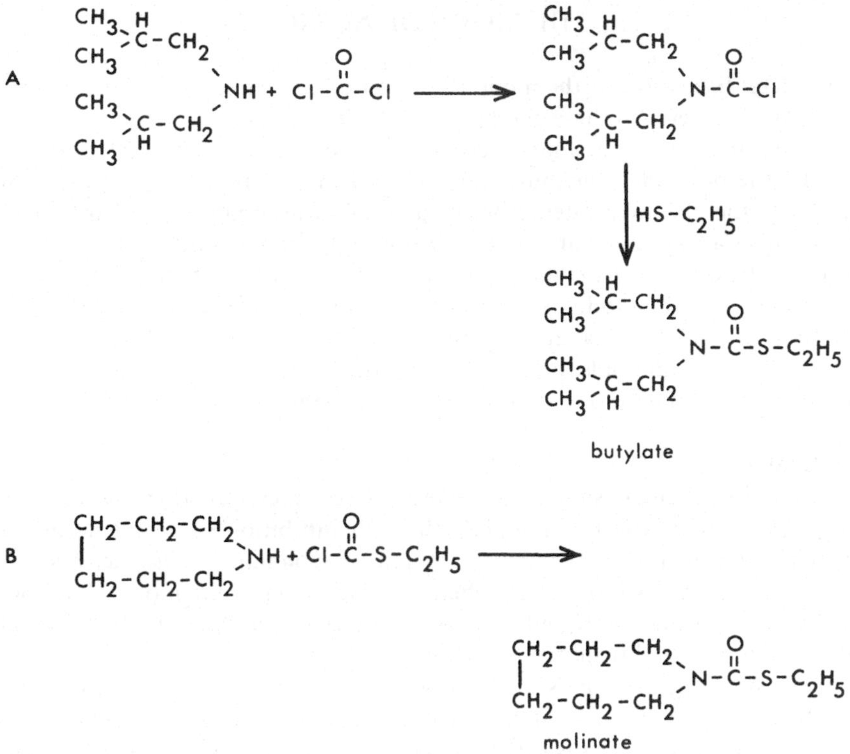

A

$$(CH_3)_2C-CH_2 \quad NH + Cl-\overset{O}{\underset{\|}{C}}-Cl \longrightarrow$$

butylate

B

$$\underset{CH_2-CH_2-CH_2}{\overset{CH_2-CH_2-CH_2}{\diagup}} NH + Cl-\overset{O}{\underset{\|}{C}}-S-C_2H_5 \longrightarrow$$

molinate

FIGURE 8. Synthesis of thiocarbamates. (A) butylate and (B) molinate.[18]

herbicide. Another common synthetic route involves the reaction of the alkyl chloro-thioformate with the appropriate alkylamine. These processes are illustrated in Figure 8 for butylate and molinate, respectively.[18]

6. Dithiocarbamates and Ethylenebisdithiocarbamates

Dithiocarbamates are prepared in a two-stage process.[15,18] The first step is common to all these fungicides: carbon disulfide is reacted with dimethylamine in sodium hydroxide to produce the sodium salt of dimethyldithiocarbamate.

$$\underset{CH_3}{\overset{CH_3}{\diagup}}NH + CS_2 \xrightarrow{NaOH} CH_3-\underset{CH_3}{\overset{S}{\underset{\|}{N-C}}}-S-Na \xrightarrow{M^+} \left[\underset{CH_3}{\overset{CH_3}{\diagup}}N-\overset{S}{\underset{\|}{C}}-S \right]_x M_y$$

sodium dimethyl-dithiocarbamate

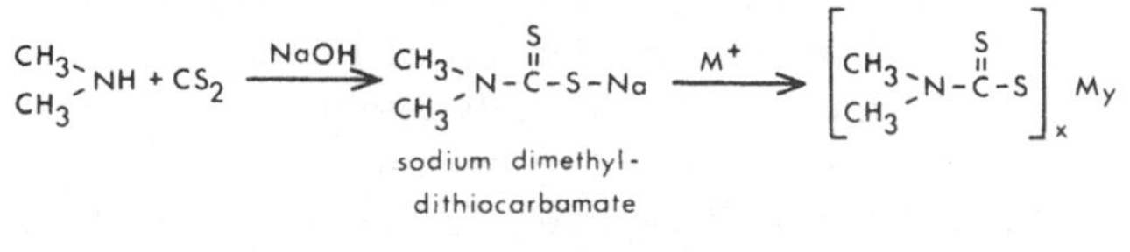

The aqueous solution of this salt is then mixed with a soluble metallic salt, such as zinc sulfate, and the resulting compound readily precipitates because of its low water solubility. Thiram is prepared by peroxide oxidation of the sodium salt.

In an analogous manner to the dithiocarbamates, the ethylenebisdithiocarbamates are prepared by precipitation of the metallic salt from water soluble sodium ethylene-bisdithiocarbamate (nabam), which is previously obtained from the reaction of carbon disulfide on ethylenediamine in the presence of sodium hydroxide.

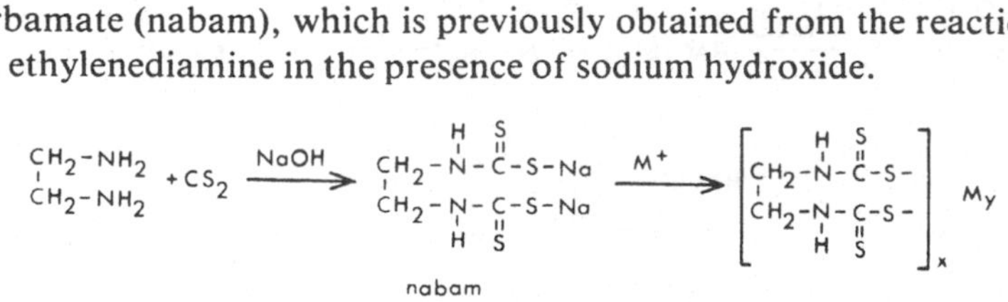

$$\underset{CH_2-NH_2}{\overset{CH_2-NH_2}{|}} + CS_2 \xrightarrow{NaOH} \underset{CH_2-N-C-S-Na}{\overset{CH_2-N-C-S-Na}{|}} \xrightarrow{M^+} \left[\cdots \right]_x M_y$$

nabam

III. MODE OF ACTION

A complete discussion of the mode of action of carbamate pesticides is beyond the scope of this chapter. All these chemicals "kill" by interfering with various biological components, and although a single crucial reaction or "biochemical lesion" may be effected by the pesticide, one must bear in mind that there is a causal chain of events leading from the chemical interaction to death of the organism.[56] Many of the exact effects in this causal chain of events are not fully understood, and observations of effects often lead to concepts in the mode of action of a particular type of pesticide. To fully appreciate the mode of action of a chemical one must be familiar with many related fields such as the anatomy, neurobiology, biochemistry and physiology of plants, insects and mammals, and fungi. This discussion follows a simple approach to explain or state the main mode of action of the carbamate pesticides.

A. Insecticides

Carbamate insecticides, similar to the o.p. insecticides, function due to their ability to disrupt the transmission of impulses through inhibition of acetylcholinesterase at certain synaptic junctions in the nervous system. Carbamate insecticides reversibly carbamylate the enzyme. A more comprehensive discussion of the mode of action of carbamate insecticides may be found in texts by O'Brien,[56] and Kuhr and Dorough.[1] The following is summarized from their works.

Transmission of nerve impulses across peripheral synapses is mediated by acetylcholine (ACh); when an impulse moving along an axon reaches a synapse, a chemical transmitter substance (ACh) is diffused across the synapse to trigger off another impulse at the postsynaptic membrane. To restore sensitivity of the synapse, the transmitter must be cleaved off the receptor to restore it to its resting potential. At cholinergic junctions (synapses utilizing ACh), the enzyme acetylcholinesterase (AChE) hydrolyzes the ACh to acetic acid and choline.

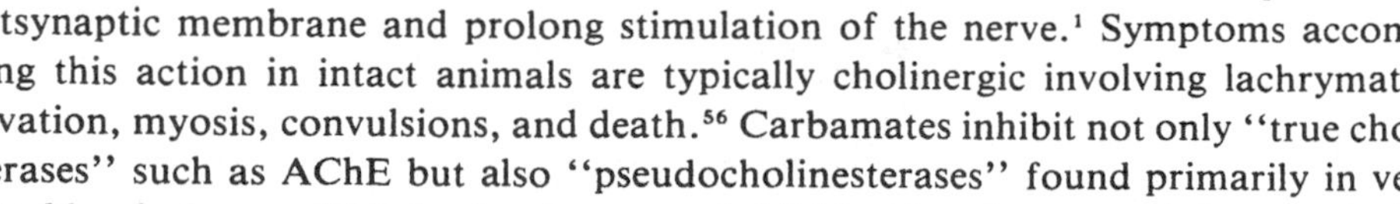

$$CH_3-\overset{\overset{\displaystyle O}{\|}}{C}-O-CH_2-CH_2-\overset{\oplus}{N}\overset{\diagup CH_3}{\underset{\diagdown CH_3}{-CH_3}} \quad \xrightarrow[H_2O]{AChE} \quad CH_3-\overset{\overset{\displaystyle O}{\|}}{C}-OH \; + \; HO-CH_2-CH_2-\overset{\oplus}{N}\overset{\diagup CH_3}{\underset{\diagdown CH_3}{-CH_3}}$$

acetylcholine   choline

Carbamate insecticides enter the synapse and react with or inhibit AChE so that the enzyme cannot cleave ACh thus causing the mediator to continue to depolarize the postsynaptic membrane and prolong stimulation of the nerve.[1] Symptoms accompanying this action in intact animals are typically cholinergic involving lachrymation, salivation, myosis, convulsions, and death.[56] Carbamates inhibit not only "true cholinesterases" such as AChE but also "pseudocholinesterases" found primarily in vertebrate blood plasma. This basis of enzyme inhibition has been applied to analysis of carbamates (see Section XII.C).

The mechanism of *N*-methyl and *N,N*-dimethylcarbamate (analogous to o.p.) insecticide inhibition of acetylcholinesterase is shown below:

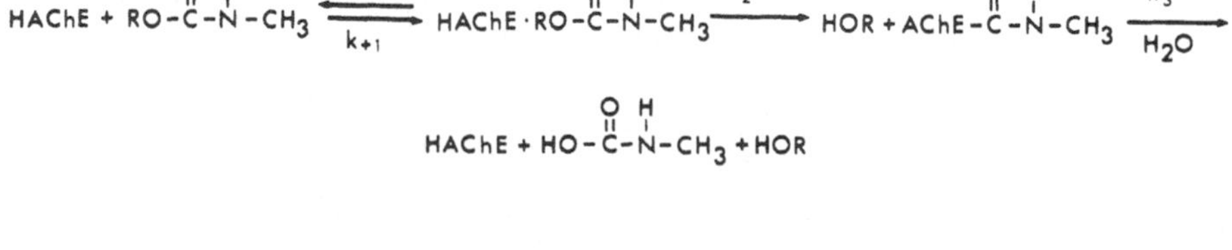

$$HAChE + RO-\overset{\overset{\displaystyle O}{\|}}{C}-\overset{\overset{\displaystyle H}{|}}{N}-CH_3 \underset{k_{+1}}{\overset{k_{-1}}{\rightleftharpoons}} HAChE\cdot RO-\overset{\overset{\displaystyle O}{\|}}{C}-\overset{\overset{\displaystyle H}{|}}{N}-CH_3 \xrightarrow{k_2} HOR + AChE-\overset{\overset{\displaystyle O}{\|}}{C}-\overset{\overset{\displaystyle H}{|}}{N}-CH_3 \xrightarrow[H_2O]{k_3}$$

$$HAChE + HO-\overset{\overset{\displaystyle O}{\|}}{C}-\overset{\overset{\displaystyle H}{|}}{N}-CH_3 + HOR$$

The three steps involved are complex formation, carbamylation of the enzyme, and finally hydrolysis (decarbamylation); each step has a rate constant associated with it that determines the activity of the specific inhibitor. Formation of the enzyme-insecticide complex is favored since the affinity constant K_a ($K_a = k_{-1}/k_1$) is small and with acetylcholine k_2 and k_3 are fast so the whole reaction proceeds forward. With the carbamates, k_2 is of the order of 1 per minute and k_3 is even slower at 0.05 per minute; K_a is exceptionally low. Therefore, small amounts of the complex and carbamylated enzyme exist under practical conditions. *N,N*-dimethylcarbamates have slightly slower carbamylation constants relative to the *N*-methyl analogs. By way of comparison, for o.p. insecticides k_2 is moderately fast (30 min^{-1}), but k_3 is extremely slow so that the phosphorylated enzyme accumulates. The major difference between the carbamate and o.p. insecticides in this inhibition is the value of k_3; since it is much quicker with the carbamates, an almost spontaneous recovery of enzyme occurs. The half-life of decarbamylation is about 30 to 40 min and the recovery from carbamate poisoning appears to occur within minutes after removal of the toxicant. The phosphorylated enzyme being less susceptible to hydrolysis shows longer recovery periods. For this reason, carbamates are often considered as "reversible" inhibitors, whereas the o.p. insecticides are considered as "irreversible" inhibitors. While in fact both types of inhibition are reversible, the rapid k_3 of the carbamate insecticides give them a marked safety factor over the o.p. insecticides.

The antidote for carbamate poisoning is atropine; atropine protects the receptor sites on the postsynaptic membrane. 2-PAM, an alternative antidote used for o.p. insecticidal poisoning, is **not** recommended for carbamate poisoning. Effects of antidotes have been reviewed.[1]

Structure-activity relationships have been studied[57] to determine chemicals with good inhibitory action and selectivity. Much of this work has to do with the goodness of the fit of the insecticide to the enzyme and electronic considerations. An alternative representation of the acetylcholinesterase inhibition by carbamate (and o.p.) insecticides is shown in Figure 9. Although the nature of the enzyme esteratic and anionic sites is not fully known it is believed to be as represented.

The carbamate molecule is attracted to the enzyme by electronic charge attraction at the esteratic site and/or hydrophobic forces at the anionic site. The stability of the complex is determined by the goodness of the fit at both the sites and the strength of hydrogen bonding. The enzyme is carbamylated via electrophilic attack by the carbamate carbonyl carbon on the serine hydroxyl group at the esteric site of the enzyme, concomitant with the release of a phenolic or enolic leaving group. The overall structure of the carbamate, particularly the δ^+ of the carbonyl created by Hammett σ inductive effects (electron-withdrawing ability) of the ring moiety, can affect the carbamylation rate (k_2). Organophosphorus insecticides show a much greater affinity for AChE because of variations in δ^+ created by P = O or P = S structure and side-chain inductive effects. Since the carbamate carbonyl will not accommodate as strong a δ^+ as will the o.p. insecticides, the σ effect will be less; however, the goodness of the fit to the lipophilic area around the anionic site may be more important to toxicity. The distance of the aromatic substituent from the carbonyl may be critical. In the last step, the carbamylated enzyme is reactivated to yield AChE and the respective methyl carbamic acid. Various carbamate structures will therefore determine insecticidal activity, and many of these factors have been reviewed by Kuhr and Dorough.[1]

B. Herbicides

Complete biochemical reactions of carbamate, thio- or dithiocarbamate herbicides are not fully understood at the present time. These herbicides are usually applied for pre-emergence control of weeds to the soil from which they may be absorbed and

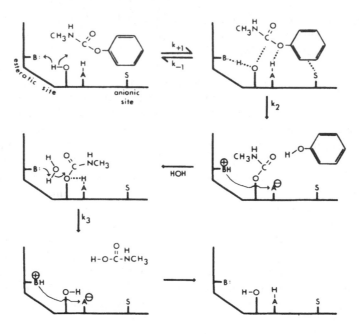

FIGURE 9.   Representation of the reaction of a *N*-methylcarbamate with acetylcholinesterase.[1,5,16,29a] B = basic group (probably histidine imidazole nitrogen), OH = serine hydroxyl, HA = acid group (possibly aromatic hydroxyl group of tyrosine), and S = anionic site (probably a carboxyl group of glutamic or aspartic acid). The esteratic and anionic sites are separated by a distance of approximately 5Å.[1] (From Kuhr, R. J. and Dorough, H. W., *Carbamate Insecticides: Chemistry, Biochemistry and Toxicology*, CRC Press, Boca Raton, Fla., 1976. With permission.)

translocated in the plants. It is believed that these herbicides are mitotic poisons and hence they alter normal plant metabolism(s). On metabolic degradation, some metabolites may form conjugates with endogenous substances for herbicidal activity. Apart from classifying these compounds as mitotic poisons, the variety of chemical structure and selectivity precludes generalizing on the mode of action of these herbicides. Many processes may be involved in the ultimate death of the plant. Certain of these herbicides have demonstrated an ability to inhibit or alter such processes as oxidative phosphorylation, photosynthesis, respiration, RNA and protein synthesis, nucleic acid metabolism, and the ATP content in tissue.

More information on this topic may be found in texts on weed science and/or herbicides.[58-61]

C. Fungicides

As with carbamate herbicides, the mode of action of the carbamate fungicides is not fully understood. It appears that the disulfide group effects its fungicidal activity through interaction with sulfhydryl enzymes and coenzymes such as in the citric acid cycle. The dithiocarbamate fungicides are respiratory inhibitors and apparently are effective through the formation of heavy metal-enzyme complexes such as with lipoic acid and pyruvic acid dehydrogenase; the nature of the heavy metal salt of the fungicide determines the activity and the enzyme system affected.

The ethylenebisdithiocarbamates (EBDCs) differ chemically from the dithiocarbamates in that they have a reactive hydrogen on the nitrogen atom; this reduces their stability and results in a different biological behavior.[62,63] The mode of action of

EBDCs is not fully understood but they appear to affect oxidation-reduction reactions. The EBDCs are decomposed into isothiocyanates[64] which may account for part of the fungicidal action. One such compound, ethylenethiuram monosulfide (ETM; EBIS;[62] DIDI[63]) shows remarkable fungitoxicity. More information on the mode of action of carbamate fungicides may be found in articles by Owens,[65] Ludwig and Thorn,[66] and Lukens.[67]

IV. ENVIRONMENTAL PERSISTENCE

Carbamates as a class of pesticides are quite labile in the environment as compared to the persistent o.c.s,[68,69] but they are somewhat more persistent than the o.p. pesticides (Figure 1, in Volume II, Chapter 2). In general, the half-lives of carbamate pesticides may be considered in terms of days or weeks. Many chemical, physical, and biological factors operating either alone or in combination may affect the persistence of pesticides. Often the carbamates form degradation products or metabolites that are also short lived but must be analyzed to prove previous carbamate exposure. Some of these metabolites may also be biologically active or toxic; for example, carbofuran forms 3-keto and 3-hydroxy carbofuran, both of which are cholinesterase inhibitors.[46] There is little evidence to suggest that the carbamates are biomagnified in the ecosystem. Although the carbamates may not often be found in routine monitoring studies on environment samples,[68,69] residues are persistent enough to be found in soil and crops after application, and in the environment following such point source events as accidental spills, land run-off, aerial spray applications, or fires in pesticide warehouses. Biological and environmental stability of carbamates has been reviewed by Kuhr and Dorough.[1] Gerakis and Sficas[70] have discussed the presence and cycling of pesticides in the ecosystem, and Haque and Freed[71] have examined environmental chemodynamics of pesticides including the carbamates.

A. Aquatic Environment
In the aquatic environment such factors as pH, temperature, ionic strength, presence of suspended solids, UV light, and microorganisms have an effect on the persistence of carbamates. Aly and El-Dib[72,73] studied these factors with respect to carbaryl, propoxur, dimetilan, and Pyrolan in water, and studies on the persistence of carbaryl and 1-naphthol in seawater have also been made;[74,75] a summary of these findings is presented below.

Carbamate insecticides are susceptible toward chemical hydrolysis in alkaline media with the N,N-dimethylcarbamates being more stable than the N-methylcarbamates. pH strongly affects the rate of hydrolysis. For example, at 20°C in water at pH 7.0, carbaryl decomposes slowly and is 99% hydrolyzed in 70 days, whereas at pH 8.0 it would be completely hydrolyzed in 9 days; propoxur is more stable in neutral waters but at pH 8.0, 99% hydrolysis occurs in 107 days. The N,N-dimethylcarbamates are expected to be stable towards hydrolysis at the pH of natural waters (Figure 10). Persistence of all carbamates would increase at colder water temperatures; carbaryl in the dark at a concentration of 10 mg/ℓ in seawater was 93% hydrolyzed in 4 days at 23°C, while at 3.5°C only 9% was hydrolyzed after 8 days. Salt content (or ionic strength) has an effect on the rate of hydrolysis since it decreases the activity of both the hydroxyl ion and the carbamate. In seawater at pH 8.0 and 20°C, 50% of carbaryl is hydrolyzed in 4 days, but in dilute salt solutions under the same conditions 50% hydrolysis occurred in 1.3 days. Most suspended solids do not readily adsorb most carbamates and thus do not have an appreciable effect on the persistence of these pesticides in water. Pyrolan is, however, adsorbed to some extent on mineral particulate rich in montmorillonite content. Carbaryl is readily adsorbed by bottom muds from which it is slowly de-

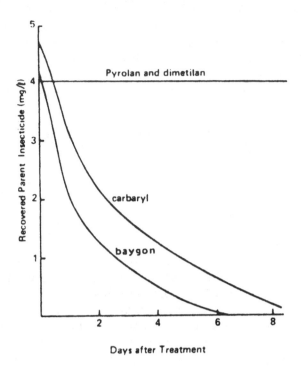

FIGURE 10.   Typical degradation of *N,N*-dimethylcarba-
mates (Pyrolan, dimetilan) and *N*-methylcarbamates (carbaryl,
propoxur [Baygon®]) in river water.[1,73]

graded. UV light has a marked effect on carbamates resulting in cleavage of the
esteratic bond,[76] but the degree of photodecomposition in deep waters may be restricted. Microorganisms can effect enzymatic hydrolysis of carbamates and result in
significant decreases in *N*-methylcarbamate concentrations;[77-79] *N,N*-dimethylcarba-
mates are more resistent to biological degradation presumably due to a lack of a specific enzymatic system since the hydrolysis step is rate limiting to their complete breakdown. Microorganisms may also become acclimatized to carbamate insecticides.[80,81]

In many of the studies on persistence of carbamates in water, the intermediate phenols could not be isolated. Most of the carbamates have short half-lives in water but
may be prone to total destruction and small amounts may persist for long periods.
Other studies have been conducted on persistence of carbamates in water[82-86] and similar results to those summarized above were found. Table 3 contains some of the half-
lives of carbamates in water.

While the carbamate pesticides are prone to hydrolysis, this does not mean they are
innocuous in aquatic systems. Faust and Gomma[89] suggested that carbamates hydrolyze slowly in water and therefore might be more persistent in natural waters than
commonly believed; this would be especially true at colder temperatures. Wolfe et
al.[31,90] showed that the alkaline hydrolysis half-life for carbamates varies dramatically
with structure (see also Section II. B). IPC and CIPC have long hydrolysis half-lives
(1×10^4 days), but their short half-life for bacterial degradation (about 3 days) indicates
that biolysis would be the most competitive degradative process for these *N*-phenyl-
carbamates. Photolysis may also be an important means of degrading carbamate pesticides. The direct photolysis half-life for carbaryl in sunlight is 6.6 days in distilled
water whereas the calculated half-lives for the direct photolysis of IPC and CIPC in a
clear water body near the surface are 254 and 121 days, respectively.[31] Hence, kinetic
data on hydrolysis, photolysis, and biolysis must be evaluated to compare the relative

Table 3
HALF-LIVES OF SOME *N*-METHYLCARBAMATES IN WATER

Compound	Water type	Temperature (°C)	pH	$t_{1/2}$	Ref.
Aminocarb	Stream	Field conditions	7.1	8.7 days	178
	Pond	Field conditions	5.5	4.4 days	178
Carbaryl	Estuarine	20	8.0	4 days	74
	Buffered	20	7.0	10.5 days	72
	Buffered	20	8.0	1.8 days	72
	Buffered	20	9.0	2.5 hr	72
	Deionized	25	9.0	173 min	86
Carbofuran	Water	25	Alkaline	8 days	110
	Paddy (field)	26—30	7.8—8.5	67 hr	92
	Pond (field)	26—30	7.8—8.5	48 hr	92
	Pond (field)	26—30	7.8—8.5	55 hr	92
	Deionized	27 ± 2	7.0	864 hr	92
	Paddy	27 ± 2	7.0	240 hr	92
	Deionized	27 ± 2	8.7	19.4 hr	92
	Paddy	27 ± 2	8.7	13.9 hr	92
	Deionized	27 ± 2	10.0	1.2 hr	92
	Paddy	27 ± 2	10.0	1.3 hr	92
Landrin®	Water	38	8.0	42 hr	114
Mexacarbate	Buffered	12—13	9.5	ca 2 days	84
	Buffered	12—13	7.4	ca 2 weeks	84
	Buffered	20	7.0	25.7 days	85
	Buffered	20	8.42	4.6 days	85
	River (nonsterile)	20	8.16—8.2	9.1 days	85
	River (sterile)	20	8.16—8.4	6.2 days	85
Propoxur	Buffered	20	8.0	16 days	72
	Buffered	20	9.0	1.6 days	72
	Buffered	20	10.0	4.2 hr	72

importance of these three processes with respect to the environmental persistence of carbamates. Wolfe et al.[90] further found that based on free energy plots, it is apparent that many carbamate pesticides will not readily hydrolyze in natural water; only in the case of *N*-substituted carbamates for which the resulting alcohols have pKa's less than 12 will hydrolysis likely be fast enough to be an important degradative pathway.

Recently, several studies have examined the persistence of carbamate pesticides during flooded rice culture.[91-96] With carbofuran, about 1 day was required for solubilization of the granular formulation, and there appeared to be a variable time source interaction (Table 4).[91,92] Rainfall appeared to wash some carbofuran from the rice foliage into the water.[91] Only small concentrations of carbofuran metabolites were found; 3-keto carbofuran appeared after a time lag indicating it was a degradation product.[91] Loss of carbofuran was pH dependent, and the hydrolysis was more rapid in paddy water than in distilled water (Table 3).[92] Hydrolysis of carbofuran appeared to be primarily chemical, but degradation of carbofuran phenol was biological.[93] Degradation of carbofuran in soil under flooded conditions was much slower compared to paddy water,[93] but was more rapid than under nonflooded conditions.[94] Carbofuran phenol was the only metabolite detected and it was probably rapidly bound to the soil.[93] Under laboratory conditions, there was evidence that carbofuran may to an appreciable extent be photodecomposed in an aquatic environment, whereas oxidative and biochemical degradation and volatilization may not be important means of dissipating residues.[91-93] Overall, it was suggested that carbofuran dissipated from plot water via an adsorptive mechanism.

Degradation of molinate in rice field water was primarily photochemical, and hy-

Table 4

CARBOFURAN RESIDUAL LEVELS IN RICE PADDY WATER
(pH 6.0-6.5) AVERAGED OVER TREATMENTS AT THE VARIOUS
SAMPLING PERIODS

		Hours following application[b](kg/ha)[c]						
Year	SNK[a]	0	24	48	96	192	384	768
1973	0.104	0.565a	0.158b	0.063c	0.032c	0.026c	0.020c	0.007c
1974	0.440	0.170b	1.080a	—	0.073c	0.501b	0.209bc	0.046c
1975	0.297	0.001d	0.787a	0.406bc	0.550b	0.270c	0.203cd	0.009d

[a]   Student-Newman-Keuls' (SNK) range test for $p = 7$ in 1973 and 1975; SNK for $p = 6$ in 1974.
[b]   Means within a year not followed by the same letter are significantly different at 5% level.
[c]   Microgram per liter concentration normalized to kilograms per hectare to account for variable plot water depths.

Adapted from Deuel, L. E., Jr., Price, J. D., Turner, F. T., and Brown, K. W., *J. Environ. Qual.*, 8(1), 25, 1979. By permission of the American Society of Agronomy, Crop Science Society of America, and Soil Science Society of America.

drolysis and microbial breakdown was demonstrated to be negligible.[95] The photoproducts disappeared rapidly under practical use conditions. Volatilization from field water was by far the major route for dissipation of molinate (Table 5). Molinate volatilizes readily from dilute organic solution, and the rate of loss was negligible at 15°C but very rapid (half-life of 1.6 days) at the typical field temperature of 28°C; molinate was found in the atmosphere above the field.[95] In western Canada, where the thiocarbamate herbicides, particularly triallate, are widely used for weed control in cereals, residues in air are extremely high.[97] In Japan, a case of human and large-scale fish intoxication from molinate polluted water has been reported.[96] Further studies on the photochemical decomposition[98] and chemodynamics[71] of the thiocarbamate herbicides and their environmental impact on aquatic systems are required.

Many of these findings relate to static field or laboratory conditions, and in dynamic systems, dilution of the carbamates must be considered the largest variable in determining their persistence. In a study on pesticides in 11 agricultural watersheds in southern Ontario,[99-101] carbofuran and EPTC were found at trace levels in only 23 and 5.7% of the samples, respectively. Total losses of pesticides from land use to water were calculated at 21 g carbofuran and 0.1 g EPTC. Small run-off losses of carbaryl and carbofuran were also found in other watershed studies.[83,102] Following a fire at a pesticide warehouse, carbaryl and bufencarb appeared to move downstream through several monitoring stations in discrete quantities presumably reflecting solubility, run-off, leaching, and movement of suspended solids.[103] In point source contaminations, such as in wells,[104] the persistence of some pesticides may be exceedingly long, and in many cases unpredictable.

1. Water Quality Objectives

The International Joint Commission has developed specific water quality objectives, which, if not exceeded, will protect the most sensitive beneficial use of the boundary waters.[105] These objectives were established to protect aquatic life or its consumers (i.e., fish, birds, and mammals), public water supply, and/or recreational use, depending upon which is the most sensitive. Therefore, with respect to the unspecified, nonpersistent pesticides where neither "no effect" nor estimated "safe" levels have

Table 5
LOSS OF MOLINATE FROM AN
ORDRAM-TREATED RICE
FIELD

Process	Estimated loss (%)
Soil adsorption and metabolism	<10
Plant uptake and metabolism	<5
Aqueous microbial metabolism	<1
Hydrolysis	<1
Photolysis	5—10
Volatilization to atmosphere	75—85

Reprinted with permission from Soderquist, C. J., Bowers, J. B., and Crosby, D. G., *J. Agric. Food Chem.*, 25, 940, 1977. Copyright 1977, American Chemical Society.

been determined for these compounds, it was recommended that protection be afforded aquatic life through the use of a 0.05 safety factor applied to the median lethal concentration in a 96-hr test for any sensitive local species. For example, the calculated value for carbofuran in water is 6.0 $\mu g/\ell$.[101] It is desirable that drinking water be free of these biocides, and Health and Welfare Canada has established an objective and acceptable limit for carbamates in raw and drinking water at a not detected level with a maximum limit of 0.100 mg/ℓ, expressed as the parathion equivalent in cholinesterase inhibition.[106]

B. Soil and Sediment

The fate of pesticides in soil is dependent on soil type, pH, organic content, adsorption, moisture content, volatilization, microbial population, air oxidation, and photodecomposition, to mention a few factors. Fate of herbicides in soils has been studied more intensively than with insecticides. Many carbamates are applied directly to the soil either as a herbicide or insecticide and thus it is important to know their persistence. Soil constitutes a reservoir for pesticides wherein they may ultimately be degraded or removed by air, water, or plant and animal uptake.[1]

Early reports on carbaryl persistence indicated that it was rapidly degraded in the soil environment with a half-life of 8 days.[107] However, more recent studies indicate that carbaryl may persist longer,[1] and in one case, 135 days was required for 95% disappearance.[83] A lag period in the degradation (25 to 116 days) indicated that carbaryl degradation is primarily microbial. Of significance to the aquatic environment is the observation that of 4 kg applied to the field, 5.77 g (0.14%) was lost during the season in run-off water (75%) and sediment (25%).

Carbofuran is often applied directly to the soil and its persistence may be more significant.[108] There is some indication that residues may build up when annual treatments are employed.[109] Persistence of carbofuran appears to be influenced by soil sterilization and soil pH.[110] In 3 to 50 weeks, 50% breakdown occurred, with degradation being 7 to 10 times faster in alkaline soil (pH 7.9) than in acid or neutral soils (pH 4.3 to 6.8). Indirect evidence suggested carbofuran was hydrolyzed to its phenol, which was slowly metabolized by microorganisms.

Caro et al.[102] found in a 2-year study that carbofuran disappeared from soil by apparent first-order kinetics, the half-life ranging from 46 to 117 days. About 5 to 10% was converted to 3-keto carbofuran, which disappeared at about the same rate as the parent compound. Persistence increased with in-furrow application, but degradation was acclerated by high soil moistures, high pH, or heavy soil texture. From 0.5 to 2% of applied carbofuran was lost in run-off mainly in the water; larger losses occurred after disturbance of the soil surface at harvest. Because of the higher water solubility of carbamates compared to o.c. pesticides, most of the pesticide run-off moved in the water rather than with the solids. Comparison between carbaryl and carbofuran in the same field indicated that carbaryl was less persistent and produced smaller losses through run-off.

The major metabolite of carbofuran in soil is 3-keto carbofuran. Soil type may to some extent affect the degree of conversion, and recent evidence suggests that a higher conversion rate occurs in muck soil.[111,112] Carbofuran phenol is also produced and exists as a bound residue.

The major isomer of bufencarb was readily metabolized by soil organisms to only one metabolite, *m*-(1-hydroxy-1-methylbutyl)-phenyl methylcarbamate, in trace amounts and both compounds degraded at about the same rate (50% loss in 1 to 2 weeks) under laboratory conditions at ambient temperature.[113] Rates of dissipation of Landrin® from nonsterile soils were primarily dependent on soil type, but organic matter was not a factor; alkaline hydrolysis was the major cause of Landrin® degradation although microorganisms were also involved.[114]

Aldicarb persistence in soil has been well studied. Aldicarb has been shown to decompose more slowly in soils than in plants, and to have a half-life of 7 to 12 days depending on soil type.[115] Measurable quantities of aldicarb were determined after 12 weeks although, over the study period, aldicarb sulfoxide was the major product determined. Degradation is dependent on soil type, moisture level, and temperature.[116] Residues of aldicarb and its transformation products do not persist in the soil after a 90-day growing season. Rainfall caused significant decline in residue concentration. Aldicarb and its oxidation products are very mobile in soil and hence are easily leached.[117] Recent studies with aldicarb, aldicarb sulfoxide, and aldicarb sulfone, individually, showed similar results.[118,119] On Long Island, N.Y., excessive use of aldicarb on potatoes resulted in its leaching to the high water table and contamination of the Upper Glacial Aquifer, a prime source of drinking water; carbofuran has also been found in the well water.[120]

Sharom et al.[121] observed that adsorption, desorption, and the mobility of insecticides in soil and water is dependent to a large extent on the nature of the adsorbents and the water solubility of the insecticide. The order of adsorption on soil for 12 pesticides was DDT > leptophos > dieldrin > endrin > ethion > chlorpyrifos > lindane > parathion > diazinon > carbaryl > carbofuran > mevinphos. Organic soil had the greatest adsorptive capacity followed by stream sediment > Beverly sandy loam > Plainfield sand. Desorption and mobility of the insecticides increased with increasing solubility of the insecticides (Figure 11), although the rate was dependent on the adsorbent. Since the carbamate insecticides such as carbaryl and carbofuran have high water solubilities (40 and 320 ppm, respectively), one would expect to find these compounds in the water system as opposed to being adsorbed on suspended particulates. Wauchope[122] noted that pesticides with solubilities of 10 ppm or higher are lost from agricultural fields mainly in the water phase of run-off and erosion control practices would have little effect on such losses. Felsot and Dahm[123] found that in general, adsorption of o.p. and carbamate insecticides from aqueous solution takes place mainly on organic matter surfaces, and although adsorption generally is reversible, desorption was not always complete.

Similarly, the aqueous solubility of a pesticide may be related to its *n*-octanol/water partition coefficient.[124] These values may be useful as a means of predicting soil adsorption,[125] biological uptake,[126] lipophilic storage,[127] and biomagnification[128-132] of the pesticide. This data in turn may be used to make analytical or managerial decisions regarding particular pesticides.

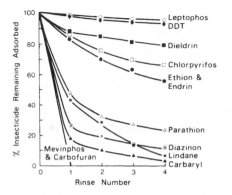

FIGURE 11. Desorption of 12 insecticides from Beverly sandy loam with four (200 mℓ) rinses of distilled water (0.2 *M* KH₂PO₄ solution for carbaryl). (Reprinted from Sharom, M. S., Miles, J. R. W., Harris, C. R., and McEwen, F. L., *Water Res.*, 14, 1095, 1980. With permission.)

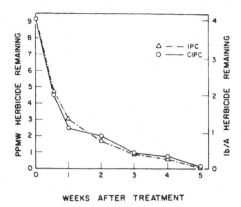

WEEKS AFTER TREATMENT

FIGURE 12. The dissipation of IPC and CIPC from soil after field application of the herbicide at 4 lb/A. (Reprinted from Parochetti, J. V. and Warren, G. F., *Weed Sci.*, 16, 13, 1968. With permission.)

Table 6
TIME FOR 50% DEGRADATION OF
THIOLCARBAMATE HERBICIDES IN
REGINA HEAVY CLAY AND WEYBURN
LOAM

	Time in weeks[a]	
Thiolcarbamate	Regina heavy clay	Weyburn loam
Pebulate	2—3	2—3
Vernolate	2—3	2—3
EPTC	4—5	4
Diallate	5—6	4
Triallate	10—12	8—10

[a] Average of four replicates.

Reprinted with permission from Smith, A. E. and Fitzpatrick, A., *J. Agric. Food Chem.*, 18, 720, 1970. Copyright 1970, American Chemical Society.

Thiocarbamates are very volatile and must be immediately incorporated into the soil to prevent losses; even so, air residues due to volatilization may be quite high.[95,97] In general, the thiocarbamates are not persistent herbicides, lasting from 1 to 3 months under most field conditions (Table 6)[36,61,133] although there is some indication that triallate may persist from one growing season to the next;[134] minimum degradation occurred during the winter months in Canada. Kaufman[133] has reviewed the degradation of carbamate herbicides in soils. Volatility is a major loss of thiocarbamates (Table 5), with these losses being reduced on dry soils. Thiocarbamates may also be photodecomposed.[98,135] Soil microorganisms contribute significantly to the disappearance of thiocarbamates through enzymatic attack at the alkyl group, the amide linkage, or the ester linkage.[133,136,137]

The *N*-phenylcarbamates, IPC and CIPC, are not persistent herbicides, and they show phytotoxic action for only 1 to 2 months (Figure 12); CIPC is an inherently more potent herbicide and hence is biologically more active.[138] These compounds resist leaching into the soil;[139] however, volatilization losses are low particularly in dry soils.[140] Soil microorganisms are responsible for most of the degradation of N-phenylcarbamates;[133,136,139,141] there is an indication that methylcarbamates are competitive inhibitors of the phenycarbamate hydrolyzing enzyme.[142]

Environmental factors conducive to increased microbial activity, e.g., high organic matter levels, increased aeration, and increased soil moisture content and temperature, tend to reduce persistence of carbamate herbicides in soils.[133] Carbamate insecticides, as well as other pesticides, are also prone to microbial degradation[136] in soil and water.[75,77-79,143-147]

There is little information on the persistence of carbamate pesticides in sediments. Carbaryl and 1-naphthol are adsorbed by estuarine mud and slowly decomposed; both compounds were adsorbed from seawater and reached a maximum concentration in the mud after 8 to 12 days and remained near this level for about 30 days.[74] Carbofuran residues have been reported in tailwater pit sediment[148] and in soil under flooded rice culture.[93,94]

C. Fish

Carbamates are toxic to fish, and fish exposure can occur after spills or land run-off. Catfish were exposed to labeled carbaryl at 0.05 or 0.25 mg/ℓ water, or fed 0.28 or 2.8 mg/kg/week for 50 days;[149] maximum retained radioactivity totalled only 9 ppb after feeding or 11 ppb after water exposure at the higher dosages. Kanazawa[87] reared Motsugo in an aquarium containing 1 mg/ℓ of the three carbamates carbaryl, 2-sec-butylphenyl N-methylcarbate (BPMC), and 3,5-dimethylphenyl N-methylcarbamate (XMC). Uptake of pesticide by the fish in 1 day was 7.5 ppm carbaryl, 3.5 ppm BPMC, and 1.4 ppm XMC (Figure 13). Carbaryl was metabolized at the fastest rate and concentrations decreased to 0.89 ppm after 7 days. Uptake of carbamates is generally much less than for the o.p. pesticides. Carbamates are metabolized rapidly in animals and show little if any propensity for storage in animal tissue[150] (see Section IV.E).

In some cases, the toxicity of carbamates to fish can be affected by the water quality parameters such as pH. It has recently been found that fresh solutions of mexacarbate at pH 9 to 9.5 are several times more toxic to fish than are fresh solutions at pH 6.5, and upon aging for several days the level of toxicity increases at both pHs.[84]

The toxicity of carbamate pesticides to fish is used as a criterion by the International Joint Commission in establishing water quality objectives (Section IV.A.1).[101,105]

D. Plants

In assessing residues in or on plants, one must consider how the pesticide was applied. Some pesticides are applied directly to the crop, whereas others are systemic. A systemic pesticide is one which is abosrbed by the plant and is translocated within plant tissues.[58,150-153] Most systemic pesticides are carried up with the transpiration stream and are termed apoplastic, but they can in fact enter membranes into plant cells. A few compounds are transported to metabolic sinks (roots, buds, shoot apices, flowers, and fruits) and they are termed symplastic. Typically, a systemic pesticide may be applied to seeds, to the foliage, or to the soil for root uptake.

Carbamate insecticides are not very stable in plants being rapidly decomposed by oxidation and conjugation (Section V).[1] Carbaryl dissipation on sprayed fruit depends on the fruit type and half-lives of 28 days on lemon peels and 42 days on oranges,[154] and 7 to 10 days on apples, peaches, and plums[155] have been reported; actual residue deposit of 1.5 to 13 ppm was dependent on variations in the surface of the fruit. In general, dissipation of carbaryl is very rapid, with most of the residue being lost within 1 to 2 weeks. In a forest field test, carbaryl residues dropped, after an application of 1 lb/A, to below the 1 ppm level within 28 days for grass, 47 days for geranium, and

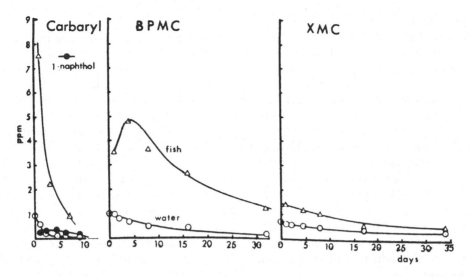

FIGURE 13. Uptake and fate of carbamate insecticides by fresh water fish, Motsugo (Δ) and its persistence in water (O). (Reprinted from Kanazawa, J., *Bull. Environ. Contam. Toxicol.*, 14, 346, 1975. With permission.)

67 days for aspen; 3.8 ppm on Douglas-fir was found after 67 days.[156] Estimated half-lives (days) were grass 8, geranium 3, aspen 8, and Douglas-fir 4.5. Following aerial forest spraying with mexacarbate, residue levels declined rapidly, in most of the plants investigated, after the first day.[157]

Residues of carbofuran on sprayed alfalfa were detected on days 0 and 7 but not after 14 days.[158] Rapid dissipation was also found in cotton and corn plants.[46] There is a rapid uptake of carbofuran by plants and translocation to leaf margins.[159] Metabolism indicated that carbofuran is converted to 3-hydroxy and 3-keto carbofuran, with the 3-keto form being hydrolytically unstable with subsequent conversion to its phenol.[160] Conjugation of the phenol and 3-hydroxy carbofuran occurs in plants. In mugho pine, translocation of carbofuran was to the needles and metabolism was slower than in other plants or animals.[161] Persistence of up to 2 years in mugho pine foliage in the field has been reported.[162]

Aldicarb is another soil-applied systemic carbamate insecticide. Aldicarb is rapidly oxidized in cotton plants to its sulfoxide which was relatively stable, but several other metabolites including oximes were also present; the total residue remaining after 21 days was about 60%.[115] In potatoes a similar metabolism occurred.[195] The fate of aldicarb was determined by the formation of the sulfone and sulfoxide metabolites and their subsequent degradation to their corresponding oximes and finally to water soluble metabolites. Foliage residues were high whereas in the tubers the levels decreased due to dilution of growth. Seldom was the parent carbamate found in these studies.[163]

In general, the carbamate pesticides are not very stable in or on plants. Dissipation may be by weathering of surface residues, dilution due to plant growth, through metabolism, or by translocation of systemics. The scope of this article does not permit a more detailed discussion of carbamate residues in plants. The above data, however, is indicative of the persistence of carbamates. Information on terminal residues are available in IUPAC reports.[164-171] Nevertheless, these facts re-emphasize that substrates must be analyzed for the parent compound as well as its metabolites to determine if toxic levels persist or if exposure to carbamates occurred.

Knowledge of environmental persistence of pesticides is particularly important to the residue analyst in selecting appropriate sample sizes and plant parts, and in assessing residue results. Based on the residue determined, one can predict the approximate time after exposure or the time required for disappearance of a pesticide.

E. Model Ecosystems

Metcalf et al.[172] devised a simple laboratory model ecosystem to facilitate the evaluation of pesticide biodegradability and magnification in food-chain organisms. These ecosystems simulate typical terrestial-aquatic environments and may include soil, water, plants, larvae, crabs, clams, snails, frogs, and fish, etc. Usually, a radiolabeled compound is introduced and, after an appropriate period of time, the radioactivity (on TLC plates) is determind in the various components in the ecosystem. Studies have been conducted with carbofuran,[173] bufencarb,[174] and carbaryl, 2-sec-butylphenyl *N*-methylcarbamate (BPMC), and 3,5-dimethylphenyl *N*-methylcarbamate (XMC).[87,175]

Carbofuran, applied to sorghum plants, reached a peak concentration in the water after 7 days whereupon it was rapidly hydrolyzed with hydroxylation of the benzofuranyl moiety.[173] Metabolites in the water were carbofuran phenol, 3-keto carbofuran, 3-hydroxy carbofuran, 3-hydroxy carbofuran phenol, and *N*-hydroxymethylcarbofuran. No parent carbofuran was found in the living organisms, but it was present in dead clams. Carbofuran was highly toxic to crabs, clams, snails, and *Daphnia* immediately after introduction of the compound, but most organisms survived restocking 20 days later.

The concentration of bufencarb insecticide reached a maximum in the water after 5 days and thereafter rapidly hydrolyzed with a half-life of less than 4 days.[174] Bufencarb was highly toxic to aquatic organisms immediately after application and metabolism was limited in the killed species. Few metabolites were found in the water. Kanazawa[87] studied the uptake and excretion of three carbamate insecticides by fish in an aquarium. Figure 13 shows the levels of these compounds in the water and fish. In a subsequent study, Kanazawa et al.[175] suggested that carbaryl and XMC are relatively persistent in the aquatic environment. Daphnids were particularly susceptible to these compounds. The soil-applied insecticides slowly desorbed into the water whereupon they were hydrolyzed. Soil (sediment) was the major repository for the insecticides which appeared primarily as bound residues.

These data show that *N*-methylcarbamates can be transported from a terrestial to an aquatic environment. Due to their rapid hydrolysis in water, they do not represent a long-term environmental problem; however, their short-term adverse toxicity is cause for concern. Rapid movement of the compounds into the aquatic system may result in death to many organisms. Furthermore, all compounds showed bioaccumulation ratios of the order of 100 to 4000, although these are not of the order of magnitude found with o.c. pesticides and PCBs. In general, the carbamate pesticides are not toxic to nontarget organisms; however, there are exceptions. In an aquatic environment *Daphnia* is particularly susceptible to *N*-methylcarbamates as are molluscs and crustaceans.[1] Fish are susceptible to higher (ppm) concentrations of the insecticides. With respect to nontarget terrestial species, bees are sensitive to the *N*-methylcarbamates, especially carbaryl (LD$_{50}$, 0.23 μg/insect).[176] Carbofuran has been shown to be toxic to earthworms.[177] From an analytical point of view, the parent pesticide may be the only compound present in a dead species; however, in moribund or other species from an affected area, several metabolites may be present.

F. Environmental Study

Even though the carbamates are generally nonpersistent, their overall effect on the environment must be considered. Such an example[178] concerns the distribution and persistence of aminocarb residue (Figure 14) in spruce foliage, forest soil, and natural waters from the Larose forest collected after an experimental aerial application of 70 g ai/ha (1 oz ai/A). The concentration at 0.6 days after application in pond and stream water samples were 2.1 and 1.9 ppb, respectively, and they disappeared through physical and biological processes to below the level of analytical sensitivity (0.1 ppb) within

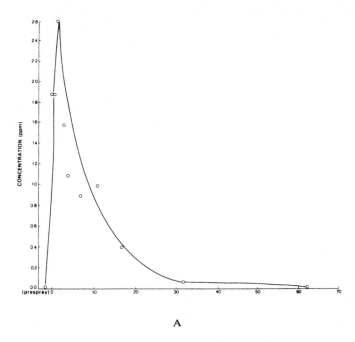

A

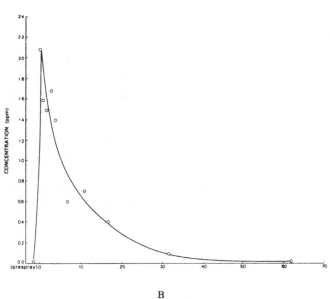

B

FIGURE 14. Dissipation of aminocarb in forest environment substrates following aerial application. (A) stream water, (B) pond water,
and (C) spruce foliage. (Adapted from Sundaram, F. M. S., Volpe,
Y., Smith, G. G., and Duffy, J. R., Report CC-X-116, Chemical
Control Research Institute, Canadian Forestry Service, Ottawa, Ontario, 1976.)

32 days. The rate of disappearance was slow and uneven in stream water due to foliar
washings by rain and run-off waters from land. The half-life in these two waters were
different and ranged from 4.4 to 8.7 days for the pond and stream water, respectively.
The initial (0.6 day) concentration of aminocarb in spruce foliage was found to be 0.7
ppm; the concentration increased to a maximum of 2.2 ppm after 4 days. After this

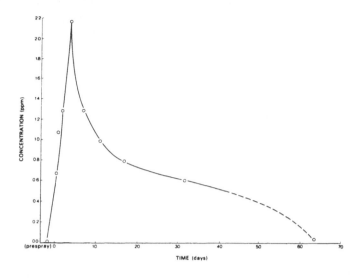

FIGURE 14C

period, the residue levels decreased exponentially with a half-life of 5.6 days and disappeared completely after 64 days probably due to physical and environment factors. No measurable amounts of aminocarb were found in the post-spray soil samples.

G. Summary

Further studies are required on long-term effects and the environmental significance of residues on various nontarget species of fauna and aquatic organisms exposed to carbamate pesticides.[179] The carbamates are generally nonpersistent (relative to the o.c. pesticides); however, there is evidence that they are persistent enough to have an effect on the aquatic environment.

Development of multi-residue methods should reflect those compounds that are in use for agricultural or forestry purposes.[99,180] Unfortunately, many multi-residue studies on environmental residues confine themselves to the o.c. and to a lesser extent the o.p. pesticides. In the few monitoring studies conducted on carbamate pesticides, there is sufficient evidence to indicate that the carbamates may occasionally be present in water[100,101] and soil,[111] and although the levels are low, significant loadings of these compounds are possible.[100,101] In most cases, these contaminations are related to agricultural run-off or forest spraying, and to point sources. Monitoring for carbamates and their metabolites may indicate likely sources of contamination, and appropriate remedial measures applied.

V. DEGRADATION AND METABOLIC PROCESSES

The chemist involved with environmental monitoring for pesticides and their metabolites is not concerned with metabolism per se, but rather with the degradation products, that is, those products that may be found during residue analysis in significant amounts. Many texts and articles have been written on the subject of pesticide metabolism, but most of these deal with the biochemical transformations rather than with environmental products. These studies are often based on in vitro or in vivo experiments using radiolabeled compounds and TLC,[181] and few relate directly to aquatic or soil systems. Albeit there are microorganisms present in these environs that could and do cause parallel reactions,[133] the variety of biochemical transformations and subsequent degradation of metabolites that occur in different species precludes many of these products being present in soil or water samples. Nevertheless, continued studies are required for these substrates to ascertain what, if any, metabolites may be present; a recent study[182] found a new degradation product of mexacarbate in fresh water. Residue methodology may have to be designed or modified to determine these metabolites, although many are short lived, nontoxic, or are present at very low concentrations. During residue analysis, the analyst often encounters many extraneous GLC peaks and often wonders whether they are, in fact, metabolites.

Ideally, a good pesticide is stable for pest control yet is biodegradable to nontoxic end products so that persistence and toxicity to the environment is not a problem. The success of the carbamates is generally due to their efficacy, short persistence during the control period, and complete breakdown to relatively innocuous end products. These end products tend to be ubiquitous materials such as phenols, anilines, mercaptans and alcohols; however, some carbamates may form fairly stable and toxic oxidation products (See Section IV). In general, metabolites retaining the carbamate moiety are toxicologically more important; hydrolysis of the carbamate ester destroys the cholinesterase activity.

Before true metabolism studies on carbamates were conducted it was assumed that hydrolysis of the carbamate ester was the main metabolic pathway.[183] Subsequent studies have shown that radiolabeled metabolites could be organo-extractable, water soluble, and unextractable,[184,185] and this fact has strong analytical implications. Furthermore, the response of a plant or animal to an xenobiotic is to convert it to a more polar compound which makes analytical extraction and cleanup difficult. Metabolism of carbamates, whether it be chemically or biochemically via enzyme catalyzed reactions, occurs through three principal routes: hydrolysis, oxidation, and conjugation. Although the metabolites formed in plants, insects, and mammals are generally similar, differences exist in their rate of formation, their conjugating moiety, and their ultimate fate within the organism.[186] In animals, hydrolysis is the main pathway although oxidative metabolism, primarily in the liver, also occurs. In plants and insects, hydrolysis is slow and oxidative metabolism tends to predominate. Conjugation of the hydrolysis and/or the oxidation products often occurs rapidly; conjugates in animals are rapidly eliminated as a detoxification process. However, plants do not readily eliminate these conjugates, some of which may be as toxic as the parent compound, and in fact, they may be stored for a considerable length of time. Persistence in plants and insects is often limited by the oxidative enzyme systems present in the organism, whereas animals usually eliminate 70 to 80% of an administered dose of carbamates as metabolites in the urine within 24 hr.[187] Differences in metabolism also occur within and between the different groups of carbamate pesticides, and as such, a variety of metabolites or degradation products must be routinely analyzed depending on the parent compound. A good example of the difference in metabolism between similar compounds is found with the oxime carbamates aldicarb[115,116,118,119,163,188-196] and thiofanox,[196-199] and methomyl,[200-202] and oxamyl.[203-205]

This discussion considers the basic metabolic routes by which all carbamates (or pesticides) are degraded rather than the specific metabolism of an individual pesticide. More details on the metabolism of carbamates may be found in the references.[1,16,56,58,59,133,136,186,187,206,207]

A. Hydrolysis

Hydrolysis of carbamate pesticides is an important detoxification process. Many carbamates are hydrolyzed at the ester or amide linkage by esterases to such products

Table 7
COMPARISON OF THE RATES OF HYDROLYSIS FOR VARIOUS AGENTS IN THE RAT (INTRAPERITONEAL)

Agent	Exhaled $C^{14}O_2$ (% original activity)	
Propoxur	31	
HRS 1422	49	
UC10854	53	
Chlorxylam	58	
Mercaptodimethur	66	⎫
Aminocarb	67	⎬ para-substituted
Mexacarbate	77	⎭
Dimetilan	49	
Isolan	74	

Adapted from Schlagbauer, B. G. L. and Schlagbauer, A. W. J., *Residue Rev., 42,* 1, 1972. With permission.

as phenols, oximes, enols, and anilines. Also formed is the corresponding acid which is spontaneously decomposed to carbon dioxide and the amine:

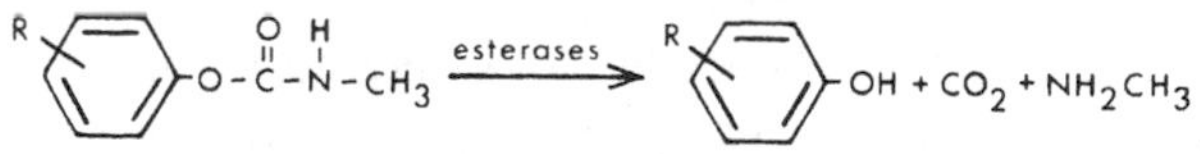

Enzymatic hydrolysis may also occur after prior oxidation of the carbamate moiety, such as with *N*-hydroxymethylcarbamate products.[1,207]

In general, *N,N*-dimethylcarbamates are enzymatically hydrolyzed in vivo more rapidly than the *N*-methylcarbamates. This trend is the reverse of that found with chemical hydrolysis (See Section II.B). IPC is very slowly hydrolyzed.[16] Differences in enzymatic hydrolysis rates depend on the structure of the carbamate insecticide and on the species contacting the chemical.[1] Carbaryl is readily hydrolyzed in rat, sheep, guinea pig, and dog; however, it is fairly resistant to hydrolysis in monkey and pig.[1,49,190] As discussed under mode of action, ring substitution affects the rate of hydrolysis, and *para*-substitution increases the rate. A typical comparison between some insecticidal carbamates is shown in Table 7.[16,209]

B. Oxidation
Kuhr and Dorough[1] showed the possible sites of oxidation on a hypothetical *N*-methylcarbamate molecule (Figure 15). Mixed-function oxidase (MFO) enzymes have been shown to be capable of catalyzing aromatic and aliphatic hydroxylation, *N*-, *O*-, or *S*-dealkylation, deamination, desaturation, expoxidation, and *N*- or *S*-oxidation. Often several of these mechanisms occur sequentially to form degradation products. Hydrolysis can also occur with those compounds retaining the carbamate moiety.

One multistep process is oxidative *N*-dealkylation. This reaction can occur at the carbamate amine (methyl or dimethyl) or at the dimethylamino ring substituent. *N*-hydroxymethyl analogs have been shown for carbaryl, propoxur, carbanolate, aminocarb, and mexacarbate, whereas *N*-dealkylation has been observed with mexacarbate, aminocarb, and dimetilan.[1] Desmethylformamido and desmethyl metabolites of pirimicarb may account for more than 50% of the residue in some plants.[210]

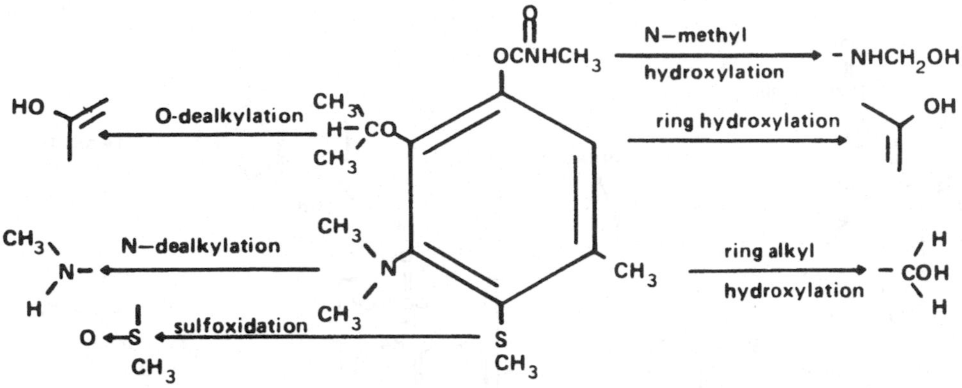

FIGURE 15.   Sites of oxidation on a hypothetical *N*-methylcarbamate. (From Kuhr, R. J. and Dorough, H. W., *Carbamate Insecticides: Chemistry, Biochemistry and Toxicology,* CRC Press, Boca Raton, Fla., 1976. With permission.)

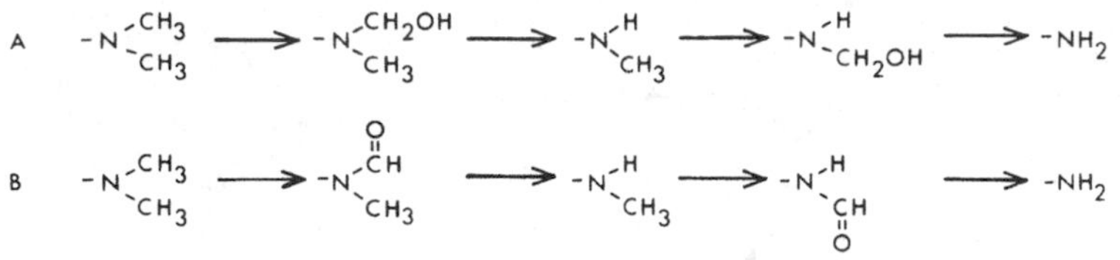

Another major oxidative route is sulfoxidation of the thio ester. This process is probably a two-step reaction with further oxidation of the sulfoxide to the sulfone:

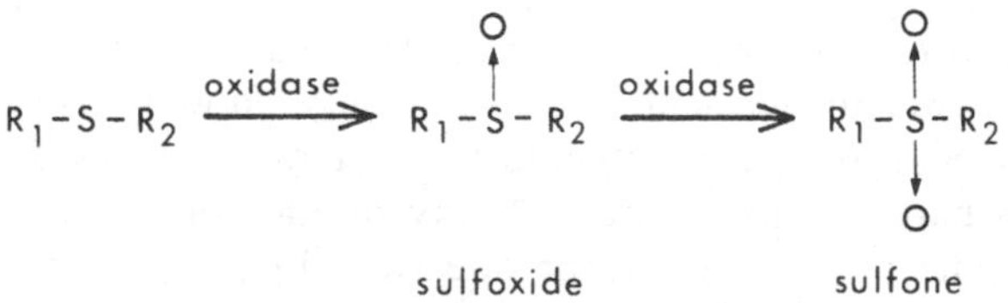

This process accounts for significant oxidative residues of aldicarb, and to a lesser extent of methiocarb. Thiocarbamate herbicides may also be oxidized to the sulfoxide.[211]

The third major oxidative metabolic route is ring and ring alkyl hydroxylation. Ring hydroxylation is important in the metabolism of carbaryl and carbofuran; this process probably proceeds through epoxidation to the hydroxy compound, then to the keto, and eventually to ring cleavage. This reaction is shown for carbaryl in Figure 16.

One very important oxidative scheme involving the ethylenebisdithiocarbamate fungicides is the degradation to the carcinogen ethylenethiourea. Many reports such as those by IUPAC[212] or Ripley et al.[213-215] and references therein demonstrate that ethylenethiourea is present in many crops following field treatment with EBDC fungicides (Section XI.C).

C. Conjugation

Conjugation refers to the process whereby pesticide metabolites are joined to other biological natural products for storage or excretion. Generally, metabolites containing such chemical moieties as $-COOH$, $-NH_2$, $-SH$, and $-OH$, formed as the result of hydrolysis or oxidation, react with natural components to form glucuronides, glycosides, sulfates, mercapturic acids, amino acid conjugates, and methylated conjugates.[1]

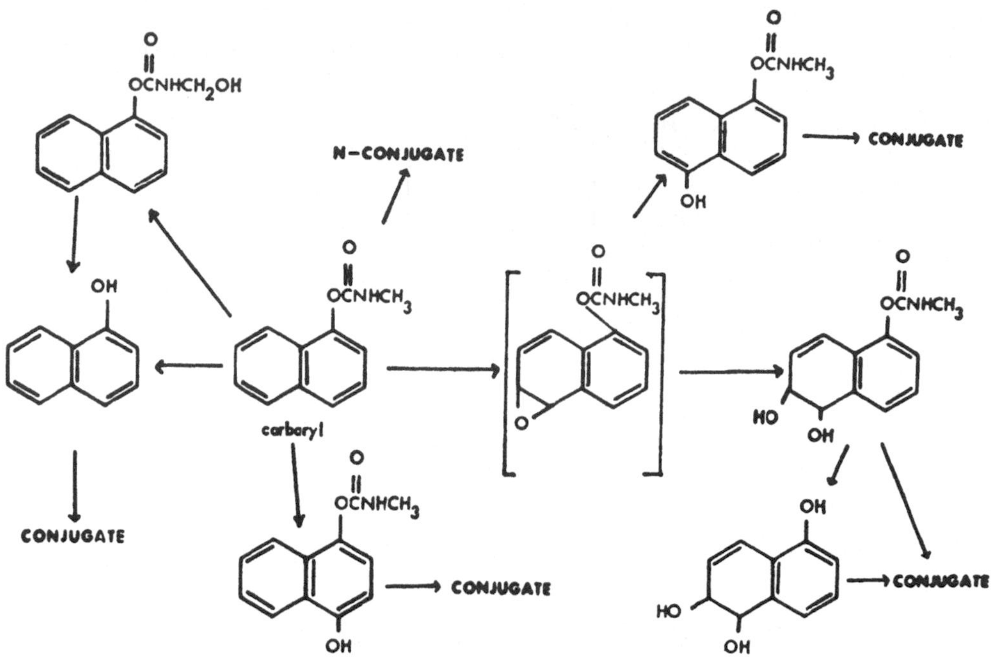

FIGURE 16.   Proposed metabolic pathway for carbaryl in animals. (Reprinted with permission from Dorough, H. W., *J. Agric. Food Chem.*, 18, 1015, 1970. Copyright 1970, American Chemical Society.)

A glycoside refers to a sugar found in nature in combination with a nonsugar component (aglycone group); glycoside conjugates are found primarily in plants and insects with mono-, di-, and trisaccharides serving as the sugar residue.[216] In animal tissues glucose is oxidized to glucuronic acid, which is conjugated with the aglycone for excretion from the body. Sulfates are more common in mammals, phosphates are found in insects, and plants may not form either. Many of the conjugates tend to be water soluble although some may be bound or unextractable; many of the aglycones, or hydroxylation products, of carbamate insecticides formed in plants, are anticholinesterase agents. Usually conjugates have a β-D-linkage and the nature of conjugation is demonstrated by release of the aglycone by enzymes such as β-glucosidase, β-glucuronidase or aryl sulfatase; some glucuronides have been determined by GC.[217,218] The variety of aglycones thus determined indicate a wide variety of oxidation-metabolism prior to conjugation.

Examples of typical conjugates and their methods of synthesis (See Section V.E) are shown in Figure 17. As stated, these conjugates are water soluble or bound to plant residues; Table 8 shows the percentage of water soluble metabolites found in bean plants 3 to 6 days after a 25-μg injected dose. Since many carbamates form conjugates in significant amounts and since they may be released to toxic moieties, analytical schemes such as those for propoxur[219,220] or carbofuran[160] must be developed to analyze for these conjugates (as the aglycone.)

D. Typical Metabolic Pathways

Typical metabolic breakdown pathways[206] are illustrated in Figures 16 and 18 to 23 for carbaryl, carbofuran, aminocarb, aldicarb, IPC, pebullate, and nabam, respectively. These representative metabolisms illustrate the variety of possible breakdown products, yet in all cases the general reactions involved are simply hydrolysis, oxidation, and conjugation.

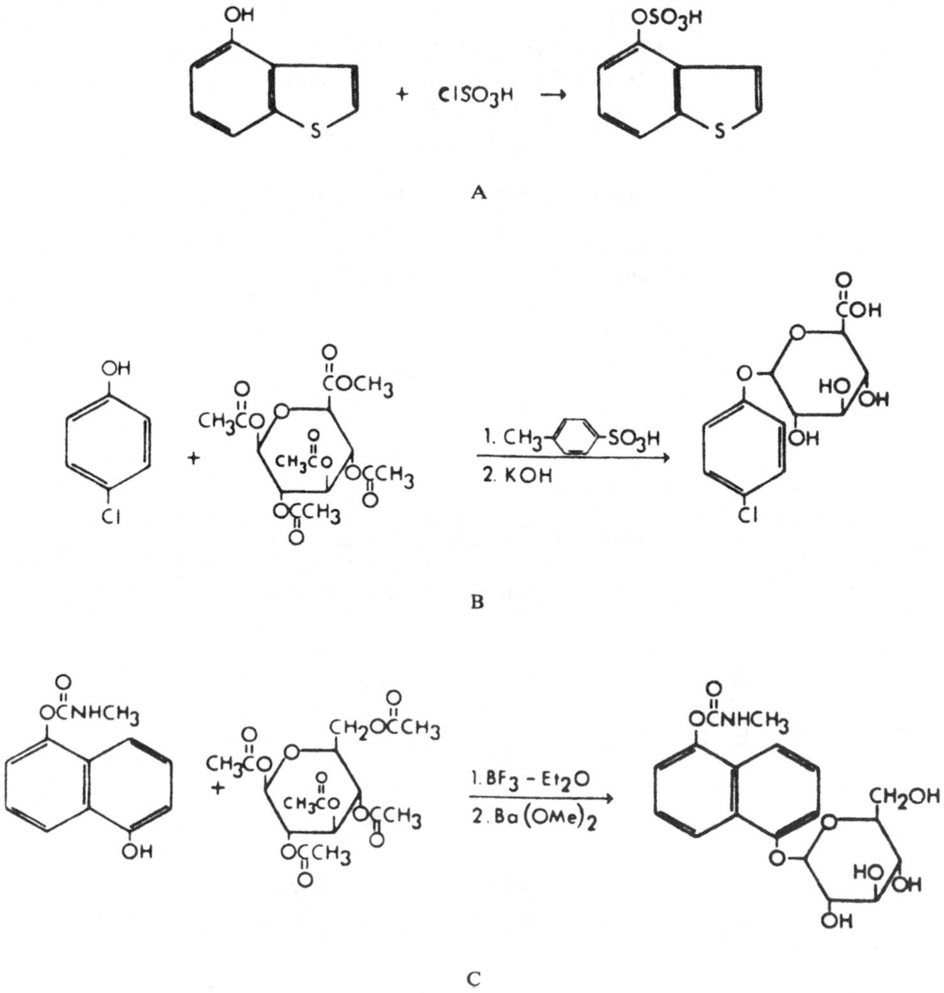

FIGURE 17. Structure and synthesis of some conjugates of carbamates or metabolites. (A) Synthesis of the sulfate of Mobam phenol,[221] (B) the glucuronide of 4-chlorophenol,[222] and (C) the glucoside of 5-hydroxy carbaryl.[223] (From Kuhr, R. J. and Dorough, H. W., *Carbamate Insecticides: Chemistry, Biochemistry and Toxicology,* CRC Press, Boca Raton, Fla., 1976. With permission.)

E. Synthesis of Carbamate Pesticide Metabolites

To determine metabolic products which are often present in minute amounts, standards of these compounds for comparison by co-chromatography are required. Residue analysts are usually concerned with those metabolites or degradation products that may be present as terminal residues, but they too require metabolite standards for comparison purposes. Usually these principal metabolites are available from the pesticide manufacturers, but occasionally the analyst is required to synthesize these standards himself. It is important in these syntheses that high-purity products be used as the "standards". The synthetic route to prepare some terminal compounds is the same route employed analytically to determine the carbamates as "total" residue (Section IX.D).

Hydrolysis products are usually available from chemical manufacturers because they are easily prepared and are also used in the manufacture of the carbamate. Less common phenols, such as that of Mobam, may be prepared synthetically.[224] Phenolic metabolites of carbamates that have undergone enzymatic or chemical hydrolysis are eas-

Table 8
PER CENT WATER-SOLUBLE METABOLITES OF CERTAIN CARBAMATE INSECTICIDES, FOUND IN PLANTS

Compound	% Water-soluble metabolites
Aminocarb	7—8
Carbanolate	45—66
Carbaryl	35—39
Dimetilan	16—25
HRS-1422	12—23
Isolan	20—26
Methiocarb	8—9
Mexacarbate	4—5
Propoxur	59—68
UC-10854	56—64

Reprinted with permission from Kuhr, R. J. and Casida, J. E., *J. Agric. Food Chem.*, 15, 814, 1967. Copyright 1967, American Chemical Society.

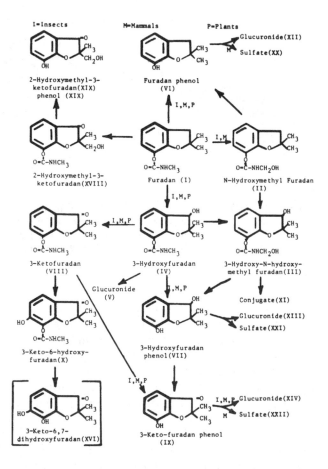

FIGURE 18.   Metabolic pathways of carbofuran (Furadan®).[206]

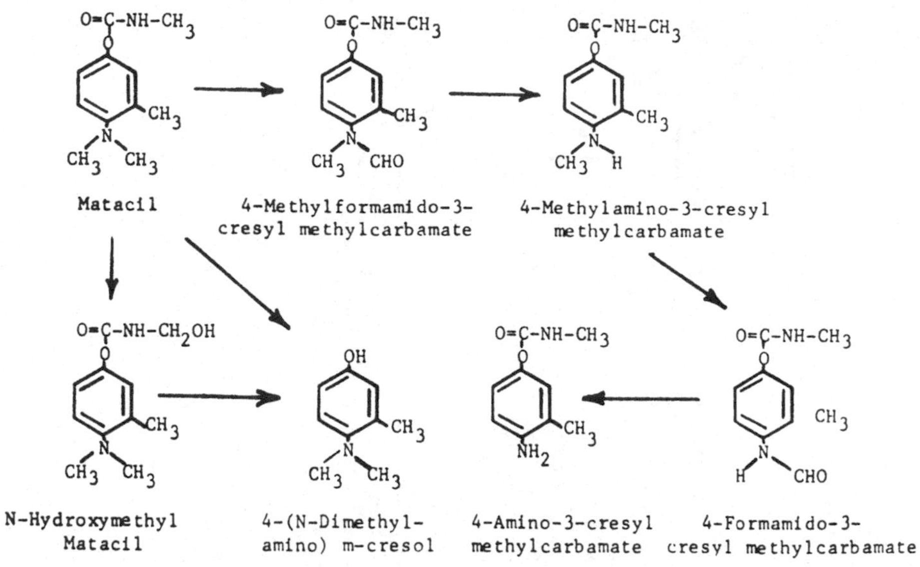

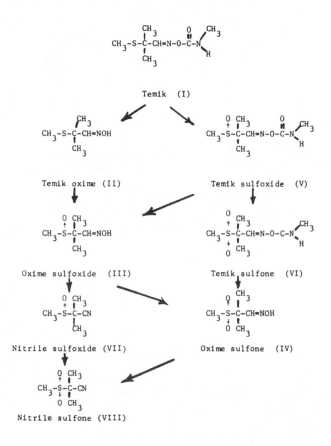

FIGURE 19. Metabolic pathways of aminocarb (Matacil®).[206]

FIGURE 20. Metabolic pathways of aldicarb (Temik®).[206]

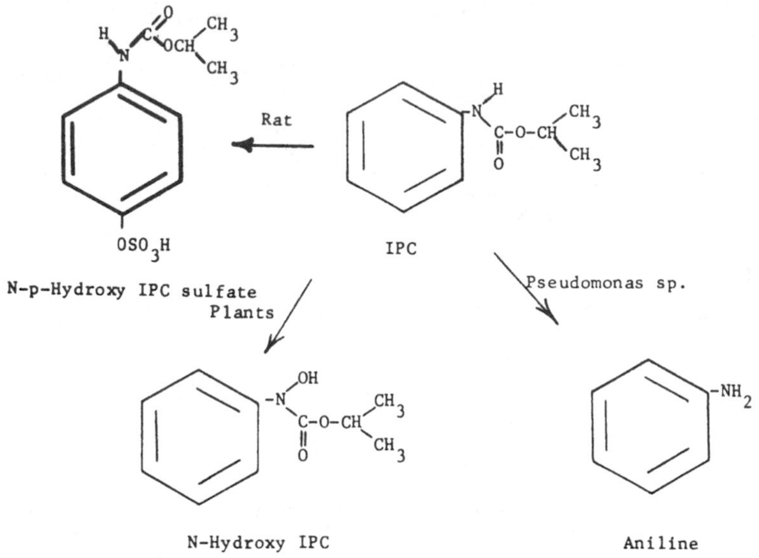

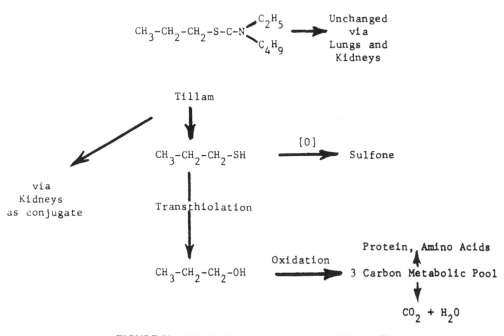

FIGURE 21.   Metabolic pathways of IPC.[206]

FIGURE 22.   Metabolic pathways of pebulate (Tillam).[206]

ily obtained by alkaline hydrolysis of the parent compound, appropriate pH adjustment of the aqueous phase, and partitioning into an organic solvent to isolate the phenol, enol, oxime, or aniline (Section X.C).

Oxidative metabolites have been prepared prior to many metabolic studies and some of the synthetic routes for preparing labeled and nonlabeled metabolites have been reviewed by Kuhr and Dorough.[1] Some of the carbamate oxidative metabolites that have been synthesized are N-hydroxymethyl;[46,48,113,225-227] ring hydroxy;[48,49,113,228-231] hydroxyalkyl[113,230,231] and carbofuran[46] derivatives; and sulfoxides and sulfones of al-

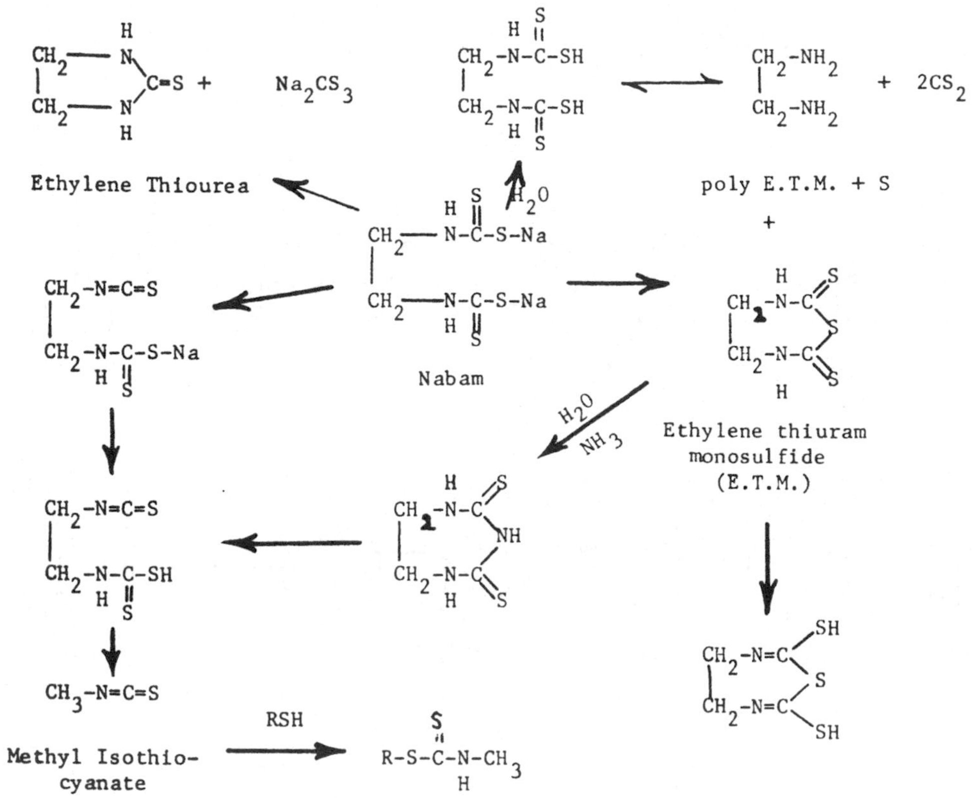

CH$_2$—N(H)
CH$_2$—N(H) C=S + Na$_2$CS$_3$

CH$_2$-N-C-SH (H, S)
CH$_2$-N-C-SH (H, S)

CH$_2$-NH$_2$
CH$_2$-NH$_2$ + 2CS$_2$

Ethylene Thiourea

poly E.T.M. + S

CH$_2$—N-C-S-Na (H, S)
CH$_2$—N-C-S-Na (H, S)
H$_2$O

N abam

CH$_2$-N=C=S
CH$_2$-N-C-S-Na (H, S)

CH$_2$-N-C (H, S)
CH$_2$-N-C (H, S)

Ethylene thiuram monosulfide (E.T.M.)

H$_2$O / NH$_3$

CH$_2$-N-C (H, S)
CH$_2$-N-C (H, S) NH

CH$_2$-N=C=S
CH$_2$-N-C-SH (H, S)

CH$_2$-N=C-SH
CH$_2$-N=C-S-SH (S)

CH$_3$-N=C=S

Methyl Isothio-cyanate

RSH → R-S-C-N-CH$_3$ (S, H)

FIGURE 23. Metabolic pathways of nabam.[206]

dicarb,[10,54,115,190,192,232] methomyl,[200] and thiocarbamates.[233] A few of the reaction schemes are illustrated in Figure 24.

Conjugates of the oxidation metabolites of the carbamates have been prepared[49,217,221-223,234-236] mainly for comparison purposes during TLC-metabolic studies. The residue analyst is not usually concerned with these conjugates per se but is interested in the aglycone that could be released. Figure 17 shows some of these products and methods of preparation.

Many of these papers provide spectroscopic identification (IR, NMR, MS) of the products and some of this data may be beneficial to the analyst attempting to confirm the identity of an unknown.

VI. RESIDUE ANALYSIS

The problems of residue analysis of pesticides are great. A pesticide analyst must have many and varied skills and, above all, he must be a good analytical chemist.[237] Carbamate pesticides, unlike the o.c. and o.p. insecticides, are not generally amenable to multi-residue GLC analysis. GLC analysis of intact N-methylcarbamates is hindered by their thermal lability and their lack of sensitivity to ECD; detection of carbamates is usually based on detection of a heteroatom using an element-specific detector. These specific detectors, however, are not specific for a pesticide, but only to a specific element and the presence of co-extractives or artifacts often makes the job of pesticide identification difficult. The selectivity of carbamate pesticides is based on the variety of substitution to the basic carbamate moiety, and unfortunately, this diversity of

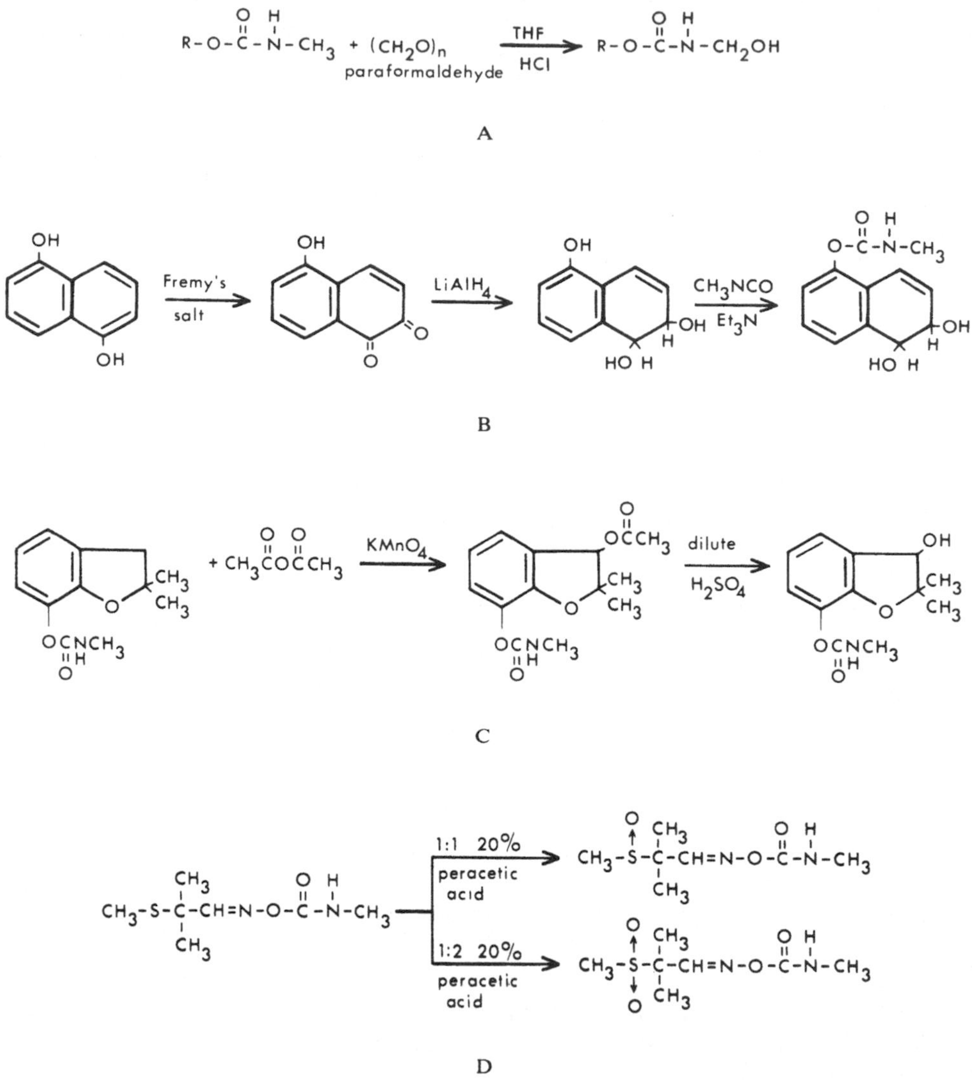

A

B

C

D

FIGURE 24. Synthesis of carbamate metabolites. (A) N-hydroxymethylcarbamates,[227] (B) ring hydrox-ylated carbaryl,[1,229] (C) 3-hydroxy carbofuran,[46] and (D) aldicarb sulfoxide and sulfone.[10] (From Kuhr, R. J. and Dorough, H. W., *Carbamate Insecticides: Chemistry, Biochemistry, and Toxicology*, CRC Press, Boca Raton, Fla., 1976. With permission.)

chemical structure and the propensity of most carbamates to be degraded to several metabolites in the environment presents numerous problems to the pesticide residue analyst.

In general, the chemist receives two types of analytical requests. The easiest analysis to handle is the request for a single compound (and its metabolites) in a specific sub-strate following field application of the pesticide. A specific, sensitive method may be developed to handle the separation, qualitation, and quantitation of the specific ap-plied compound(s). Often this procedure is not applicable to other carbamates or sub-strates without changes in the methodology. On the other hand, the analyst often re-ceives a request for the analysis of "carbamate pesticides". This situation arises in monitoring or regulatory laboratories wherein a variety of substrates must be checked for many possible carbamates and their metabolites. At the present time there is no multi-residue method available to determine all carbamate residues, although several procedures allow multi-residue determination of some carbamates.

The history of carbamate methodologies parallels the development of procedures and instrumentation for other pesticide classes, and also the increase in use of carbamate pesticides and requests for their analysis. Originally many methods were based on spectroscopic or TLC techniques with their inherent difficulties, particularly with respect to sensitivity. The development of GC (Section IX) as a routine analytical technique represented a major step forward for all pesticide analysis and the use of sensitive detectors such as EC or AFID, and specific detectors such as the FPD, electrolytic conductivity, or tunable N-P allowed for better quantitation of small amounts of pesticides. Although not all carbamates can be analyzed by GC, this technique still remains one of the most widely used. The advent of high speed liquid chromatography in the last few years represented another significant step forward. With this technique the heat sensitive carbamates can easily be analyzed and multi-residue methodologies are now appearing in the literature.

The main steps involved in a residue method are extraction, cleanup, and analysis and identification. Each step in the procedure may influence the next; for example, a relatively clean water sample may not require cleanup whereas plant or animal extracts may require two or more cleanup steps. Usually the method of determination or detection (Sections IX to XII) influences the degree to which co-extractives or interferences need to be removed. Extraction methods may have to be modified depending on the goal of the analysis, i.e., whether metabolites, particularly conjugates, are to be analyzed. Tables 9 to 15 summarize extracting solvents, cleanup procedures, and the determination method used to analyze various carbamates in different substrates. Few multi-residue procedures have been published, but the similarity in the methodologies indicates that many carbamates may be extracted and cleaned up under the same conditions. The one aspect of carbamate pesticide analysis limiting multi-residue applications is the determination step. Hence, it is not surprising that determination of carbamate pesticides and their metabolites remains the focus for many residue analysts.

Several reviews on the state-of-the-art in carbamate analysis have previously been reported[1,238-250] as well as literature reviews.[251-253] Some specific methods are contained in available analytical manuals and treatises.[13,20,254-257] Additional comments are found in periodic reports by AOAC[258,259] and IUPAC.[260] Among the published methodology for carbamates, only a few refer to water and a paucity of information is available for sediments and biota. The large majority of the methods are for plants and foods; however, the technology developed for these substrates is useful for environmental samples. Therefore the following review and evaluation of the major aspects of carbamate analysis published in the literature will be presented to form background information for the discussion on environmental samples.

A. Sampling and Sample Preservation

Sampling techniques and sample preservation are important criteria in any monitoring program. Although sampling and analysis are distinct parts of the program, they are interdependent since each controls certain aspects of the other.[261] An analytical result can be no more valid than the samples or sampling scheme used.[262] Good sampling procedures should be practiced[248,261-264] and the following precautions may be taken to ensure that the sample received and analyzed represents the field situation.

Samples should be analyzed as soon as possible after collection to avoid any biological or chemical changes to the pesticides; in the interim the sample should be frozen or in the case of water stored at a low temperature (about 4°C). Since many of the *N*-methylcarbamates are prone to hydrolysis at the pH of natural waters, the addition of acid (to about pH 2) is recommended. Photochemical changes are also possible and the samples should be stored in the dark. In an effort to reduce many of the problems associated with transit and storage of samples, some laboratories recommend field extraction, or adsorption of the pesticides onto resin columns (Section VII.A.1).

Table 9

EXTRACTION, CLEANUP, AND DETERMINATION OF *N*-METHYLCARBAMATES FROM WATER

Compound(s)[a]	Substrate	Extracting solvent	Cleanup	Determination	Ref.
Carbaryl	Water, estuarine	CH$_2$Cl$_2$	(Florisil)® (a)Florisil	Colorimetric (a)HPLC-UV	Johnson and Stansbury[107] (a) Pieper[156]
Carbaryl, propoxur	Neutral water	CH$_2$Cl$_2$		TLC	Abbott et al.[307]
Carbaryl, Mobam	Buffered water	pH 2, ether		TLC	MacNeil et al.[30]
Carbaryl, propoxur	Tap water	H$^+$, CH$_2$Cl$_2$ or CHCl$_3$		TLC	El-Dib[308]
Carbaryl, aminocarb, mexacarbate methiocarb, propoxur	River water	CHCl$_3$		TLC	Eichelberger and Lichtenburg[88]
Carbaryl	Run-off	H$^+$, CHCl$_3$		GLC-derivatization	Caro et al.[83]
Bufencarb	Water	H$^+$, CH$_2$Cl$_2$		GLC-direct	Tucker[309]
Aminocarb	Stream, pond	pH 7.5 with NaHCO$_3$, toluene	(Acid-base partition)	GLC-derivatization	Sundaram et al.[178]
Carbofuran	Run-off	0.25 *N* with HCl, CH$_2$Cl$_2$		GLC-derivatization	Caro et al.[102]
	Tail water	EtAc-hexane (1:1), CHCl$_3$	Celite® 545-MgO-Norit® charcoal	GLC-direct	Kadoum and Mock[148]
Carbofuran, 3-OH, 3-K	Rice paddy water	CH$_2$Cl$_2$-ether (3:1)		GLC-direct	Seiber et al.[92]
	Water	CH$_2$Cl$_2$	Nuchar-Attaclay-Florisil®	GLC-direct	Cook[310]
	Rice paddy water	CH$_2$Cl$_2$		GLC-derivatization	Deuel et al.[91]
Carbaryl, 1-naphthol	Tap and natural water	H$^+$, CHCl$_3$	Water, base partition	Colorimetric	Aly[311]
	Water	pH 4, CH$_2$Cl$_2$	XAD-8	GLC-derivatization	Nagasawa et al.[312]
Carbaryl, butacarb, methiocarb, propoxur	River	Ce(SO$_4$)$_2$, Na$_2$SO$_4$, CHCl$_3$		GLC-derivatization	Cohen et al.[313]
Carbaryl, carbofuran, phenols	Deionized and river water	H$^+$, CH$_2$Cl$_2$		GLC-derivatization	Seiber et al.[314]

Carbaryl, carbofuran, 3-OH, 3-K, methiocarb, Mobam, propoxur, bufencarb, bendiocarb, phenols	River, stream, lake, tap water	pH 3—4, Na₂SO₄, CH₂Cl₂	Derivatization, silica gel	GLC-derivatization	Coburn et al.[27], Frank et al.[100,101], Ripley[315]
Carbaryl, carbofuran, methiocarb, propoxur, bufencarb, aminocarb, mexacarbate	Distilled	Na₂SO₄, CH₂Cl₂	Silica gel	GLC-derivatization	Thompson et al.[278]
Carbaryl, carbofuran, propoxur, methiocarb, aminocarb, mexacarbate, pirimicarb, methomyl	Natural (neutralized)	XAD-2, eluted with EtAc	None	GLC (N-P) direct	Sundaram et al.[306]

Note: Main or original method is listed first. Optional parameters are in parentheses — i.e., (Florisil®) = this may be used if required but is not essential. (a), (b) designate procedures where additional authors made a change in pesticide or method — i.e., for carbaryl determination, (a) Pieper (Reference 156) used the Johnson and Stansbury method, but with Florisil® and HPLC-UV instead of colorimetric determination.

a Compounds in series determined simultaneously.

Table 10
EXTRACTION, CLEANUP, AND DETERMINATION OF *N*-METHYLCARBAMATES FROM SOILS AND SEDIMENTS

Compound(s)[a]	Substrate	Extracting solvent	Cleanup	Determination	Ref.
Carbaryl	Soil	Acetone	(Coagulation), Florisil®	Colorimetric	Johnson and Stansbury[107]
	Fortified soil	Acetone	H₂O, CHCl₃, extraction, coagulation, pH 12 buffer, CCl₄ extraction	GLC-derivatization	Gutenmann and Lisk[216]
	Soil	Benzene	Coagulation	IR	Kuhr et al.[317]
	Soil	Acetone-EtOH-H₂O (1:1:1)	CH₂Cl₂ extraction	GLC-derivatization	Caro et al.[83]
	Soil and sediment	Acetone-H₂O (1:1)	CH₂Cl₂ extraction, Florisil®	HPLC-UV	Pieper[156]

Table 10 (continued)
EXTRACTION, CLEANUP, AND DETERMINATION OF N-METHYLCARBAMATES FROM SOILS AND SEDIMENTS

Compound(s)[a]	Substrate	Extracting solvent	Cleanup	Determination	Ref.
Carbaryl, 1-naphthol	Mud (sediment)	CH₂Cl₂	Florisil®	Colorimetric	Karinen et al.[74]
Bufencarb	Soil	CH₂Cl₂	Silicic acid-Nuchar-Attaclay	GLC-direct	Tucker[309]
Landrin®	Fortified soil	Acetone	CH₂Cl₂ extraction	GLC-derivatization	Asai et al.[114]
Aminocarb	Soil	CH₃CN	Acid-base partition, ether, CHCl₃ partition, pH 7.5 with NaHCO₃, benzene extraction	GLC-derivatization	Sundaram et al.[178]
Carbofuran	Soil	MeOH-CH₂Cl₂ (2:5)	Alumina	GLC-direct	Williams et al.[109]
	Fortified flooded soil	CHCl₃-ether (1:1)	TLC	Colorimetric	Venkateswarlu et al.[94,318]
	Sediment	Acetone, EtAc	Celite 545®-MgO-Norit® charcoal	GLC-direct	Kadoum and Mock[148]
(a) Carbofuran (b) 3-Keto carbofuran	Fortified soil	0.25 N HCl	CH₂Cl₂ extraction	GLC-direct	(a) Siddaramappa et al.[93] (b) Ragab et al.[319]
Carbofuran, 3-OH,3-K	Soil	CHCl₃	None	GLC-derivatization	Miles and Harris[111]
	Fortified soil	0.25 N HCl	CH₂Cl₂ extraction, alumina	GLC-derivatization	Butler and McDonough[320]
	Soil	0.25 N HCl	CH₂Cl₂ extraction, Nuchar®-Attaclay	GLC-direct	Cook[310]
Carbofuran, 3-OH, 3-K, propoxur, Landrin®, bufencarb, methiocarb, carbaryl, mexacarbate	Soils	Acid ammonium acetate	CH₂Cl₂ extraction	GLC-derivatization	Caro et al.[102,321] Coburn et al.[277] Holland[322]

Note: Main or original method is listed first. Optional parameters are in parentheses. (a), (b) designate procedures where additional author(s) made a change in pesticide or method. See Note, Table 9 for additional information.

[a] Compounds in series determined simultaneously.

Table 11

EXTRACTION, CLEANUP, AND DETERMINATION OF *N*-METHYLCARBAMATES FROM ANIMAL, BIRD OR FISH TISSUE

Compound(s)[a]	Substrate	Extracting solvent	Cleanup	Determination	Ref.
Carbaryl	Bird tissue	CH_2Cl_2	Coagulation	Colorimetric	Kurtz and Studholme[325] AOAC[255]
Carbaryl, 1-naphthol	Animal tissue (a) Poultry	CH_2Cl_2	Base partition, CH_3CN-Skellysolve F partition, Florisil®	Colorimetric	Claborn et al.[326] (a) Johnson et al.[327]
Carbofuran, 3-OH, 3-K	Animal tissue, milk	Acetone	Acidification, freezing out, CH_2Cl_2 extraction, Nu-char®-Attaclay-silica gel	GLC-direct	Cook[310]
Carbofuran	Animal tissue	0.25 *N*HCl	CH_2Cl_2 partition, Florisil®, CH_3CN-pet. ether partition, Florisil®	GLC-derivatiza-tion	Wong and Fisher[328]
Carbofuran phenols	Milk, muscle, liver, kidney	Acetone	Acidification, freezing out, CH_2Cl_2-EtAc (3:1) extrac-tion, ethoxylation, base partition, DNPE derivati-zation, (Nuchar ® C-190N)	GLC-derivatiza-tion	Cook et al.[329]
	Eggs, poultry	0.25 *N*HCl	CH_2Cl_2-ether (3:1) extrac-tion, ethoxylation, base partition, DNPE derivati-zation, alumina	GLC-derivatiza-tion	Cook et al.[329]
Propoxur, *o*-OH	Animal tissue	CH_3CN	Hexane partition, H⁺, $CHCl_3$ extraction, Florisil®	GLC-derivatiza-tion	Stanley and Thornton[220] Anderson[330]
Aminocarb	Fish	CH_3CN	H⁺, ether, $CHCl_3$ partition, pH 7.5 with $NaHCO_3$, ben-zene extraction, H_2O wash	GLC-derivatiza-tion	Sundaram[331]
Carbaryl, BPMC, XMC	Fish	CH_3CN	Hexane partition, Florisil®	GLC-derivatiza-tion	Kanazawa[87]

Table 11 (continued)
EXTRACTION, CLEANUP, AND DETERMINATION OF *N*-METHYLCARBAMATES FROM ANIMAL, BIRD OR FISH TISSUE

Compound(s)[a]	Substrate	Extracting solvent	Cleanup	Determination	Ref.
Carbaryl	Bees	CH_2Cl_2	Florisil®	(a) Fluorescence (GLC)	(a) Argauer et al.[332]
				(b) TLC-EI Colorimetric	(b) Winterlin et al.[333] Morse et al.[334]
		$CHCl_3$ Benzene	Alumina CH_3CN-pet. ether partition, Florisil®, base partition	Colorimetric	Johnson and Stansbury[107,335]
		$CHCl_3$	CH_3CN-hexane partition, Florisil®	GLC–derivatization	Butler and McDonough[336]
	Bees, chicken, trout	Acetone	H_2O, $CHCl_3$ extraction, coagulation, pH 12 buffer, CCl_4 extraction	GLC–derivatization	Gutenmann and Lisk[316]
1-Naphthol	Urine	H^+, reflux, OH^-, benzene extraction	Derivatization, silica gel	GLC–derivatization	Shafik et al.[337]
Banol and phenol	Milk	Acetone	CH_2Cl_2 extraction, Florisil®, base partition	GLC–derivatization	Argauer[338]
Carbaryl	Milk	Pentane, ether	CH_3CN-hexane partition	GLC–derivatization	Butler and McDonough[336]
Carbofuran and phenol Mobam and phenol	Milk	Acetone	CH_2Cl_2 extraction, alumina, hydrolysis, steam distillation	GLC–derivatization	(a) Bowman and Beroza[339]
					(b) Bowman and Beroza[340]
Mobam	Milk	CH_2Cl_2-hexane (3:1)	CH_3CN-pet. ether partition	Colorimetric	Chasar and Lucchesi[341]
Propoxur, *o*-OH	Milk	Acetone	$CHCl_3$ extraction, CH_3CN-hexane partition	GLC–derivatization	Stanley and Thornton[220] Anderson[330]

Note: Main or original method is listed first. Optional parameters are in parentheses. (a), (b) designate procedures where additional author(s) made a change in pesticide or method. See Note, Table 9 for additional information.

[a] Compounds in series determined simultaneously.

Table 12
EXTRACTION, CLEANUP, AND DETERMINATION OF N-METHYLCARBAMATES FROM PLANT TISSUE

Compound(s)[a]	Substrate	Extracting solvent	Cleanup	Determination	Ref.
Carbaryl	Fruits and vegetables	$CHCl_3$	Coagulation (alumina-Celite®)	Colorimetric	Miskus et al.[342] Elessawi and El-Refai[343]
		CH_2Cl_2	Coagulation, CH_2Cl_2 extraction, (Florisil®, NaOH wash, CH_3CN-pet. ether partition, hydrolysis, acid-base partitioning, coagulation)	Colorimetric	Johnson[344,345] Johnson and Stansbury[107] Benson and Finocchiaro[346] AOAC[255]
	Grain	MeOH	None	Colorimetric	Rangaswamy and Majumder[347]
	Fruits and vegetables	CH_2Cl_2	Coagulation	TLC	Finocchiaro and Benson[348]
		CH_2Cl_2	Channel layer chromatography	TLC	Palmer and Benson[349] Faucheux[353]
	Broccoli, lettuce, potato, apple	CH_3CN	Hexane-H_2O partition	TLC-EI	Wales et al.[350]
		CH_2Cl_2	Coagulation	Polarography	Gajan et al.[351]
	Fruits and vegetables	CH_3CN	Coagulation, CH_2Cl_2 extraction, Florisil®	Polarography (TLC, colorimetric)	Porter et al.[352]
	Beans, tomatoes	CH_2Cl_2	Hydrolysis to 1-naphthol	Fluorescence	Argauer and Webb[354]
	Apples, beans, corn, broccoli Cotton	Acetone	$CHCl_3$ extraction, coagulation, pH 12 buffer, CCl_4 extraction	GLC-derivatization	Gutenmann and Lisk[316] (a) Ware at al.[355]
	Green beans	CH_2Cl_2	Florisil®	GLC-derivatization	Ralls and Cortes[356]
	Snap beans	CH_2Cl_2	Coagulation	GLC-derivatization	Van Middelem et al.[357]
	Corn, potato	Acetone	Hexane-CH_2Cl_2 (1:1)extraction, Florisil®, (dansylation)	HPLC-UV or fluorescence	Lawrence and Leduc[358]

Table 12 (continued)
EXTRACTION, CLEANUP, AND DETERMINATION OF N-METHYLCARBAMATES FROM PLANT TISSUE

Compound(s)[a]	Substrate	Extracting solvent	Cleanup	Determination	Ref.
	Cotton	Acetone or CH_2Cl_2	None	HPLC-UV	Ware et al.[359]
	Pear, rice	MeOH	CH_2Cl_2 extraction, coagulation, (alumina), hydrolysis	HPLC-fluorescence	Kojima et al.[360]
	Forest foliage	$CHCl_3$	CH_3CN-hexane partition, Florisil®	HPLC-UV	Pieper[156]
Mexacarbate	Forest foliage	0.5 NH_2SO_4	Ether + $CHCl_3$ wash, pH 7.3 with $NaHCO_3$, benzene extraction	GLC-direct	Pieper and Miskus[157]
Aminocarb	Forest foliage	CH_3CN	H^+, ether, $CHCl_3$ wash, pH 7.5 with $NaHCO_3$, benzene extraction	GLC-derivatization	Sundaram et al.[178]
Landrin®	Corn	CH_3CN	Hexane partition, hexane-ether (1:1) extraction, alumina-Solka-Floc-Darko	GLC-derivatization	Lau and Marxmiller[361]
Banol	Plant tissue	MeOH	Pet. ether, ether extraction, CH_3CN partition, alumina	GLC-derivatization	Boyack[362]
		MeOH, alkaline hydrolysis	Steam distillation, pentane extraction, acid-base partition	GLC-direct (phenol)	Boyack[362]
Banol, phenol	Grass, apples, cucumbers, tomatoes	CH_2Cl_2	Florisil®, base partition	GLC-derivatization	Argauer[338]
Banol, methiocarb	Alfalfa, apple	Acetone	Coagulation, $CHCl_3$ extraction, hydrolysis	GLC-direct (phenol)	Bache and Lisk[363]
Methiocarb and metabolites	Apples, pears, corn	Buffer, acetone, 2 N HCl	CH_2Cl_2 extraction, silica gel, alumina	GLC-direct (phenol)	Bowman and Beroza[364]
Methiocarb, SO,SO₂	Blueberries	Buffer, acetone	$CHCl_3$ extraction, silica gel	GLC-derivatization	Greenhalgh et al.[365]
Mobam	Fruits and vegetables	CH_2Cl_2	Coagulation,(Florisil®, CH_3CN-pet. ether partition)	Colorimetric	Chasar and Lucchesi[341]
	Lettuce	Methanol	None	GLC-derivatization	Moye[366]

Compound	Substrate	Solvent	Cleanup	Method	Reference
Mobam, phenol	Grass	CHCl₃	Alumina	GLC-direct (phenol)	Bowman and Beroza[340]
Propoxur	Plant tissue	Acetone, CHCl₃	pH 5, CHCl₃ extraction, hexane-CH₃CN partition, Florisil®	GLC-derivatization	Stanley et al.[219] Anderson[330]
Propoxur, o-OH, N-CH₂OH			pH 5, CHCl₃ partition, enzyme hydrolysis, CHCl₃ extraction, silica gel	GLC-derivatization	Stanley et al.[219] Anderson[330]
Carbofuran, 3-OH, 3-K	Fruits and vegetables (a) Tobacco (b) Lettuce	0.25 *N* HCl	CH₂Cl₂ extraction, CH₃CN-hexane partition, Nuchar®-Attaclay-(silica gel, silicic acid, Florisil®)	GLC-direct (b)GLC-derivatization	Cook et al.[160,310] Cassil et al.[367] (a) Hawk et al.[368] (b) Van Middelem et al.[369]
Carbofuran phenols	Plant tissue	0.25 *N* HCl	CH₂Cl₂-EtAc(3:1) extraction, ethoxylation, NaOH partition, DNPE derivatization, Nuchar C-19ON	GLC-derivatization	Cook et al.[329]
Carbofuran, 3-OH, 3-K	Cucumbers, lettuce, tomatoes, potato	0.25 *N* HCl	CH₂Cl₂ extraction, alumina	GLC-derivatization	Butler and Mc-Donough[320]
	Small fruits (a) Carrots	0.25 *N* HCl, CH₂Cl₂ blend of solid residue	CH₂Cl₂ extraction, alumina-silica gel, (Florisil®)	GLC-direct	Williams and Brown[370] (a) Finlayson et al.[371]
Carbofuran, 3-OH, 3-K, phenols	Alfalfa	0.25 *N* HCl	CH₂Cl₂, ether extraction, Florisil®, (TLC)	GLC-derivatization	Archer[372]
	Strawberries	0.25 *N* HCl	CH₂Cl₂, ether extraction, EtAc-hexane (1:1) extraction, silica gel	TLC	Archer et al.[373]
Carbofuran, phenol	Corn	CHCl₃	Alumina, hydrolysis, steam distillation	GLC-derivatization	Bowman and Beroza[339]
Carbofuran	Grass	Acid ammonium chloride	Neutralization, CH₂Cl₂ extraction	GLC-derivatization	Holland[322]
Carbofuran, 3-K, 3-OH (nonconjugated)	Corn, potato, turnip, wheat	Acetone	CH₂Cl₂ extraction, Florisil®	HPLC-UV	Lawrence and Leduc[374]
(a) 3-OH (conjugated)		(a) 0.25 *N* HCl-CH₂Cl₂(solids)		(a) HPLC-UV, GLC-direct, GLC-derivatization	(a) Lawrence et al.[375]

Table 12 (continued)
EXTRACTION, CLEANUP, AND DETERMINATION OF *N*-METHYLCARBAMATES FROM PLANT TISSUE

Compound(s)[a]	Substrate	Extracting solvent	Cleanup	Determination	Ref.
Carbaryl, methiocarb, propoxur	Lettuce and apples	CH_2Cl_2	Steam distillation	TLC-El (GLC-derivatization)	Ernst et al.[376]
	Apples	Toluene-hexane (3:1)	Florisil®	GLC-derivatization	Johansson[377]
	Pears, lettuce, apples	Acetone	$CHCl_3$ extraction, coagulation, $Ce(SO_4)_4$, $CHCl_3$ extraction	GLC-derivatization	Cohen et al.[313]
Carbaryl, Mobam, carbofuran	Apples, potatoes, sugarbeets, grass	$CHCl_3$	Coagulation, $CHCl_3$ extraction, Florisil® (alumina, Norit A®)	GLC-derivatization	Butler and McDonough[378]
Carbaryl, carbanolate, Landrin, propoxur, (bufencarb, carbofuran, methiocarb, Mobam, promecarb, and others)	Vegetable crops	CH_3CN	Pet. ether partition, H_2O, CH_2Cl_3 extraction, coagulation, CH_2Cl_2 extraction	GLC-derivatization	Holden[379,380] AOAC[255]

Note: Main or original method is listed first. Optional parameters are in parentheses. (a), (b) designate procedures where additional author(s) made a change in pesticide or method. See Note, Table 9 for additional information.

a Compounds in series determined simultaneously.

Table 13
EXTRACTION, CLEANUP, AND DETERMINATION OF OXIME CARBAMATES

Compound	Substrate	Extracting solvent	Cleanup	Determination[a]	Ref.
Aldicarb	Fruit and vegetables	$CHCl_3$	Coagulation, $CHCl_3$ extraction, base partition, hydrolysis	Colorimetric	Johnson and Stansbury[390]
	Citrus	Acetone	$CHCl_3$ extraction, Florisil®, hydrolysis	Colorimetric	Meagher et al.[391] Hendrickson and Meagher[392]

Compound	Substrate	Extraction	Cleanup	Method	Reference
Aldicarb and metabolites[b]	Oranges, apples, sugar beets, potatoes	$CHCl_3$	Nuchar®-Al_2O_3-Florisil®-MgO-Hyflo Supercel®, Na_2SO_4-Celite®, CH_3CN extraction, Florisil®-Nuchar®	GLC-FPD (S)-direct and oxidized	Maitlen et al.[393]
	Sugar beets	$CHCl_3$	Acid-base hydrolysis-partition, $CHCl_3$ extraction, silica	GLC-MCD (S)	Beckman et al.[394]
	Oranges, potatoes, carrots, corn, green beans, silage	Acetone-CH_2Cl_2 (1:1)	Florisil®, oxidation	GLC-FPD (S)-sulfone	Carey and Helrich[395]
Total toxic aldicarb[b]	Apples, potatoes, cucumbers, alfalfa, cottonseed	$CHCl_3$	Na_2SO_4-Celite®, CH_3CN extraction, oxidation, Florisil®	GLC-FPD (S)-sulfone	Maitlen et al.[396]
	Tomato (a) Soybeans	Acetone-$CHCl_3$ (4:1) + peracetic acid	$NaHCO_3$, $CHCl_3$ extraction, Florisil®	GLC-FPD (S)-sulfone	Lindquist et al.[397] (a) Krueger and Mason[398]
	Soil, cottonseed, lint	Acetone-H_2O (1:1)	Oxidation, $CHCl_3$ extraction, Florisil®	GLC-FPD (S)-sulfone	Woodham et al.[399]
	Grass and weeds Birds and mammals	Acetone-H_2O (1:1) Acetone	Oxidation, $CHCl_3$ extraction, Florisil®	GLC-FPD (S)-sulfone	Woodham et al.[400]
Methomyl	Plant and animal tissue, soil	Ethyl acetate	H_2O partition, H^+, hexane partition, $CHCl_3$ extraction, OH^- hydrolysis	GLC-MCD (S)-oxime	Pease and Kirkland[401]
	Tobacco			(a) GLC-derivatization	(a) Tappan et al.[402]
	Tomatoes	Ethyl acetate	H_2O, OH^- hydrolysis, H^+, ethyl acetate extraction	GLC-FPD (S)-oxime	Krueger et al.[403]
	Tobacco	Ethyl acetate	H_2O, H^+, hexane partition, $CHCl_3$ extraction, OH^- hydrolysis	GLC-FPD (S)-oxime	Fung[404]
	Tobacco	$CHCl_3$-MeOH (9:1)	OH^-, ether partition, hydrolysis, H^+, CH_2Cl_2, Florisil®	GLC-FPD (S)-oxime	Leidy et al.[405]
	Tobacco	2.5% benzene-CH_2Cl_2	Coagulation, CH_2Cl_2 extraction, (Florisil®)	GLC-FPD (S)	Reeves and Woodham[406]
	Soil, sediment, water	CH_2Cl_2	Florisil®	GLC-FPD (S)	Reeves and Woodham[406]

Table 13 (continued)
EXTRACTION, CLEANUP, AND DETERMINATION OF OXIME CARBAMATES

Compound	Substrate	Extracting solvent	Cleanup	Determination[a]	Ref.
	Rape seed, oils, meals	Ethyl acetate	H+, H2O, hexane partition, CHCl3 extraction	TLC-EI	Mendoza[407]
	Foods	Acetone-benzene-0.1 NH_2SO_4 (19:1:1)	Low-temperature cleanup[408,409]	TLC-EI	McLeod et al.[410]
	Soil, rye grass	Acid ammonium chloride	Na2CO3, neutralization, CH2Cl2 extraction	GLC-derivatization	Holland[322]
Thiofanox-total[c]	Soil, plant	Acetone	Oxidation, H+, CHCl3 extraction, Florisil®	GLC-FPD (S)-sulfone	Chin et al.[411]
	Water	0.2 N HCl, CHCl3 extraction	None		
Individual metabolites of thiofanox	Cotton seed	CH2Cl2	(Cyclohexane partition)		
	Potato	Acetone	10% NaCl-cyclohexane partition, oxidation	GLC-FPD (S)-sulfone	Chin et al.[411]
Oxamyl	Plant and animal tissue, soil	Ethyl acetate	H2O, hexane partition, pH 12, CHCl3 partition, hydrolysis, CHCl3 partition, NaCl, ethyl acetate-MeOH (9:1) extraction	GLC-FPS (S)-oxime	Holt and Pease[412]
	Milk	Hexane partition, ethyl acetate			
	Soils, wheat, potato, water	Ethyl acetate, H2O	H2O, hexane, CHCl3 partition, OH- hydrolysis	Colorimetric	Singhal et al.[413]
	Citrus leaves	MeOH	None	HPLC-UV	Davis et al.[414]
Oxamyl and methomyl	Vegetables	Ethyl acetate	H2O, hexane partition, NaCl, CHCl3 extraction	HPLC-UV	Thean et al.[415]
	Vegetables	Acetone	Base hydrolysis, acidification, CHCl3 extraction, alumina	GLC-derivatization	Chapman and Harris[534]

Note: Main or original method is listed first. Optional parameters are in parentheses. (a), (b) designate procedures where additional author(s) made a change in pesticide or method. See Note, Table 9 for additional information.

a   Compound injected (see Section IX.D).
b   Total toxic aldicarb = aldicarb and sulfoxide and sulfone.
c   Total thiofanox = thiofanox and sulfoxide and sulfone and sulfone oxime.

Table 14

EXTRACTION, CLEANUP, AND DETERMINATION OF *N*-PHENYLCARBAMATES

Compounds[a]	Substrate	Extracting solvent/ method	Cleanup	Determination	Ref.
IPC	Lettuce	CH_2Cl_2	H+ hydrolysis, steam distillation	Colorimetric	Bissinger and Fredenburg[416]
CIPC	Soil and crops (a) Potato	CH_2Cl_2	H+ hydrolysis, steam distillation	Colorimetric	Gard and Rudd[417] (a) Gard[418]
IPC IPC, CIPC	Food crops	Base hydrolysis, distillation	None	Colorimetric	Montgomery[419] (a) Ferguson and Gard[420]
CIPC	Potatoes	CH_2Cl_2	Aluminum oxide column	IR	Ferguson et al.[421]
Barban	Potatoes	Ethylene dichloride	Celite®-Na_2SO_4-Attaclay, base, distillation	Colorimetric	Riden and Hopkins[422]
CIPC	Potatoes	Acetone	Hexane extraction	GLC-derivatization	Gutenmann and Lisk[423]
CIPC IPC Barban	Water	(a) CH_2Cl_2 (b) H+, $CHCl_3$ or CH_2Cl_2 (c) pH 2, ethyl ether	(Alumina)	TLC	(a) Abbott et al.[307] (b) El Dib[308] (c) MacNeil et al.[30]
CIPC, IPC, Barban	Water (a) Soil and plant	$CHCl_3$ (a) Acetone	TLC (a) $CHCl_3$ extraction, hydrolysis, acid-base partition, TLC	GLC-derivatization	Cohen and Wheals[424]
CIPC	Fruits and vegetables	Ethanol or ethanol-H_2O	Pet. ether extraction, MgO-cellulose	GLC-FID	Onley and Yip[425]
CIPC IPC Swep	Foods	Ethanol	$CHCl_3$ extraction	GLC-derivatization	Lawrence[426] Lawrence and Laver[427]

Note: Main or original method is listed first. Optional parameters are in parentheses. (a), (b) designate procedures where additional author(s) made a change in pesticide or method. See Note, Table 9 for additional information.

[a] Compounds in series determined simultaneously.

Table 15
EXTRACTION, CLEANUP, AND DETERMINATION OF THIOCARBAMATE HERBICIDES

Compound(s)[a]	Substrate	Extracting solvent/method	Cleanup	Determination	Ref.
EPTC	Crops and soils	(a) Hexane (b) H+, steam distillation	H+, steam distillation	Colorimetric	Batchelder and Patchett[428] Patchett et al.[37,38]
	Soil	Steam distillation	Pet. ether extraction	GLC-FID	Hughes and Freed[429]
	Sugar beets	H+, H2O, steam distillation	Hydrolysis, acid-base partition	GLC-derivatization	Crosby and Bowers[430]
EPTC, Pebulate Vernolate	Soil	Steam distillation	H+, iso-octane extraction	Colorimetric	Koren et al.[31] Smith and Fitzpatrick[36]
EPTC, diallate	Corn	Acetone	CH2Cl2-pet. ether extraction, Florisil®	GLC-CCD(N), GLC-derivatization	Lawrence[432]
Diallate, triallate	Soil	Benzene-iso-propanol (2:1)	Na2CO3 partition, Nuchar®-Attaclay	GLC-ECD	Smith[36,433,434]
Triallate	Soil (a) Straw and grain	iso-octane-iso-propanol (2:1)	H2O partition (a) H2O partition, Nuchar®-Attaclay	GLC-ECD	McKone and Hance[435]
Molinate	Water	H+, hexane	Hydrolysis, acid-base partition	GLC-derivatization	Crosby and Bowers[430]
Molinate Metabolites	Water	H+, hexane (a) CH2Cl2	None	GLC-FID (a) GLC-derivatization	Soderquist et al.[95]
Vernolate and metabolites	Soil Soil	H2O, distillation Iso-propanol-H2O-hexane	H+, CH2Cl2 extraction None	GLC-FPD (S), TLC	Hermanson et al.[436-438]
Pebulate	Tobacco	H+, steam distillation	H+, iso-octane extraction, alumina, silica gel	GLC-FID	Long and Thompson[439]
EPTC, molinate, cycloate, butylate, pebulate, vernolate, diallate	Fruit and vegetables (a) Soybean, alfalfa	Ethanol (a) Ethanol-water (3:2)	Pet. ether extraction, MgO-cellulose	GLC-AFID, GLC-FPD (S)	Onley and Yip[425]
	Water	Iso-octane	None	GLC-FPD (S)	Frank et al.[100,101]

EPTC, butylate, pebulate, vernolate, cycloate, molinate	Crops and soils	H_2O, steam distillation	H^+, *iso*-octane extraction, (silica gel)	GLC-FPD (S), GLC-CCD (N)	Ja et al.[440]
Thiocarbamates	Water	$CHCl_3$	Ion-exchange, LiChroprep® Si-60	HPLC-UV	Schulten and Stoeber[441]

Note: Main or original method is listed first. Optional parameters are in parentheses. (a), (b) designate procedures where additional author(s) made a change in pesticide or method. See Note, Table 9 for additional information.

ᵃ Compounds in series determined simultaneously.

Frequently, the analyst finds that he can not analyze samples as quickly as desired. Hence, it may be desirable to extract the sample upon receipt (or in the field) and store the extract pending analysis. The stability of a pesticide during any type of storage is often questionable, and it is often beneficial to examine the stability of a standard or field-incurred residue in a sample or extract under various conditions of time, temperature, and in different substrates. Some initial work has been done in this area (Table 16).[265] The stability of carbofuran[266] and carbaryl[154,267,268] appears to be more dependent on the substrate than on the storage temperature. On the other hand, methomyl stability is related to the storage temperature and it appears to be most stable under freezing conditions.[265,269] EBDC fungicides tend to decompose immediately after maceration of substrates and storage by freezing is preferred over refrigeration.[270,271] CIPC and IPC showed significant losses during storage.[272]

Clean glass bottles with Teflon® or aluminum lined tops should be used for water samples. Polyethylene bags or glass jars are recommended for other substrates. Interferences may be obtained with other storage containers such as plastic. Often pesticides adsorb on the storage vessel; with water samples, it is recommended that the bottle be rinsed with the extracting solvent which is then added to the sample for extraction.

VII. EXTRACTION

A. *N*-Methylcarbamates

The primary concern in the development of a suitable residue method is to account for the parent compound and for those toxic metabolites which, as shown from metabolic studies, might be formed from the parent compound.[219] Often not only the cholinesterase inhibiting compounds have to be determined, but also some metabolites (phenols) that could indicate previous exposure to a pesticide. In the case of carbofuran, phenolic and carbamate residues must be determined and the concentrations included as part of the tolerance for the parent material.[273] While many analysts consider the determination step (Sections IX to XII) to be the most critical part of the analysis, many losses can be attributed to incomplete extraction of the pesticide from the sample matrix or during partitioning of sample extracts. These losses may lead to the reporting of erroneous and misleading data.

Metabolic studies have shown that residues in sample matrices may be organo-extractable, water-soluble, and unextractable. Suitable extraction procedures must be developed to recover all the compounds; unfortunately, the nature of the metabolites from the various *N*-methylcarbamates makes the development of a multi-residue procedure difficult. Parent compounds are extracted directly into an organic solvent while the water-soluble metabolites are usually extracted after hydrolysis to their organo-extractable aglycone form. Methylene chloride is the most widely used solvent for the extraction of *N*-methylcarbamates probably due to its excellent solubility towards the carbamates, and its low bp which allows facile concentration (Section VII.G). However, the sample matrix may be a factor in favor of other solvents; benzene is better than methylene chloride for extracting carbaryl in bees due to its solubility towards beeswax.[107] Extraction efficiencies may be determined by monitoring radiolabeled compounds, but few laboratories have this capability and thus they must rely on recoveries determined from fortification studies. Analytical aspects of fortification recoveries have been discussed.[274-276]

1. Water

Methylene chloride or chloroform are used almost exclusively for the extraction of *N*-methylcarbamates from water (Table 9). Few comparative studies on the extraction of carbamates from water have been conducted. Coburn et al.[277] found higher recov-

Table 16
STABILITY OF SOME CARBAMATES ON AND IN VARIOUS SUBSTRATES AND IN THEIR EXTRACTIVES SOLUTIONS STORED UNDER COLD CONDITIONS FOR VARIABLE PERIODS

Pesticide	Substrate	Extractives solvent	Storage temperature (°C)	Storage period	Apparent loss[a] of insecticide (%)	Ref.
Carbaryl	Lemons	Methylene chloride	10	8 months	nd	154
	Tomatoes	None	12.7	1 week	nd	267
	Green beans	None	7	11 days	20	268
	Beans	Methylene chloride	4	3 months	40	265
Methomyl	Corn	Ethyl acetate	4	3 months	85	265
	Lettuce	Ethyl acetate	4	1 month	25	265
	Corn fodder	None	−15	4 months	nd	265
	Corn silage	None	−15	4 months	nd	265
	Corn	None	4.5	1 month	65	265
	Corn	none	−17	1 month	nd	265
	Corn	None	−36	4 months	nd	265
	Tomatoes	None	4.5	1 month	nd	265
	Tomatoes	None	−17	1 month	nd	265
	Tomatoes	None	−36	4 months	nd	265
Maneb	Kale	None	5	3 days	15	270
	Kale	None	−15	3 days	nd	270
Zineb	Kale	None	5	3 days	10	270
	Kale	None	−15	3 days	nd	270
IPC	Apples	None	0	5 months	28	272
	Apples	None	10	4 months	52	272
	Plums	None	0	6 weeks	27	272
	Plums	None	10	3 weeks	23	272
	Tomatoes	None	0	1 month	12	272
	Tomatoes	None	10	6 weeks	29	272
	Tomatoes	None	20	1 month	63	272
CIPC	Apples	None	0	5 months	33	272
	Apples	None	10	4 months	48	272
	Plums	None	0	6 weeks	28	272
	Plums	None	10	3 weeks	37	272
	Tomatoes	None	0	1 month	9	272
	Tomatoes	None	10	6 weeks	35	272
	Tomatoes	None	20	1 month	56	272

[a] nd, not detectable.

Adapted from Kawar, N. S., DeBatista, G. C., and Gunther, F. A., *Residue Rev.*, 48, 45, 1973. With permission.

eries with methylene chloride compared to benzene or ethyl acetate. In a multi-class residue study, methylene chloride is generally acceptable,[278] and ethyl acetate-hexane followed by chloroform has also been used.[148]

The distribution of a compound between two immiscible liquid phases can be expressed in terms of its partition coefficient; however, a more convenient term is the p-value which represents the fractional amount of compound present in the nonpolar or less polar phase after partitioning between two phases of equal volume.[23,279] p-Values can also be used to select a solvent system for partitioning cleanup and for determining the number of partitioning steps required to quantitatively extract a compound. Theoretical considerations on the extraction of pesticides from water have been discussed.[274,280-283]

Apart from the choice of solvent, other considerations include pH, ionic strength,

Table 17
RECOVERY (%) OF CARBAMATE
PESTICIDES FROM NATURAL
WATERS[a,b]

Compound	pH 6.8	pH 3—4	pH 3—4 + 10 g Na₂SO₄[c]
Propoxur	41	59	88
Carbofuran	45	71	97
3-Ketocarbofuran	72	97	93
Metmercapturon	92	80	87
Carbaryl	86	98	98
Mobam	93	95	95

[a] One microgram of each pesticide was added to 1 ℓ water sample and pH conditions were altered with addition of 50% H_2SO_4.

[b] Recovery values are the average of 2 determinations on 2 different GLC columns.

[c] The sodium sulfate was added to the samples after acidification.

Reprinted from Coburn, J. A., Ripley, B. D., and Chau, A. S. Y., *J. Assoc. Off. Anal. Chem.*, 59, 188, 1976. With permission.

turbidity of the sample, water-solvent ratios, number of extractions, solute type and concentration, and the method of extraction (separatory funnel vs. vortex stirring). Coburn et al.[277] found that increased recoveries were obtained with acid conditions and in the presence of a salt (Table 17). Acidification of samples also minimizes hydrolytic losses. The addition of a salt aids in the extraction by salting out the solvent from the water and also reduces emulsion problems. Generally, several extractions in a separatory funnel are employed, and a 10:1 or less sample to solvent ratio is used.

Analytical adsorption onto various columns has been employed as a sampling/preservation and/or extraction technique; this technique has been primarily applied to the o.c. pesticides and pollutants in air and water samples. Such columns as polyurethane foam,[284-289] Amberlite® polymeric resins (Rohm & Haas Co.),[290-301] and HPLC Seppaks® (Waters Associates, Inc., Milford Mass.)[302-305] have been used; the use of Seppaks® for extraction/concentration of residues has also been termed "trace enrichment". Generally, a large volume of sample (in the field or laboratory) is passed through the column with the organic contaminant being adsorbed onto the column matrix. Subsequently (in the laboratory), the column is eluted with an appropriate solvent such as diethyl ether or ethyl acetate to desorb and collect the residue. Usually only a small volume of eluant is required and after concentration a potential 100- to 1000-fold concentration of the pollutant has been achieved. Recoveries of the studied o.c. and o.p. pesticides and PCBs have been very good; mirex has been a notable exception.

Recently, Sundaram et al.[306] showed that an XAD-2 column could be used to extract 7 N-methylcarbamates and pirimicarb from water samples. One liter of water was percolated through 2.5 g of Amberlite® XAD-2. The carbamates were desorbed from the column with 40 mℓ of ethyl acetate and analyzed by direct GLC using an N-P detector. Best recoveries (particularly for the N-methylaminophenyl N-methylcarbamates) were obtained at neutral pHs; depressed recoveries for aminocarb and its phenol were observed at acidic pH values. Unfortunately, poor reproducibility and recovery for aminocarb phenol were also found with natural water samples. Low recovery was also obtained with the water soluble methomyl. The XAD-2 resin column could be regenerated with methanol and water with no signs of deterioration. Aminocarb and its phenol were stable once extracted and adsorbed onto the column for up to 14 days; this technique, therefore, could also be used as a collection, storage, and transportation procedure thus reducing sample preservation problems (Section VI.A).

a. Typical Extraction Methods for Aqueous Samples
i. Separatory Funnel Method (Coburn et al.,[277] Frank et al.,[100,101] and Ripley[315])

Placed a measured volume of sample (typically 0.1 to 1.5 ℓ) in a separatory funnel. Rinse sample container with methylene chloride and add to separatory funnel. If sample was not preserved with acid, adjust pH to 2 to 4 with 50% H_2SO_4. Add 10 g Na_2SO_4 or 10 mℓ saturated aqueous Na_2SO_4. Extract sample with methylene chloride (about ¼ to 1/10 volume of sample) by shaking separatory funnel for 1 to 2 min. Let phases separate and drain CH_2Cl_2 into another separatory funnel if phenols are to be removed (Section VIII.A.1). Alternatively, dry extract through Na_2SO_4 into a round bottom flask for concentration (Section VII.G) and re-extract sample with CH_2Cl_2 adding this to previous extract.

If emulsions are a problem, they may be broken in one of the following ways. (1) Salt out organic by adding more Na_2SO_4 or NaCl saturated aqueous solution, or about 1 tsp of Na_2SO_4 or NaCl, or a pinch of sodium lauryl sulfate. Gently swirl mixture until phases separate. (2) Transfer emulsion to another separatory funnel, add some water, and use salting out Procedure 1. (3) Transfer emulsion to centrifuge bottles and centrifuge at high speed for 5 min.

If aminophenyl N-methylcarbamates are to be included, neutralize and adjust sample pH to 7.3 to 7.5 with $NaHCO_3$, and re-extract with methylene chloride (or benzene) and add to previous extracts or analyze separately.

ii. XAD-2 Resin Column (Sundaram et al.,[306] Ripley[315])

Preparation of column — Slurry 2.5 g Amberlite® XAD-2 in about 50 mℓ distilled water. Decant water and fines; repeat several times. Transfer resin-water slurry to 1.1-cm in diameter × 35-cm glass chromatographic column containing a glass wool plug and reservoir. Wash column with 2 × 20 mℓ methanol followed by 3 × 50 mℓ distilled water. Maintain water level 1 cm above resin bed when not being used. After use, the resin may be regenerated by washing with methanol and water as described above.

Sorption and desorption — Neutralize sample, if required, with aqueous $NaHCO_3$ solution. Percolate sample through column by gravitation. Aspirate column with vacuum to remove residual water. Rinse sample container with 2 × 10 mℓ ethyl acetate and transfer to column. Elute column with a total of 40 mℓ ethyl acetate and dry through Na_2SO_4 with 2 × 10 mℓ ethyl acetate rinses. Combine eluates for concentration or removal of phenols.

2. Soils and Sediment
When analyzing soils and sediments, consideration must be given to the soil type and water content (field capacity). Sediments and river bottom muds usually have a higher water content and contain large amounts of decomposing organic matter and miscellaneous debris as compared to field soils.[323] Some pesticides are strongly adsorbed to some soil types and in general, lower recoveries of pesticides are obtained with soils high in organic matter.

Usually soils are extracted as received (moist) or after addition of some water to approach field capacity; some methods call for air-dried soil prior to extraction, but this may cause some losses due to volatilization. Samples are sieved to remove stones and twigs, and a moisture content is determined on a subsample so that residue data can be compared on a common (dried) basis. Chiba[324] reviewed some factors affecting the extraction of o.c. pesticides fom soils, and many of the considerations are pertinent to the carbamates. Many of the factors are interrelated and the individual influence is often difficult to ascertain. Among the factors that may influence the recovery of pesticides from soils and/or affect the data reported are sampling technique; extraction equipment and sample manipulation; choice of extracting solvent(s); time and temperature of the extraction; soil factors including moisture content, soil type, age of residues, and presence of microorganisms; the pesticide factors of class and chemical structure, volatility, concentration, and formulation; presence of co-extractives; and, the validity of fortification techniques.[324] With the carbamates, one must also consider the nature of possible metabolites and the method of determining the recovered residue. For o.c. pesticides, direct GLC is fairly reproducible and provides a convenient tool to monitor the recovery. However, with the carbamates there is a propensity for lower results due to the extra manipulative steps required for their determination. If one were to use labeled compounds, recovery of all the radioactivity does not in itself mean that the compound has been recovered 100%. It is possible that a residue or metabolite may be bound or conjugated and is not readily available in its aglycone form, or a metabolic change may have occurred and the radioactivity is associated with a further metabolite or breakdown product such as carbon dioxide or methylisocyanate.

Many studies on only the parent *N*-methylcarbamates in soil matrices use an organic solvent or solvent mixture for extraction (Table 10). Generally, the soils are mechanically shaken or less often blended with the solvent. Water-miscible solvents such as methanol, acetone, acetonitrile, water-ethanol-acetone (1:1:1), and water-acetone (1:1) have been employed followed by extraction of the aqueous/organic system with the solvent of choice (usually methylene chloride) for the compound under examination. 1-Naphthol was not extracted with the acetone-ethanol-water solution.[83] Water-immiscible solvents such as methylene chloride, chloroform, benzene, 2.5% methanol in methylene chloride, and chloroform-diethyl ether (1:1) have also been used to extract *N*-methylcarbamates from soils. The soil to solvent ratios vary from 1:1 to 1:6; generally, sufficient solvent is added to completely wet the soil and produce a slurry for the extraction step. Recoveries from fortified samples are usually greater than 70%, but they are very dependent on the soil type. Additional cleanup steps may be required depending on the soil characteristics and the method of determination.

One would expect the ease of extracting adsorbed insecticides from soils and sediments to be related to their water solubilities. Therefore, the more water-soluble carbamates should be easily desorbed from these particulates relative to DDT (Figure 11). However, the nature of the adsorbent is very important, and care must be exercised when analyzing high organic soils as pesticides are much more strongly adsorbed to these matrices relative to sandy soils.

Several procedures have been developed to extract bound or conjugated residues of carbofuran, and other *N*-methylcarbamates. Cook et al.[160] refluxed samples in 0.25 *N* HCl to hydrolyze water-soluble conjugated residue forms of carbofuran in corn plant to the organo-extractable aglycone forms (see Section VII. A. 4). This principle has been applied to soils fortified with carbofuran and its toxic metabolites.[93,319,320] Methylene chloride was used to extract the pesticides from the filtered acid solution.

Caro et al.[321] used acid ammonium acetate (pH 2.5, 0.5 *M*) to digest soils at 60°C for 1 hr; after filtering, methylene chloride was used to extract the residues. Recoveries varied depending on the soil type and the pesticide; carbofuran phenol was not extracted with the acetate solution. This procedure was found to be better than digestion with hot dilute HCl, and it was felt that the acetate was a more effective reagent for

Table 18
RECOVERY OF *N*-METHYLCARBAMATE INSECTICIDES BY ACID AMMONIUM ACETATE EXTRACTION OF 2 SOILS FORTIFIED AT 1 PPM

Insecticide		Insecticide recovery (%)	
Common or trade name	*N*-methylcarbamate substituent	Evesboro sandy loam	Muskingum silt loam
Landrin®	2,3,5- and 3,4,5-trimethylphenyl-	72.1	75.6
Propoxur	2-*iso*-propoxyphenyl-	66.3	53.1
Bufencarb	3-methylbutyl- and 3-ethylpropylphenyl	62.8	76.4
Methiocarb	4-methylthio-3,5-xylyl-	44.4	58.0
Carbaryl	1-naphthyl-	29.0	34.2
Mexacarbate	4-dimethylamino-3,5-xylyl	0.0	0.0

Adapted from Caro, J. H., Glotfelty, D. E., Freeman, H. P., and Taylor, A. W., *J. Assoc. Off. Anal. Chem.*, 56, 1319, 1973. With permission.

rupturing the stronger adsorption bonds by which weathered pesticide residues are bound to soil particles. Recoveries of several carbamates using this procedure are shown in Table 18.

3. Animal, Bird, and Fish Tissues

In developing a procedure for the extraction of *N*-methylcarbamates from tissue samples, consideration must be made of metabolic products. Hydrolysis of carbamates is more predominant in animal tissue than plant tissue, and the metabolites may be conjugated as glucuronides rather than as glycosides; these conjugated residues are effectively released by acid digestion.[220] Special consideration must also be given to the presence of fats and oils in the sample matrix that may interfere with the extraction, partitioning, and determinative steps.

In general, the *N*-methylcarbamates are extracted from the tissue by either of two methods: blending with water-miscible solvents followed by acid digestion, or direct acid digestion of the tissue matrix (Table 11). Traditionally, carbaryl residues have been extracted directly by blending with methylene chloride. A typical extraction and cleanup procedure for propoxur residues[220] is shown in Figure 25. Although this scheme is probably applicable to other *N*-methylcarbamates, its application in a multi-residue procedure has not been demonstrated. To our knowledge, no multi-residue procedure for *N*-methylcarbamates and their metabolites has been published. Filter aids such as sodium sulfate or Hyflo Super Cel® are often employed. Acetonitrile or acetone have been used to extract the carbamates from tissue samples with the fats and oil then being removed by selective partitioning with hexane or by freezing out (see Section VIII). After partitioning between the aqueous and organic solvent (CH_2Cl_2 or $CHCl_3$), the aqueous solution is usually then hydrolyzed to release any conjugates. Tissue samples have also been directly digested with hot HCl, and the hydrolyzed carbamates are then extracted with methylene chloride. Carbaryl in bees has been extracted with chloroform or methylene chloride although benzene may be a better solvent because of the beeswax.[107] Usually a 5:1 solvent to sample ratio is used and in most cases a subsequent column cleanup with Florisil® is required.

4. Plant Tissue

In general, many of the procedures used to extract the *N*-methylcarbamates are based on extraction procedures for the o.c. pesticides.[381,382] *N*-methylcarbamates are

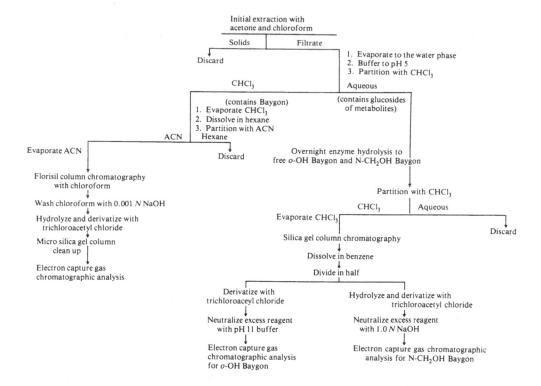

FIGURE 25.   Schematic outline of the methodology for the analysis of propoxur (Baygon®) and its metabolites in plant tissue. (Reprinted with permission from Stanley, C. W., Thornton, J. S., and Katague,
D. B., *J. Agric. Food Chem.*, 20, 1265, 1972. Copyright 1972, American Chemical Society.)

generally extracted from a plant matrix by blending with a suitable solvent (Table 12).
Blending homogenizes the sample and allows extraction of both surface and translocated residues. Sodium sulfate is often added to prevent troublesome emulsions; appropriate filter aids are frequently employed. The tendency of carbamates to hydrolyze
during analysis may be a problem with acidic substrates. The addition of a Na_2HPO_4-
citric acid buffer[364,365] prior to maceration and solvent extraction is beneficial in inhibiting hydrolysis although this problem was not completely eliminated.[364]

Water-immiscible solvents, of which methylene chloride or chloroform are the most
common, are often used to extract parent compounds. Water-miscible solvents, such
as acetone, acetonitrile, and methanol have also been used to extract the carbamates;
residues are subsequently extracted from the aqueous solvent mixture with methylene
chloride, or partitioned and/or cleaned up with hexane, ether, or a coagulation procedure.

Storherr[383] felt the Porter et al.[352] acetonitrile extraction procedure for carbaryl was
superior to and should replace methylene chloride as the AOAC recommended procedure. The current AOAC method,[255] based on the work of Holden,[379,380] involves acetonitrile extraction. Watts[384] evaluated blending with ethyl acetate or acetonitrile as
well as exhaustive Soxhlet extraction using 10% methanol in chloroform for their relative efficiencies in removal of ^{14}C-labeled carbaryl from bean plants. All three extraction procedures were found to be equally effective for the extraction of carbaryl following 2 days of simulated weathering.

Wheeler et al.[385] compared the extraction efficiency of ^{14}C-carbaryl from mustard
greens and radishes following field application of the compound which was suspended
in a commercial carbaryl formulation. Recoveries were compared between a blend-

leach and a repetitive blending followed by Soxhlet extraction; solvents used were methanol, acetonitrile, and acetone. There appeared to be an effect due to the formulation since lower recoveries were observed in this study relative to those using pure compounds.[384,386] Ware[359] observed a similar effect of formulation on extraction efficiency. All the extractable radioactivity was recovered by blending; however, the amount of extractable ^{14}C decreased with longer harvest intervals. The first blend recovered most of the radioactivity. Methanol was generally the best solvent while the blend-Soxhlet procedure was superior to the blend-leach method. Acid hydrolysis of extracted tissue released 40 to 50% of residual ^{14}C, and a higher recovery was evidenced at longer post-application intervals.

Conventional extraction procedures using organic solvents are not effective in removing conjugated carbamate residues. Metabolic studies with ^{14}C-carbofuran indicated that conjugation occurred after oxidation at the 3-position of the benzofuran ring and/or at the 7-position after hydrolysis resulting in the formation of water-soluble compounds (glycosides). Cook et al.[160] demonstrated that the conjugated forms of carbofuran could be quantitatively hydrolyzed to the aglycone form using hot acid digestion. Acid hydrolysis converts the water-soluble conjugates to the organo-extractable aglycones without destruction of the compounds. This procedure has gained wide acceptance for the analysis of carbofuran residues and is also used for the extraction of other carbamate conjugated residues. Samples are refluxed in 0.25 N HCl for 1 hr and, after cooling, the aqueous solution is extracted with methylene chloride. Alternatively, after sample extraction with water-immiscible solvents, the filter cake can be hydrolyzed with acid.[364] Additional comments on hydrolysis of conjugated residues are in another section (see Section X.C).

Williams and Brown[370] found that waxy fruits held up significant amounts of both carbofuran and 3-hydroxy carbofuran even after acid digestion and thus an extra extraction of the filtered solids with methylene chloride was necessary. They also found that the addition of mercuric chloride to the aqueous extract prior to reflux eliminated mercaptans in crucifer crops and also effectively reduced emulsions during the subsequent methylene chloride extraction of the acid solution.

Van Middelem and Peplow[386] compared the extraction recoveries of ^{14}C-carbofuran and its metabolites using five different procedures: acid digestion,[160] acetonitrile, methanol, and ethyl acetate blending, and 24-hr Soxhlet extraction using methanol. The relative efficiencies of the extraction procedures are shown in Table 19. Satisfactory recoveries were obtained with the acid digestion, with both methanol extraction procedures, and also with the acetonitrile blending, whereas depressed recoveries were obtained with ethyl acetate. While the solvent extraction procedures removed the ^{14}C-compounds, subsequent acid hydrolysis would still be required to release the conjugates to their aglycone form for determination.

Since the carbamates are stable in acid solution, either the acid digestion-extraction or aqueous-acid partitioning-hydrolysis may be employed to release conjugated residues. No reference was found on the applicability of these procedures to multi-carbamate residue methodology, although the procedures have been used individually on several N-methylcarbamates. A typical procedure is schematically outlined in Figure 26 for the extraction and subsequent partitioning and cleanup steps for propoxur and its metabolites in plant tissue.[219]

B. Aminophenyl *N*-Methylcarbamates

In several cases where a procedure has been extended to a quasi-multi-residue application, low or incomplete recoveries were observed for the aminophenyl N-methylcarbamates (aminocarb, mexacarbate).[277,278,321,379] In most cases, these compounds or their phenolic hydrolysis products were not extracted under the described conditions,

Table 19
PERCENT[a] EXTRACTION EFFICIENCY OF ^{14}C[b] FROM CABBAGE LEAVES BY VARIOUS EXTRACTION PROCEDURES

Extraction fractions	Acid digestion	Soxhlet (methanol)	Methanol[c]	Acetonitrile[c]	Ethyl acetate[c]
First extraction	90.2	88.8	71.2	65.4	32.7
Second extraction			17.6	13.3	19.3
Residual tissues	9.8	11.2	11.2	21.3	48.0

a   Calculated from comparing individual fractions to fraction total. Determined from combustion and scintillation counting. Each value represents the average of at least two replications.

b ^{14}C-Carbofuran added to soil 35 days prior to leaf sampling.

c   Extraction by blending for 5 min each time.

Reprinted with permission from Van Middelem, C. H. and Peplow, A. J., *J. Agric. Food Chem.*, 21, 100, 1973. Copyright 1973, American Chemical Society.

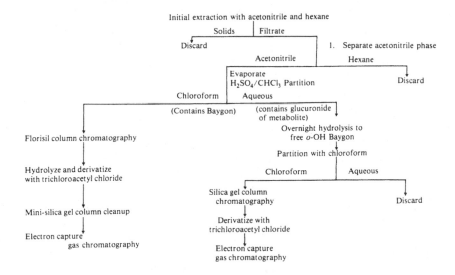

FIGURE 26.   Schematic outline of the methodology for the analysis of propoxur (Baygon®) and its metabolites in animal tissue. (Reprinted with permission from Stanley, C. W. and Thornton, J. S., *J. Agric. Food Chem.*, 20, 1269, 1972. Copyright 1972, American Chemical Society.)

and therefore special analytical considerations must be made in those situations wherein these compounds must be analyzed.

The parent compounds demonstrate properties of amines whereas the phenols show amphoteric properties. Under acidic conditions, the water-soluble amine salt of the parent compound would be formed thus precluding organic extraction; mexacarbate in forest foliage extracts has been shown to be stable in 0.5 N H_2SO_4 at 0°C for at least 2 months.[157] Under basic conditions hydrolysis of the carbamate ester would occur. Hence, it is not surprising that extraction into organic solution is usually carried out under neutral conditions; sodium bicarbonate is employed for the neutralization.

Marquardt and Luce[387,388] used these principles for the extraction and cleanup of mexacarbate in peaches and undelinted cotton prior to hydrolysis and colorimetric determination of the 4-dimethylamino-3,5-xylenol with luteoarsenotungstic acid (see

Section XII.E). Studies concerned specifically with the aminophenyl N-methylcarbamates have made use of this partitioning cleanup method for the determination of mexacarbate in forest foliage[157] and water,[85] and aminocarb residues in fish,[331] spruce foliage, soil, stream and pond water.[178] Samples may be extracted with methanol, acetonitrile, or 0.5 N H₂SO₄ as water-miscible solvents for watery substrates and with benzene for oily or waxy substrates.[388] After neutralization (pH 7.3 to 7.5) of the aqueous sample or extract with sodium bicarbonate, solvents such as benzene or toluene are used to extract the compound. Two concurrent determinations, one for the phenol and one for hydrolysis, allows calculation of the concentration of the parent compound as well as of the hydrolysis product. Any free phenol or parent compound may alternatively be partitioned from the organic extract with 2 N H₂SO₄.

C. Oxime Carbamates

Several procedures are available for the extraction, cleanup, and determination of oxime carbamates (Table 13). Depending on the sample material, varying amounts of oxime carbamate metabolites (see Section IV) may be present in the sample and the methodologies reflect the analysis for these compounds. Extraction of oxime carbamates is carried out using acetone, chloroform, methylene chloride, and most commonly ethyl acetate. Following extraction with ethyl acetate, water is added and the organic solvent removed by evaporation; the aqueous phase is either directly extracted or oxidized and extracted with chloroform. Water samples have been extracted directly with methylene chloride,[406] with chloroform after acidification,[411] or with ethyl acetate.[413]

D. N-Phenylcarbamates

Analytical methods for determining residues of N-phenylcarbamates (Table 14) fall into two general categories: (1) those involving hydrolysis, either directly on the sample or after extraction into a solvent such as methylene chloride and (2) those involving direct extraction and quantitation without hydrolysis.[52,53] Hydrolyzed samples yield the aniline which is steam distilled from the sample. N-Phenylcarbamates are usually extracted directly from neutral or acidified water samples using a polar solvent.

E. Thiocarbamate Herbicides

The thiocarbamate herbicides are usually directly extracted from solid and aqueous samples using hydrocarbon solvents such as *iso*-octane, hexane, and benzene (Table 15); isopropanol is also often added as a co-solvent to facilitate extraction. Steam distillation of acidified samples has been frequently employed with the distillate being subsequently extracted with *iso*-octane. Steam distillation allows extraction and cleanup of samples in one step. Onley and Yip[425] extracted seven thiocarbamates from food crops using ethanol or aqueous ethanol with subsequent partitioning of the aqueous phase with petroleum ether; adequate recoveries were obtained after two such extractions. Molinate was extracted from water samples with hexane; however, several of its polar metabolites required methylene chloride for extraction.[95]

With soil samples, particular care must be taken prior to analysis to prevent drying out of the sample as loss of the volatile thiocarbamates could occur; there is a large loss of EPTC from moist soils due to vaporization.[442,443] Air drying of soil samples to constant weight at 20°C caused loss of triallate,[435] while lower triallate recoveries were noted from fortified dry soils; furthermore, a certain minimum amount of moisture appeared to be necessary for good extraction of the residues.[434] As suggested by Smith,[434] extraction should be conducted on soils adjusted to have a moisture content at or above the wilting point.

Extraction of thiocarbamates from water with *iso*-octane has successfully been used in a multi-residue monitoring study.[100,101]

F. Dithiocarbamates and EBDC Fungicides

These compounds are not normally extracted prior to their determination by the CS_2 evolution technique (Section XII.E.1). These compounds are insoluble in most organic solvents; however, if it is required, they may be stripped from plant tissue; chloroform is satisfactory for ferbam, ziram, and thiram; water is used for nabam; and 1 N disodium ethylenediamine tetraacetic acid is necessary for maneb and zineb.[444] Obviously, this stripping method is unsatisfactory for multi-residue application. Once extracted, these compounds tend to hydrolyze and hence immediate analysis is imperative.

G. Concentration of Sample Extracts

Concentration of sample extracts is a potential source of large losses or variability in recovery of compounds during any analytical procedure although some obvious precautions may alleviate this situation (See Volume I, Chapter 2). There are several procedures available for concentration of sample extracts[254,255] including rotary vacuum or flash evaporation, Kuderna-Danish or micro-column evaporation, special evaporative equipment, or evaporation on the bench or in a steam bath using the atmosphere or a stream of inert gas. Most laboratories today routinely use the rotary vacuum evaporation method having previously determined the optimum parameters for this step.[445,446]

For example, early recommended methods for carbaryl called for the concentration of methylene chloride extracts using a special evaporative condenser.[255,344,345] Recovery studies indicated that this time-consuming step could be replaced by evaporation on a steam bath provided that ethylene glycol or other glycols were added to the extract.[346] Subsequent methods employed a 60°C water bath and air evaporation,[342] a Kuderna-Danish set-up,[348] and finally rotary vacuum evaporation became the norm.

In general, evaporation losses may be attributed to volatilization, thermal breakdown, and the manipulative steps. Most investigators agree that evaporation to dryness, especially in the absence of plant extractives or fat, must be avoided. Since the carbamate pesticides are more volatile than other classes of pesticides, evaporation to dryness should be avoided. To aid in this requirement, a "keeper", usually a high bp liquid, is added in small volume (0.5 to 2 mℓ) so that the solvent extract may be evaporated leaving the pesticide in the keeper. Common keepers are *iso*-octane, glycols, acetophenone, and GLC liquid phases such as OV-101 or DC-200.

To prevent volatility losses or thermal decomposition, particularly with the heat labile N-methylcarbamates, a water bath temperature of 30 to 60°C is employed with the rotary vacuum evaporator; particular care must be observed with methomyl and its oxime, which require a lower temperature of about 30°C. Furthermore, it is advisable to use a lower bp solvent (such as methylene chloride) for the extraction steps to eliminate the need of the higher temperatures required to evaporate chloroform or acetonitrile. With the Kuderna-Danish concentration, petroleum ether is facily evaporated whereas higher boiling solvents require high block temperatures, longer times, and are prone to bumping losses. To ensure quantitative results, it should be observed that bumping is prevented and that the concentrated extracts are transferred quantitatively for the subsequent analytical step.

VIII. CLEANUP OF SAMPLE EXTRACTS

Cleanup of sample extracts is an important step in any analysis. For example, coextractives may mask an entire chromatogram making qualitation and quantitation impossible, or an artifact may produce a GLC peak similar to that of a pesticide thus causing misidentification. The degree of sample cleanup required is dependent upon the sample type and the nature of the subsequent determination step. Water extracts are generally clean as compared to the co-extractives encountered from animal or plant tissue extracts. Few interferences are usually encountered with specific detectors whereas derivatization of an extract containing many co-extractives with functional groups similar to the one being derivatized on the carbamate could produce numerous peaks during EC examination (Sections X and XI). Quite often more than one type of cleanup operation is required to obtain "clean" extracts for spectroscopic or chromatographic analysis.

Cleanup procedures may be considered as separation of components of interest from those not required for a specific analysis. Obviously the removal of extraneous natural material is ideal, but cleanup may also entail removal or separation of pesticide metabolites, other classes of pesticides or pollutants, as well as fractionation of compounds of interest which can serve as partial confirmation of identity. Many of the methods used for the cleanup of the *N*-methylcarbamates are an extension of those developed for the analysis of organochlorines in fatty and nonfatty foods, and for the colorimetric determination of carbaryl. Cleanup of samples for pesticide residue analysis has been discussed.[447] Tables 9 to 15 show the cleanup procedures used with the different types of carbamate pesticides.

Liquid-liquid partitioning may be used to clean up sample extracts due to the differential solubility of compounds between two liquid phases. It has been demonstrated that in a petroleum ether (or hexane)-acetonitrile solvent system, non-ionic lipids and fats partition into the hydrocarbon and the more polar pesticides partition into the acetonitrile.[448] Johnson and Stansbury[107] discussed the applicability of this procedure to carbaryl samples, and presently this procedure is widely used for all the *N*-methylcarbamates.

Three to four volumes of water is added to the acetonitrile, which is then extracted with a polar solvent such as methylene chloride; carbaryl was not quantitatively recovered from aqueous acetonitrile with petroleum ether.[352] Alternatively, the acetonitrile is evaporated and the carbamate is dissolved in an appropriate solvent for the subsequent analytical step; however, due to the low volatility of acetonitrile, this step may be time consuming. Phenols and other acidic compounds may be removed from organic extracts by shaking briefly with dilute alkali solutions (see Section VIII.A).

Plant waxes and other water-insoluble extractives may be precipitated from sample extracts using a coagulation solution.[107,346] Typically an aqueous solution of 0.1% ammonium chloride and 0.2% phosphoric acid is added to the extract in warm acetone or methanol. After standing at rt, a filter aid is added and the slurry is filtered with vacuum, and the pesticides are extracted with methylene chloride. Since the coagulation solution is very acidic, neutralization is required prior to extraction of aminophenyl *N*-methylcarbamates.[389] Waxes from bee extracts have also been precipitated using methanol,[334] and oils from animal tissue have been frozen out of an acid solution.[310]

Liquid-solid adsorption chromatography is based on the reversible adsorption of solutes from a mobile liquid phase onto a solid stationary phase. Sample extracts are placed on the adsorbent and preferentially removed with eluants of different polarity. Morley[449] has reviewed the application of column cleanup to pesticide residues. Many of the common adsorbents such as Florisil®, alumina, silica gel, and various carbons have been used to cleanup the *N*-methylcarbamates. Judicious choice of eluants allows separation of the carbamates from co-extractives and metabolites; extraneous material may be eluted in a wash fraction prior to elution of the carbamates or they may be left adsorbed on the column. One problem common to all column chromatography is the quality of the adsorbent particularly since its activity is dependent on water content.[450] Adsorbents have been used as received, after activation, or after partial deactivation with water. Strict quality control must be exercised in the laboratory to monitor each new batch of adsorbent; most investigators caution that elution patterns should be confirmed in each laboratory. It may be for this reason that few standardized methods exist particularly for the cleanup of the *N*-methylcarbamates. For example, Florisil® was found to be satisfactory for the cleanup of extracts containing carbaryl.[107] Carbaryl is eluted with water-saturated methylene chloride leaving most of the extraneous material on the column. The water content of the Florisil®, however, influences the recovery with the optimum degree of deactivation required depending on the investigator or more precisely, the quality of the adsorbent used. Carbaryl could not be recovered from PR grade Florisil®, which is commercially activated at 1400°F, but the standard grade (activated at 1200°F) was found to be satisfactory although there was some variability in recovery, and with some batches of Florisil® no recovery of carbaryl was obtained.[336] Generally the adsorbents are activated in the laboratory by heating overnight at 300°C and then small batches are deactivated with water and equilibrated prior to use. Elution patterns must be checked with each new batch to ensure quantitative recoveries.

Florisil®[451] (synthetic magnesium silicate) has been widely used for cleanup of pesticide extracts particularly those containing lipid material. Initially analysts hoped to use the standardized Mills Florisil® cleanup[254,255,452,453] that many laboratories use for the o.c.s. Unfortunately, the nonpolar solvents used to elute the o.c. and o.p. pesticides did not elute the more polar *N*-methylcarbamate pesticides; carbamates can be eluted with methylene chloride, chloroform, or polar mixtures of hydrocarbons with chlorinated hydrocarbons or ethyl acetate. Use of these polar eluants tend to co-elute much of the co-extractive material. Several methods are available in the literature for the cleanup of *N*-methylcarbamates on Florisil®, but no standard procedure has been adopted and most investigators use widely varying conditions.

Alumina (Al₂O₃) has also been used widely as an adsorbent, but again no standard conditions have been adopted. Both activated and up to 10% deactivated alumina have been employed; basic alumina is not recommended as the carbamates may hydrolyze on this column. Alumina is more efficient than Florisil® or silica gel in removing organic acids, pigments, and other materials from water, although silica gel may be successful in eliminating pigments that are not removed by alumina or Florisil®.[454] Carbamates are eluted with chloroform or mixtures of hydrocarbons and other polar solvents. Separation of phenols from the parent carbamates can also be accomplished on alumina (see Section VIII.A).

Silica gel (SiO₂·xH₂O) is more inert than alumina and has been used to clean up extracts, separate pesticides by class, and for fractionation and cleanup of derivatives. For cleanup purposes, the carbamates have been eluted with hexane or ether. Although other adsorbents do not satisfactorily separate o.c., o.p., and carbamate pesticides, 20% deactivated silica gel has shown applicability in separating pesticides in multi-class, multi-residue studies.[278,455] A micro-column of Grade I deactivated silica gel is eluted sequentially with hexane, 60% benzene in hexane, and 5% acetonitrile in benzene; *N*-methylcarbamates are almost entirely eluted in fraction three. Micro-silica gel columns have also been used to clean up and/or fractionate chloroacetate,[337] dinitrophenyl-methylamine,[456] and pentafluorobenzyl[240,277,457] derivatives of *N*-methylcarbamates prior to GLC analysis.

Carbon is a nonpolar adsorbent which is widely used for column chromatography due its ability to remove organic and color matter from extracts. Use of Norit A® charcoal is preferred since low recoveries of carbamates were found with Nuchar® C-190;[378] however, the latter has been used for carbofuran cleanup in potatoes and some tissues.[329] Few investigators use the carbon directly; however, it is often incorporated with other adsorbents for cleanup purposes. Workers at Niagara Chemical Division of FMC Corporation[160,310,367,368] and others[369,378,458] have used various combinations of

Nuchar-Attaclay® with Florisil®, silica gel, silicic acid, or alumina to clean up carbofuran in a wide variety of substrates. The elution system consisted of methylene chloride followed by ethyl acetate-hexane mixtures. Other adsorbent mixtures such as Celite®-MgO-Norit® charcoal have been used.[148] Generally the various adsorbents are sandwiched together in one column. Carbon has also been directly added to decolorize sample extracts.

TLC has been used to clean up and separate pesticides and metabolites. Magallona[243] has discussed some separations of carbamates by TLC. Although carbamates can be separated by TLC (see Section XII.B), the applicability of this technique is usually limited to metabolic studies using radiolabeled compounds. Compounds may be easily isolated and subjected to subsequent analysis. Steam distillation, often in conjunction with hydrolysis, has been used to extract and clean up samples in one step prior to thiocarbamate and N-phenylcarbamate analysis (Tables 14 and 15).

A. Separation of the Parent Carbamates from Their Phenols

In such analytical schemes as derivatization of the carbamate phenols or direct GLC of the carbamates and determination of the chromatographed phenol, the metabolically formed carbamate phenols must first be separated from the parent compound. If this is not done, high residue results for the carbamate could occur since the phenolic metabolite would contribute to the GLC peak. If only a qualitative measure of carbamate exposure is required, then this separation would be unnecessary. Often the separation of the metabolites from the parent molecule also allows removal of co-extractives such as natural phenols which could interfere with the analysis.

As an alternative to pre-separation of the phenol, the extract may be split into two fractions; one would be hydrolyzed and derivatized while the other would be derivatized only.[314] After quantitation of both fractions, the amount of carbamate can be calculated by subtraction of the phenol portion from the total (hydrolyzed) value.

Johnson and Stansbury[335] discussed the separation of 1-naphthol from carbaryl by liquid-liquid partitioning. The organic solution containing both compounds is shaken with dilute aqueous NaOH. 1-Naphthol is transferred to the base layer as sodium naphthoxide and upon acidification the regenerated 1-naphthol may be extracted with methylene chloride. This procedure has gained wide acceptance for the ease with which pigments, phenolics, and other acidic compounds can be separated from the parent pesticide. Since some hydrolysis can occur during the partitioning with base, it is important to use a dilute base solution and to keep the contact time to a minimum. Typical alkaline solutions that have been used include 0.25 N NaOH and 0.1 M K_2CO_3. The latter being a weaker base is probably the better choice since it would minimize the chance of hydrolysis of the parent carbamates. With aminocarb and mexacarbate, their phenols are partitioned out of the organic solution with 2 N H_2SO_4 and then are re-extracted from the aqueous phase after neutralization with bicarbonate.

Bowman and Beroza demonstrated that carbofuran,[339] Mobam,[340] and methiocarb[364] could be separated from their more polar phenols by liquid chromatography on deactivated alumina. Extracts were eluted first with chloroform to obtain the carbamate and then with methanol or 2% acetic acid in chloroform to obtain the more polar phenols. Adsorption chromatography with Florisil® and various mixtures of n-hexane and anhydrous ethyl ether yielded separation of radiosynthesized N-methylcarbamates from their respective phenols.[27]

Free, naturally occurring phenols, as well as the carbamate phenols, can be eliminated from sample extracts by selective oxidation.[313] Oxidation by potassium permanganate, ferric chloride, potassium dichromate, nitric acid, hydrogen peroxide, ammonium persulfate, and ceric sulfate has been studied, and only ceric sulfate removed the phenols completely with no loss of the carbamates studied. Satisfactory results were obtained when 20% acetone was present in the aqueous phase. After oxidation with 50 mg ceric sulfate in 4 *N* H₂SO₄, for 15 min, the carbamates were extracted with chloroform for subsequent determination.

1. Typical Isolation of Carbamate Phenols (Coburn et al.,[277] Ripley[315])

Wash the combined methylene chloride (or organic extract) from Section VII.A.1.a with 75 to 100 m*l* of 0.1 *M* K₂CO₃. Transfer CH₂Cl₂ to round-bottom flask (drying with Na₂SO₄ if not proceeding to hydrolysis) and concentrate (Section VII.G). Acidify K₂CO₃ partitioning solution and extract phenols with anhydrous ethyl ether or methylene chloride (Section X.C and X.C.1). Neutralize with aqueous NaHCO₃ and extract with methylene chloride or benzene if amino phenols are required.

It is imperative that a dilute base solution be used for this partitioning and contact time should not exceed 5 min. Exceeding these conditions may result in hydrolysis of the parent *N*-methylcarbamate.

IX. GAS CHROMATOGRAPHY (GC)

Two themes pervade the literature concerning residue analysis of carbamate pesticides and in particular that of *N*-methylcarbamate pesticides. One opinion among analysts is that direct GC of *N*-methylcarbamate pesticides is difficult because of their tendency to break down to their corresponding phenols on GC columns. Alternatively, thermally stable derivatives of either the intact carbamate or the hydrolysis products may be prepared, but many workers feel that these additional steps make the residue procedure time consuming and concomitantly increases the chances of pesticide loss. Some brief introductory comments may help to clarify these opinions.

The on-column decomposition of *N*-methylcarbamates is well documented in the literature; however, much of the work was accomplished early in the history of carbamate analysis and many papers refer specifically to carbaryl. The development of better GLC column supports and liquid phases has allowed for direct determination of several carbamates. One of the most widely used application of direct GLC analysis is for carbofuran using a nitrogen detector. The main problem associated with direct GLC analysis of carbamate insecticides is that few compounds or their metabolites can be determined under the same GLC conditions and therefore multi-residue analysis is not readily applicable. Another drawback is the wide range of sensitivities exhibited by carbamates during direct GLC analysis. The statement that carbamates are not amenable to direct GLC analysis must be qualified: some *N*-methylcarbamates can be analyzed by this technique whereas the thiocarbamate herbicides are ideally suited to direct multi-residue GLC analysis, but the dithiocarbamate fungicides almost always decompose.

Chemical derivatization was once used primarily as a confirmatory technique for pesticide residues. Today this technique has several advantages for the analysis of carbamates including the improvement of thermal stability, increasing the sensitivity, and the possibility of multi-residue analysis. Some examples of the increase in sensitivity of carbamates after derivatization are shown in Figure 27; in most cases carbaryl cannot be determined by GC without derivatization to a stable compound. Derivatization is admittedly time consuming and sometimes tedious. The judicious choice of a derivatization procedure may shorten the analysis time if some of the preliminary cleanup and partitioning steps are incorporated into the derivatization procedure. While it is true that each additional step in the method increases the chances of loss of the compound, these steps may also allow for cleanup of sample backgrounds, provide additional qualitative confirmation of residue identity, and produce more stable compounds and hence more reproducible results.

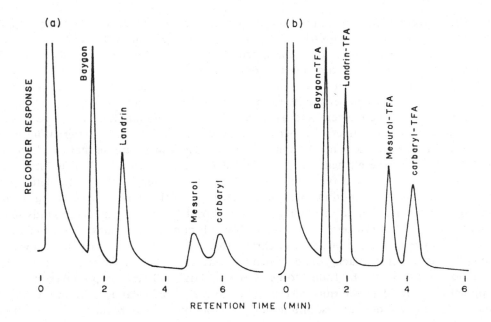

FIGURE 27. Gas chromatograms of a mixture of four *N*-methylcarbamates (a) before derivatization with trifluoroacetic anhydride, 200 ng each injected, and (b) after derivatization, 100 ng each injected. Operating parameters: column, 4-ft glass column packed with 3% SE-30; temperature, 180°C; gas flow, nitrogen carrier, 20 ml/min; AFID. (Reprinted with permission from Seiber, J. N., *J. Agric. Food Chem.*, 20, 443, 1972. Copyright 1972, American Chemical Society.)

The final selection of an analytical procedure lies with the analyst and unfortunately personal preference and prejudice often play an important role in the selection. Frequently a method that works well in one laboratory may not work well in another; although this is not a tremendous deficiency in carbamate methodologies, it does point out that no one method has yet been developed for carbamates that all analysts agree upon. Other considerations in selecting a method include available time for analysis and development of a method, the number of samples, required sensitivity, quality of results, confidence in results, reproducibility, ease of manipulation, and available chemicals and instrumentation. Much discussion still continues among analysts concerning the methods of analysis for carbamate pesticides and much of the discussion still pertains to the final method of detection.

A. Gas Chromatographic Detectors

The commonly used detectors in GC (Volume I, Chapter 2 and Volume II, Chapter 1) are electron capture (ECD),[459] flame ionization (FID),[460,461] alkali flame ionization (AFID),[462-464] flame photometric (FPD),[465] microcoulometric (MCD),[466,467] Coulson conductivity (CCD),[468,469] Hall electrolytic conductivity (Hall),[470] and various nitrogen-phosphorus detectors (N-P).[471-473] All these detectors have been employed for the detection of carbamates although each may exhibit some problems with selectivity or sensitivity. Details on the theory of operation, optimization techniques, and pesticide performance evaluations have been reported.[238,240,243,244,250,254,257,367,474-496]

These detectors may be classified as specific or nonspecific, and each type has a role in the quantitation and qualitation of pesticide residues; for example, a nonspecific detector (FID, ECD) may be employed to screen samples prior to analysis using a specific detector. With nonspecific detectors a rigorous cleanup of samples may be required prior to detection of residues. ECDs are widely used in pesticide laboratories, but they have limited application to direct GC of carbamates since few of these compounds contain electron capturing substituents; moreover, the wide variation in electron affinity characteristics of organonitrogen compounds makes screening at a selected sensitivity or detection level impractical with ECD.[492] On the other hand, following derivatization with electron capturing moieties, ECD is widely used and in fact it provides excellent sensitivity to most derivatives. Specific detectors are usually sensitive to only one element and as such less cleanup of samples may be required. Nevertheless, co-extractives or other pesticides containing the same heteroatom may cause spurious responses and caution must be exercised in assuming that the detector is "specific" for the element or compound(s) under investigation. Furthermore, an overabundance of certain types of compounds may interfere with specific detectors; excess phosphorus compounds may be detected on AFID or N-P detectors tuned to nitrogen, or cross-channel interference from phosphorus or sulfur compounds may affect specific FPD responses. Analysis of controls and fortified substrates often provides the analyst with a "background chromatogram"; with water, variations in water quality may affect the background. When one detector fails because of excessive backgrounds, another may be found to be effective. Trial and error approaches must be applied and consideration must also be given to the use of cleanup procedures, alternative detectors or derivatives, or alternative methodologies to obtain a good detection system for the pesticide and substrate being analyzed.

For the direct analysis of carbamates, nitrogen-specific detectors are widely used since all these compounds contain a nitrogen atom; however, when thermal degradation of *N*-methylcarbamates occurs the carbamate moiety is lost and hence the specificity of the detector to most of these compounds. Since FID responds to all carbon-containing compounds, discrimination between a *N*-methylcarbamate and its phenol might be difficult if they had similar retention times. The sulfur detector (FPD or conductivity detectors in the oxidative mode) has applicability to the thiocarbamates and other sulfur-containing compounds such as methiocarb, aldicarb, and methomyl; it is important to bear in mind that the FPD in the sulfur mode (394-nm filter) is nonlinear in response to sulfur compounds.[465,493] After derivatization many of the detectors may be used either nonspecifically, such as the ECD for electron-capturing derivatives, or specifically, such as the phosphorus detector for phosphoryl derivatives, the nitrogen detector for dinitrophenyl ether derivatives, or the electrolytic conductivity detector in the reductive (halogen) mode for halogenated derivatives. Confirmation of residues is usually achieved by analysis on several GC columns of different polarity and with different detectors; often two or more detectors may be used with the same instrument.[494,495]

Aue[496] has discussed the use and choice of GC detectors for the analysis of pesticides. These detectors may be changed chemically, modified, or tuned to achieve the desired specificity or sensitivity required in the analysis. Often tuning the detector for one compound may affect the response to another compound. One final comment as expounded by Westlake and Gunther[475] bears repeating:

The development of detectors responsive to less-than-microgram quantities had led to a revolution in pesticide determinations resulting in a wealth of data that would not otherwise have been obtained. Unfortunately, gas chromatography has also created more misinformation and misunderstanding than all other analytical procedures combined, primarily because operators without the training and experience necessary to evaluate their data intelligently and to recognize the pitfalls inherent in this outwardly simple method seized upon the gas chromatograph as a quick and easy way to analyze for pesticide residues, usually mistaking quantitation for determination. And those who are least qualified are, more often than not, the most certain of their results.

B. Direct GLC Determination of *N*-Methylcarbamates

GLC is by far the most commonly used analytical technique for the determination of pesticides. Since gas chromatographs are widely available in most laboratories analyzing for o.c. and o.p. pesticides, it was only natural to extend this powerful analytical tool to the analysis of carbamate pesticides. Practically, the easiest method of determining these compounds is by direct injection of the intact compound using conventional GLC parameters. Unfortunately, only some carbamates can be analyzed in this way due to their thermal instability. Direct GLC causes most *N*-methylcarbamates to decompose in various amounts to their respective phenols:

$$\text{Ph-O-}\overset{\overset{\text{O}}{\|}}{\text{C}}\text{-}\overset{\overset{\text{H}}{|}}{\text{N}}\text{-CH}_3 \overset{\Delta}{\longrightarrow} \text{Ph-OH + CH}_3\text{NCO}\uparrow \qquad (3)$$

As substitution to the nitrogen atom increases, such as with the *N,N*-dimethylcarbamates or the *N*-methyl-*N*-phenylcarbamates, the molecules are greatly stabilized so that direct GLC analysis of these intact carbamates is a general practice. Greatest thermal instability seems to occur with those *N*-methylcarbamates having few or no substituents on the phenyl ring (carbaryl, carbanolate) whereas those carbamates containing heteroatoms on the ring are more stable (carbofuran, mexacarbate).[497] Zielinski and Fishbein[498-500] studied the behavior of various substituted carbamates under different GLC conditions. They found that the presence of an aromatic moiety with *N*-substituted carbamates engendered the occurrence of degradation phenomena; however, lower GLC column temperatures minimized the thermal decomposition. Carbaryl was particularly prone to thermal cleavage to 1-naphthol. Typical percent breakdown of *N*-methylcarbamates on two columns at different temperatures is shown in Table 20.

The basis of GC separation is the distribution of a compound between two phases. In GLC, a sample is carried through the system by an inert carrier gas, and the components in the sample are separated on the GLC column due to their differential partitioning between the carrier gas and a nonvolatile solvent (stationary liquid phase) supported on an inert solid support.[501] GLC liquid phase coatings range from nonpolar (SE-30, OV-1, DC-200) through intermediate polarity (QF-1, OV-17), to polar (OV-225, Carbowax® 20M). Additional details on GLC stationary and liquid phases are available in most chromatographic supply catalogs. In general, the nonpolar liquid phases have less affinity for compounds and as such the compounds are eluted from the column quickly thus having a short retention time. Since the *N*-methylcarbamates are thermally unstable, the use of a short column and a nonpolar liquid phase allows these compounds a minimum of column residency time and thus minimizes the possibility of on-column breakdown.

For example, on a 4-ft, 3% OV-17 column at 180°C, a 4 to 66% breakdown of *N*-methylcarbamates to their phenols was observed.[497] On polar columns such as Carbowax® 20M, Apiezon N, and QF-1, most carbamates decomposed to their phenolic moieties.[502] Using a 3% OV-17 column (2 ft) at carefully controlled temperatures, certain carbamates such as aminocarb at a column temperature of 180 to 190°C can be analyzed without decomposition;[503] however, when these temperatures were exceeded decomposition occurred as indicated by multiple peaks.[504]

One of the most widely studied *N*-methylcarbamates has been carbaryl since its introduction in 1961 as an ''alternate type of pesticide'' to the o.c.s. Several investigators[27,505-509] have shown that carbaryl decomposes during direct GLC analysis. Riva and Carisano[510] were the first to determine carbaryl directly when they employed a short, 1-m column coated with 0.5 to 2% SE-30, resilanized *in situ* with HMDS operating at 160°C and an AFID. Lewis and Paris[511] were able to determine carbaryl directly using an ECD and a 0.3-m column coated with 3% SE-30 at an optimum column temperature of 145°C. Weyer[512] determined carbaryl in formulations directly on a 6-

Table 20

PERCENT BREAKDOWN OF SOME
CARBAMATES ON TWO GLC COLUMNS AND
AT THREE DIFFERENT TEMPERATURES[497]

	OV-17 (°C)			SE-30 + Carbowax® 20M (°C)		
Compound	180	190	205	190	205	215
Aminocarb	7.7	50.6	21.9	29.9	29.9	38.2
Bayer 42696	8.8	50.6	63.4	6.9	68.8	36.1
Bayer 50282	11.3	56.3	74.3	—	—	—
Carbanolate	35.3	57.7	68.9	—	—	—
Carbaryl	66.8	80.0	88.9	—	—	—
HRS 1422	4.8	8.2	8.5	38.8	11.5	12.3
HRS 9485	9.4	16.3	15.9	23.9	68.5	40.7
Methiocarb	66.1	90.6	88.3	—	—	—
3-Methylphenyl *N*-methylcarbamate	4.0	46.3	10.8	88.9	22.5	—
Mexacarbate	8.6	15.1	18.6	55.8	35.8	15.0
Phenyl *N*-methylcarbamate	—	—	—	70.8	—	—
Propoxur	13.5	21.9	17.7	26.1	44.2	41.2

Adapted from Magallona, E. D., *Residue Rev.*, 56, 1, 1975. With permission.

ft, 3% SE-30 column at 190°C using a FID with accuracy that compared to the saponification method;[50] although on-column thermal decomposition was not a problem, 1-naphthol was present in all chromatograms indicating that either technical carbaryl contained some 1-naphthol initially or a slight but reproducible decomposition was occurring. Lewis and Paris[511] found all high purity carbaryl samples contained 1-naphthol, and this fact has also been demonstrated through other means such as HPLC.[513]

Although some nonpolar columns are suitable for the GC of several *N*-methylcarbamates directly, often the column resolution is not sufficient to distinguish between other carbamates and/or their metabolites.[497] In one study with mexacarbate by direct GLC using a CCD, it was found that the three major breakdown products gave peaks with about the same retention time which could not be resolved from each other,[157] whereas Greenhalgh et al.[514] using a FPD (sulfur mode) found that while methiocarb could be determined by direct GLC, its sulfoxide and sulfone metabolites showed poor GC characteristics. As such, the direct GLC technique may be partially used as a qualitative check for some carbamates as a prelude to quantitative analysis by other techniques.

Upon direct injection of carbamates into a GC, the presence of a breakdown product may be indicated by peak tailing, low response, or the presence of several peaks in the chromatogram.[492] The phenols usually have a considerably shorter retention time on most columns; however their presence may not be detected with some detectors. Usually the identity of these extraneous peaks in the chromatogram can be determined by comparison with standards of the metabolites or breakdown products, or the GLC effluent may be analyzed by alternate spectral means. This confirmation must, however, be unequivocal. For example, during the analysis of mexacarbate, identical IR spectra were observed before and after injection of the compound into the GC, but on TLC analysis of the GLC effluent, a slight (<10%) decomposition to the phenol was evidenced.[505] Mass spectral analysis is convenient when coupled with GC,[508] although with electron impact the molecular ions are often weak and the mass spectrum may closely resemble that of the phenol (see Section XII.F).

It is possible to chromatograph some *N*-methylcarbamates by direct GLC if special care is taken in column preparation and maintenance. The columns successfully used have consisted of nonpolar liquid phases on silanized supports. Some studies have shown that polar columns may be used but often poor chromatography results with no advantage in separation.[367,492]

Carbofuran has successfully been chromatographed directly and work by FMC investigators[160,310,367] has elucidated some techniques that aid in the GC of carbamates. To prevent thermal or catalytic breakdown of compounds on metal, an inert GC system is essential. Metal transfer lines should be as short as possible and all-glass systems are best with quartz columns and injectors being superior to metal, Pyrex®, or borosilicate glass, while if necessary metal injection ports should be coated with L-31 silicone (a glassy polymer stable up to 400°C);[367] glass wool in the injector or on-column injection[512] may also reduce thermal decomposition.

Different liquid phases did not affect the direct GLC of carbofuran; however, the largest variable appeared to be the solid support, and special preparation of Chromosorb W is recommended.[367] Pre-silanized supports often did not provide sufficient deactivation to prevent decomposition of carbamates and it was found necessary to employ *in situ* silanization during column conditioning and periodically thereafter to restore column performance. A well-conditioned, deactivated column produces the best results during direct GLC of carbamates. Typically, DMCS, HMDS, TMCS, or Silyl 8® are good for silanization, and pesticides such as naled or mevinphos, or some crop extracts also appear to condition columns probably due to hydrogen bonding of active sites on the silica solid support. It may be pointed out that columns so treated with silylating agents may not be applicable for o.p. analysis with a FPD,[240,257] or with H-bonding type phases such as glycerols, Carbowax,®, or any phase which depends upon reactive hydrogens (such as −OH, −SH, −COOH, −NH₂ or −NH) for its function.[515]

A small piece of glass wool in the injection port provides a larger vaporizing surface and an easily replaceable trap for the removal of some crop volatiles. It is recommended that this glass wool be silanized and periodically replaced to maintain chromatographic performance.[160,492,516] A loose septum, of course, can also lead to low results or misidentification of pesticide identity due to variable retention times.

"Priming" or the injection of large quantities of compounds onto the column is often necessary to ensure equilibrium of the system prior to quantification of data.[160,367,511,517] A well-aged and used column is generally superior to a new column,[367,497,511,518] however, contamination due to extraneous material deposited in the system will cause degradation of some compounds.[492,502] Even those compounds that are shown to undergo minimal thermal degradation in a pure form (standard solutions) may show more severe degradation in a GLC system used for routine analysis of samples. Since the amount and quite often the type of co-extractives varies from sample to sample or day to day, the degree of degradation is unpredictable and requires constant monitoring of the GLC system by injection of standards between samples.[511,512] This degradation is one disadvantage of analyzing intact *N*-methylcarbamates particularly with automated injectors and data processors or integrators used in many laboratories.

Aue et al.[519] developed a highly deactivated column packing composed of Chromosorb W surface modified with Carbowax® 20M that allowed the analysis of thermally unstable compounds, such as the *N*-methylcarbamates. Lorah and Hemphill[516] used the "Aue" packing to determine carbaryl, promecarb, methiocarb, and mexacarbate by direct GLC; sharp, nontailing peaks, linear calibration curves, and good recov-

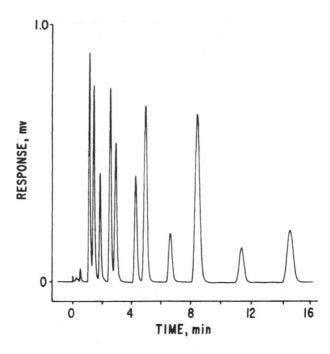

FIGURE 28.   Direct GC of a carbamate pesticide mixture on an Ultra-Bond® column. Operating parameters: column, 6 ft × 2 mm i.d. glass column packed with 0.65% OV-17 + 0.5% OV-210 on Ultra-Bond®; temperature, 170°C; gas flow, helium carrier, 25 ml/min; Hall electrolytic conductivity detector at sensitivity 10 × 8. Elution order of compounds: IPC, diallate, triallate, Meobal®, 3,4,5-Landrin®, carbofuran, mexacarbate, SWEP, dimetilan, methiocarb, and carbaryl. (Reprinted from Hall, R. C. and Harris, D. E., *J. Chromatogr.*, 169, 245, 1979. With permission.)

eries in fortified samples were obtained. The addition of a 1% solution of formic acid prior to direct GLC analysis of several carbamates on DC-200 suppressed decomposition and improved linearity;[520] however, carbaryl decomposed under these conditions.

Recently, Hall and Harris[521] examined the direct GC behavior of 32 carbamate pesticides on Carbowax® 20M surface-modified supports, with and without additional liquid phase coatings. Excellent results were obtained for 24 of the carbamates (Table 21) using several columns. Poor results were obtained with aldicarb, aldicarb sulfoxide, aldicarb sulfone, Karbutylate, methomyl, phenmedipham, and thiophanate. Pyramat exhibited some peak tailing, and some decomposition was indicated for carbaryl, methiocarb, and Meobal®; since carbaryl has a considerably longer retention time, this may have contributed to the approximate 50% degradation. Best performance with respect to peak symmetry and sensitivity was achieved on the Ultra-Bond®, 3% OV-101, 1% OV-17, and the mixed phase OV-210 + OV-17 columns (Figure 28). The other columns are still useful, however, for confirmatory multi-column analysis. In general, it appears that a moderate column temperature (185°C) should be employed and short retention times are beneficial. Hence, the Ultra-Bond® column without coating, or with a light load (1%) of polar or nonpolar stationary phase may be used for multi-residue determination of various carbamate pesticides by direct GLC.

Although many initial problems were encountered during the analysis of intact *N*-methylcarbamates, recent developments have indicated that under carefully controlled conditions these compounds may be determined intact. Retention time data for the direct analysis of intact *N*-methylcarbamates on a variety of columns are available

Table 21

RELATIVE GLC RETENTION INDICES FOR CARBAMATE PESTICIDES ON
ULTRA-BOND® WITH DIFFERENT LIQUID PHASES
(RRT, CARBOFURAN = 1.00)

				Column (170°C)		
Compound	Ultra-Bond®	3% OV-101	1% OV-17	1% Carbowax® 20M	1% OV-210	0.5% OV-210 + 0.65% OV-17
EPTC	0.87[a]	0.20	0.08	0.07	—	—
Butylate	1.00[a]	0.25	0.09	0.07	—	—
Pebulate	1.14[a]	0.25	0.12	0.09	—	—
Vernolate	1.24[a]	0.28	0.12	0.08	—	—
IPC	0.19[b]	0.31	0.19	0.22	—	0.22
Diallate	0.20[b]	0.67	0.31	0.21	0.32	0.28
Meobal	0.33	0.59	0.42	0.52	0.56	0.50
CDEC	0.34	0.66	0.40	0.30	0.40	0.37
Pyramat	0.35	0.62	0.43	0.29	0.39	0.36
Triallate	0.53	1.01	0.48	0.26	0.39	0.38
Propoxur	0.55	0.55	0.48	0.53	0.63	0.52
2,3,5-Landrin®	0.60	0.69	0.51	0.58	0.65	0.58
CIPC	0.61	0.66	0.45	0.59	0.56	0.55
Bufencarb	0.78	1.04	0.72	0.71	0.75	0.71
Terbutol	0.82	1.47	0.91	0.66	0.82	0.78
3,4,5-Landrin®	0.85	0.94	0.78	0.85	0.88	0.85
Benthiocarb	0.85	1.80	1.26	0.82	0.75	1.02
Aminocarb	0.93	1.07	0.89	0.95	1.02	0.94
Mexacarbate	0.98	1.32	0.98	0.94	1.02	0.96
Carbofuran	1.00	1.00	1.00	1.00	1.00	1.00
SWEP	1.36	1.19	0.97	1.47	1.19	1.28
Dimetilan	1.37	1.79	1.93	1.38	1.86	1.64
Methiocarb	2.10	2.25	2.13	2.28	1.96	2.20
Carbaryl	2.75	2.48	2.41	3.10	2.81	2.82

[a] Column temperature = 120°C, relative retention time to butalate.
[b] Column temperature = 150°C.

Adapted from Hall, R. C. and Harris, D.E., *J. Chromatogr.*, 169, 245, 1979. With permission.

(Tables 21 and 22).[492,497,502,516,521] It should be pointed out, however, that without extensive cleanup of samples (see Section VIII), many other compounds such as o.c. and o.p. pesticides, herbicides, and also many co-extractives may present a problem with peak identification.

Some prediction of retention time of other carbamates may be made based on empirical observation of known carbamate behavior.[497] For example, longer retention times were observed on OV-17 with thio derivatives than with ring N-substituted methylcarbamates, and the absence of a methyl group on the five position of the phenyl ring tended to decrease retention time relative to 5-methyl compounds; on SE-30 or Carbowax® 20M columns, oxygen substitution resulted in shorter retention times relative to ring N-substituted compounds.

Another problem associated with many of these discussed procedures for determining the N-methylcarbamates by direct GLC is that they pertain only to the determination of standards, and in fact, few are actually used to analyze samples for residues. Thus the problems of degradation are generally not emphasized since co-extractives were not building up in the system from repeated routine analysis. Carbofuran is the general exception with residues being reported in crops,[160,368,370,371] soils,[109,148,319] and water[148,517] using either a Coulson, MCD, or ECD.

Table 22
GLC DATA FOR CARBAMATE PESTICIDES ON 10% DC-200 COLUMN AT 180°C. RELATIVE RETENTION TIMES (RRT, PARATHION = 1.00) AND RESPONSE TO COULSON ELECTROLYTIC CONDUCTIVITY DETECTOR (N-MODE)

Compound	RRT	ng for 50% FSD
Aldicarb	<0.05	1[a]
Aldicarb sulfoxide	<0.05	1[a]
Aldicarb sulfone	<0.05	1[a]
EPTC	0.13	10
Butylate	0.17	10
Vernolate	0.17	10
Pebulate	0.18	20
Isolan	0.21	20
Molinate	0.23	20
Propoxur	0.27	20
Cycloate	0.31	20
CIPC	0.31	25
Landrin® (2,3,5)	0.32	50
Diallate	0.37, 0.39	50
CDEC	0.37	25
Methomyl	0.38	ca 200[b]
Carbofuran	0.39	35
Landrin® (3,4,5)	0.40	25
Landrin® (2 isomers)	0.32, <u>0.40</u>	50[c]
Carbanolate	0.41	35
Aminocarb	0.45	20
Bufencarb	0.35, 0.43, <u>0.47</u>, 0.53	50
Mexacarbate	0.56	25
Triallate	0.62	50
3-Hydroxycarbofuran	0.67	100
Carbaryl	0.72	75
Mobam	0.72	100
Methiocarb	0.85	100
Ziram	Footnote d	—

[a]   These compounds elute too quickly under the described conditions. They may be chromatographed using a column temperature of 140°C.

[b]   Peak tailing occurs.

[c]   Minor peak is about 25% of major peak. Major peak RT is underscored.

[d]   No response to a 200-ng injection.

Adapted from Laski, R. R. and Watts, R. R., *J. Assoc. Off. Anal. Chem.*, 56, 328, 1973. With permission.

In summary, for the direct analysis of *N*-methylcarbamates by GLC, it is recommended that a short column with a relatively nonpolar liquid phase such as 3% OV-17 or 5% DC-200 on a high-performance Chromosorb W previously conditioned with a silylating agent like Silyl 8® and primed with the carbamates in question, or the Ultra-Bond® series of column packings, be tested first. For general application, detectors such as FID, AFID, conductivity or N-P should be considered depending on availability and the objective of the analysis. Since carbaryl is the most susceptible carbamate to thermal decomposition it may be used as a test compound for column performance.[492] Fused silica capillary columns with SP-2100 or SE-54 phases are showing a lot of promise for the direct analysis of N-methylcarbamates and many other pesticides in a multi-residue type screening although little data are yet in the literature.

C. Direct GLC of Carbamate Phenols

In general, phenols chromatograph poorly[507] with the more polar phenols tending to tail badly during direct GLC analysis.[522,523] Some monohydroxy phenols have been shown to tail on several GLC columns, and the direct aqueous injection for GLC analysis of water for trace quantities of phenols is limited by the relative insensitivity and also the nonselectivity of FID.[524] The more sensitive ECD is not generally applicable unless the phenols are substituted with strong electron capturing moieties, such as halogens or nitro groups. Most methods that sucessfully determine N-methylcarbamates directly as their phenol rely on detection of a heteroatom using selective detectors. Direct GC of the phenols is not a general practice; however, after derivatization to their ethers, these representative molecules are widely determined (Section X.B).

As discussed in Section IX.B, direct GLC analysis of N-methylcarbamates often results in thermal decomposition (Table 20) of the molecule to methylisocyanate and the respective phenol (Equation 3). The phenols tend to elute quicker than the intact carbamate, and often in the studies of direct GLC of carbamates, the phenols were observed.[27,498,502,505,509]

Bowman and Beroza[364] determined methiocarb and five of its metabolites (all containing a sulfur atom) in crops as their respective phenols using a FPD in the sulfur mode (Figure 29). Extracts were first separated by liquid-column chromatography on silica gel to obtain the sulfide, sulfoxide, and sulfone, then further fractionated into carbamate and phenol portions; the carbamates were hydrolyzed to their phenols and all fractions were analyzed on DC-200. Recoveries were about 80% and sensitivities were about 0.01 ppm. Boyack[362] suggested that carbanolate, as its phenol, could be determined in crops using a FID.

Bowman and Beroza[525] described an apparatus that combined a GC containing a DC-200 column with a spectrophotofluorometer by means of a flowing liquid interface. Excitation and emission wavelengths were set to determine specific compounds such as 1-naphthol after injection of carbaryl in the GC. Bache and Lisk[526] described a detector that measures sensitive sulfur or chlorine emission after excitation of an eluted compound in a low-pressure, microwave-powered helium plasma. Carbanolate and methiocarb after alkaline hydrolysis have been determined in fortified apple extracts using this detector.[363]

The chemical changes in a compound during GC is referred to as "reaction gas chromatography"[527] (see Section X.D). The thermal degradation of N-methylcarbamates during GLC is one example of this process. Another example is the use of a reaction zone in the injection port of a GC to facilitate this breakdown. Ebing[509] injected concentrated ammonium hydroxide together with a solution of the carbamate in CS₂ and determined methiocarb as its phenol using FID. Beroza and Bowman[340] demonstrated that a phosphoric acid plug placed at the front of a GLC column could quantitatively hydrolyze the injected compound to its phenol. Mobam and its hydrolysis product could be determined in grass and milk using this method. The reaction appears to be practically instantaneous since the liberated sulfur-containing phenol migrated through the column with no appreciable difference in retention time from that of the phenol similarly injected.

In summary, direct GLC analysis of carbamate phenols has only limited application to general routine analysis; analysis of phenols or carbamates hydrolyzed to their phenols are more conveniently analyzed by chemical derivatization. The FID which is responsive to all the phenols is nonselective and generally not sensitive enough for residue applications. Specific detectors are useful for those phenols containing a heteroatom

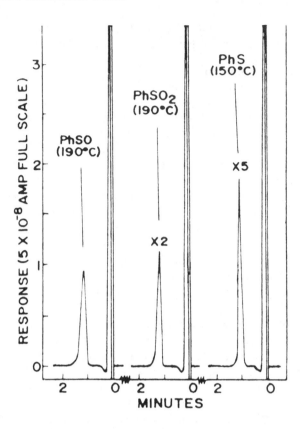

FIGURE 29.   Gas chromatograms of the three methiocarb phenol metabolite standards (25 ng injected in 25 μ*l* chloroform). Operating parameters: column, 70 cm × 4 mm i.d. glass column packed with 10% DC-200; FPD (sulfur mode); column temperatures as shown. Peak identity: PhSO-methiocarb sulfoxide phenol: PhSO$_2$-methiocarb sulphone phenol; and PhS-methiocarb phenol. (Reprinted from Bowman, M. C. and Beroza, M., *J. Assoc. Off. Anal. Chem.*, 52, 1054, 1969. With permission.)

although this aspect restricts multi-residue application. Other less common detection systems may be used, but these applications would appear limited to degradation or disappearance studies of specific compounds.

D. Direct GLC of Other Carbamates

The statement that "carbamates" may or may not be gas chromatographed directly must be qualified. As discussed above, the *N*-methylcarbamates as a whole are generally not amenable to direct GLC analysis; on the other hand, a method was published as early as 1961[429] on the determination of ethyl *N,N*-di-*n*-propylthiocarbamate (EPTC) residues in soil using GLC.[238] The statement must be further qualified as to whether the carbamate is eluted from the GC intact (same compound as injected) or whether a fragment is eluted and detected. Many situations occur wherein a representative species is eluted which may be qualitated and quantitated as the parent molecule. From a practical point of view, this may be acceptable provided the conversion is quantitative, since the same premise is used in many other analytical techniques (derivatization, colorimetry, polarography, etc.). Compounds like dicofol (an o.c.) also decomposes on GLC columns and may cause confusion in qualitation,[528,529] yet GC remains the most common method of analysis.[254] Alternatively, parent pesticides and their metabolites may be converted chemically to one representative compound for quantitation by GC.[530] In this manner, a "total" residue determination may be made and also this technique may facilitate obtaining residue data on some metabolites that might not be chromatographable or resolvable on some GLC columns.

Systematic studies on the GC of carbamate derivatives by Zielinski and Fishbein[498-500] showed that N,N-dimethyl-, N-phenyl-, and N-alkylcarbamates were thermally stable relative to phenyl N-alkylcarbamates. Amine hydrogens of urethane derivatives were found to be important molecular sites influencing separation in GC. Empirical studies of N-methyl and N,N-dimethylcarbamates[505,509,531] confirmed that the increased N-substitution stabilized the molecule, and hence direct GLC of the N,N-dimethylcarbamates was possible and practical. Retention time data for dimetilan, isolan, and Pyrolan® were determined by Eberle and Gunther;[505] these N,N-dimethylcarbamates elute intact and separate from their faster eluting enols. The recommended procedure for pirimicarb, a more important N,N-dimethylcarbamate today, and its metabolites is by direct GLC.[532]

The N-phenylcarbamates (IPC, CIPC) were not originally analyzed by direct GLC because N-phenylcarbamates, particularly those with electronegative groups on the ring, are prone to molecular rupture at GLC analytical temperatures;[500] in this study,[500] both IPC and CIPC were chromatographed although it was not specified if they eluted intact and secondly, compounds such as CIPC were found to lack sensitivity for direct electron affinity residue determination without prior cleanup and concentration.[423] Thus, these compounds were determined via hydrolysis/colorimetry or chemical derivatization in many early studies (Table 14).

In later years, Peck and Harkiss[508] showed that both IPC and CIPC were eluted intact from four GLC columns; identical mass spectra were obtained before and after GC. Retention time data for these compounds have also been reported by Laski and Watts[492] using a Coulson (nitrogen mode) detector (Table 22).

Oxime carbamates (methomyl, aldicarb, thiofanox, oxamyl) and their oxidative metabolites are prone to thermal decomposition to their oximes[210] or nitrile[190] analogs which are hard to resolve[393] during direct GLC analysis despite the low analytical temperatures (100 to 140°C) employed:

$$
R_1-\underset{\underset{R_2}{|}}{C}=N-O-\underset{\underset{H}{|}}{\overset{\overset{O}{||}}{C}}-N-CH_3 \quad
\begin{cases}
\triangle \;\rightarrow\; R_1-\underset{\underset{R_2}{|}}{C}=N-OH \quad \text{oxime} \\
\triangle \;\text{or} \\
\;\searrow\; R_1-C\equiv N \quad \text{nitrile}
\end{cases}
$$

Reeves and Woodham[406] described a procedure for the direct GLC determination of methomyl and its sulfoxide and sulfone although the latter two compounds may not be important in methomyl metabolism (Section V); the column used was 10% DC-200 and detection was by FPD in the sulfur mode. Williams[533] determined methomyl directly on a OV-101/OV-210 column using a nitrogen-specific MCD. No confirmation of the eluted species was provided in these two studies. Braun et al.[210] reported that under direct GLC analysis, methomyl and its oxime eluted at identical retention times. On-column conversion may be more critical at low residue concentrations.

Direct determination of methomyl is generally unsatisfactory since the methomyl peak tends to tail badly under isothermal GLC conditions and this leads to reduced sensitivites.[534] Fung[404] reported that methomyl oxime was three times as sensitive as the parent compound under identical GLC conditions. Originally, methomyl methods, described by Pease and Kirkland,[401] involved chemical hydrolysis of methomyl to its oxime prior to GLC analysis:

$$CH_3-S-\overset{\overset{\displaystyle CH_3}{|}}{C}=N-O-\overset{\overset{\displaystyle O}{||}}{C}-\overset{\overset{\displaystyle H}{|}}{N}-CH_3 \xrightarrow{\ \ OH^-\ \ } CH_3-S-\overset{\overset{\displaystyle CH_3}{|}}{C}=N-OH$$

Analysis of the oxime was by temperature-programmed GLC on 10% FFAP and detection was by microcoulometric-sulfur titration. Fung,[404] using this hydrolysis procedure, FPD(S), and GC-MS, confirmed that methomyl oxime was eluted from the column.

Residues of methomyl have been reported using direct analysis[210,535] and by the hydrolysis procedure.[403,536-538] One common feature was reported for both procedures: the GLC system required priming with concentrated standards to equilibrate the column. Furthermore, interferences were found with some substrates using specific sulfur detection although the use of other detectors such as N-specific[533] or AFID[539] alleviated the problem and facilitated analysis. Oxamyl is also analyzed by hydrolysis to its oxime for GLC analysis.[412]

Knaak et al.[190] reported that aldicarb and its metabolites, several of which are toxicologically important, were not thermally stable during direct GLC analysis: aldicarb formed 2-methyl-2-(methylthio)propionitrile, the sulfone and sulfone oxime formed 2-methyl-2(methylsulfonyl)propionitrile, and the sulfoxide and sulfoxide oxime formed methacrylonitrile. Maitlen et al.[393] analyzed aldicarb and its metabolites directly by GLC using a FPD(S), however, the aldicarb peak because of its short retention time, could not be resolved from the solvent front on several columns and under different GLC conditions. It was postulated that extensive pyrolysis of aldicarb occurred since the "aldicarb" peak had a shorter retention time than 2-methyl-2-(methylthio)-propionitrile. Since aldicarb sulfone is relatively stable and has a longer retention time than the other compounds, analysis was based on peracetic acid oxidation of the aldicarb to the sulfone after liquid chromatographic fractionation of the metabolites on Florisil®; the sulfoxide could be determined directly or after oxidation to the sulfone. Individual metabolite concentrations could be determined by subtraction of before and after oxidation data. Variations in the oxidation,[395] cleanup, and fractionation procedures[163,394,395] have been reported for the determination of individual metabolites. Carey and Helrich[395] note that the aldicarb sulfone is detected as the thermal degradation product—sulfone nitrile.

Subsequently, Maitlen et al.[396] described a rapid procedure to determine "total" aldicarb residues for the three toxic compounds; the whole extract is oxidized with peracetic acid to aldicarb sulfone and aldicarb sulfone oxime which are then separated on Florisil® followed by GLC analysis of the sulfones by FPD-sulfur detection. Total aldicarb residues have been reported in a variety of substrates.[397-400,540]

Analogous to aldicarb, thiofanox shows similar chemistry and metabolizes to several toxicologically important compounds. Chin et al.[411] described a procedure that allows determination of total thiofanox or some individual metabolites following column chromatography on Florisil® and oxidation with peracetic acid; aldicarb does not interfere because of its relative insensitivity and different GLC retention time.

In general, oxime carbamates may be determined by direct GLC under carefully controlled, low-temperature column conditions. Total residue results are obtainable after oxidation of metabolites to the sulfone although even this compound may not elute intact from the GC; usually the eluted fragment is sufficiently reproducible to facilitate analysis. Priming the column aids in reducing thermal decomposition.

Thiocarbamate herbicides are an excellent example of carbamate compounds that may be analyzed directly by GLC as first shown by Hughes and Freed.[429] Many papers report on the analysis of individual thiocarbamates;[35,36,95,434-436,439,492,541-544] however, these compounds are ideally suited to multi-residue monitoring.[425,440] All the thiocarbamates are quickly and easily resolved and analyzed isothermally by direct GLC (Figure 30). Screening of water samples has been accomplished using this technique.[100,101]

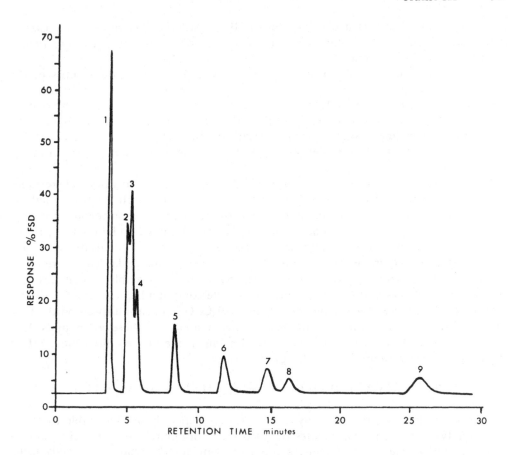

FIGURE 30. Gas chromatogram of a thiocarbamate mixture. Peak identity: (1) EPTC, (2) butylate, (3) vernolate, (4) pebulate, (5) cycloate, (6) molinate, (7 and 8) diallate, and (9) triallate. Column, 6 ft × 4 mm i.d. packed with 3.6% OV-101 + 5.5% OV-210 on 80 to 100 mesh Chromosorb W, AW, DMCS treated; temperature, 140°C; carrier gas flow (nitrogen), 60 ml/min; FPD, sulfur mode.

Generally all these compounds are detected with a FPD (sulfur mode) although ECD or N-detectors (AFID, conductivity, N-P) may be used with less sensitivity and selectivity. GLC analysis of some thiocarbamate metabolites may require other detectors.[95,211]

Ethylenebisdithio and dithiocarbamate fungicides are nonvolatile and as such are not amenable to direct GLC analysis (Section XI.C). Zielinski and Fishbein[545] injected these compounds directly onto a 4% QF-1 column at 180°C and found thermal decomposition to ethylenethiourea (ETU) and minor amounts of ethylenethiuram monosulfide and other components; quantitation was based on detection of ETU using a FID. Laski and Watts[492] reported no nitrogen response to 200 ng of ziram on 10% DC-200. Various fragments of these fungicides such as carbon disulfide or hydrogen sulfide have been analyzed by GLC[546-548] as may be derivatives of parent compounds[549] or their hydrolysis products.[550] Colorimetric determination of evolved carbon disulfide remains the method of choice for the analysis of these compounds (see Section XII.E.1).

In general, only the *N,N*-dimethyl, *N*-phenyl, and thiocarbamate pesticides are routinely analyzed by direct GLC. Under carefully controlled conditions, other carbamates may be analyzed in this manner; however, the eluting species is often a pyrolytic fragment. Oxidation of aldicarb or thiofanox and their metabolites to the corresponding sulfone avoids problems of analyzing many different compounds and reduces the number of compounds that pyrolyze during GLC. Various detectors may be used to detect these compounds depending on the sensitivity and selectivity required as well as the presence of heteroatoms.

GC-MS (see Section XII.F) is a facile technique in determining the identity of the GC eluting peak(s). Care must be taken in interpretation of the mass spectrum since compounds like the *N*-methylcarbamates have almost identical EI spectra to their phenols. Moreover, in confirming residues in samples, authentic standards must be compared; for example, single-ion monitoring for the molecular ion might cause no response for the pesticide as in the case of oxime carbamates where the oxime or nitrile is actually eluted.

GC is generally recognized as the best method of analysis for many pesticide residues because of its sensitivity and its value as a means of identification.[551] Unfortunately, GC is not always possible for many types of compounds; some carbamates fall into this area. Although other carbamates may be chromatographed under controlled conditions, routine analytical laboratories may not always be able to dedicate and maintain a gas chromatograph ideally suited for only these compounds. Therefore other techniques must be considered. For example, the thermally unstable carbamates may be chemically derivatized to more stable or volatile compounds prior to GC, or they may be analyzed at ambient temperatures by HPLC. One must not dwell on the failures of direct GC, but rather accept the challenge in analyzing these compounds. It is not impossible to analyze carbamate pesticides, but it may take a little more effort than experienced with some other classes of pesticides.

X. DERIVATIZATION OF *N*-METHYLCARBAMATES

As discussed above, the *N*-methylcarbamates are generally not amenable to direct GLC analysis. Not only carbamates but also such compounds as fatty acids, amino acids, carbohydrates, steroids, and some drugs can not the analyzed directly using conventional GLC parameters. Much effort and ingenuity has gone into the preparation of thermally stable and volatile derivatives of these compounds. Derivative chemistry may be defined as the preparation of a chemical derivative selected or specifically designed to improve the chromatographic analysis.[552]

Many pesticide residue analysts consider derivatization to be an unnecessary step in an analytical methodology that adds considerable time and possible losses to the procedure. Reasons for derivatizing *N*-methylcarbamates, apart from their thermolability during direct GLC analysis, include increasing detector sensitivity (particularly with the ECD), increasing volatility, obtaining better chromatographic separations, extending a procedure to multi-residue analysis, use for confirmatory purposes, and enhancing stability of the compounds. Derivatization is also a convenient way to increase the molecular weight of a compound for mass spectrometry. For example, derivatization of methylamine with fluorodinitrobenzene increases the molecular weight from 31 to 265; similarly, the amount of material available for detection is increased.[430]

No one derivative or derivatization procedure is perfect for all uses and may be recommended as a panacea against the failure of direct GC; however, any of the described methods may provide a solution to the specific problem an analyst may have. Often methods do not work well on an interlaboratory or even intersubstrate basis but when a procedure has been found that works well, it should become an essential feature in the analytical arsenal. Derivatives and the reactions by which they are formed thus become extremely important considerations in any chromatographic procedure. Derivatization ranks equally with sample selection, preprocessing, and chromatography as an important part of the overall analytical technique.[553]

Derivatization reactions must not be thought of as capricious per se, since many factors may dictate the response to a specific derivatization procedure. A procedure may in fact be synthetically sound and work on a preparative scale; however, during the micro-reactions involved in residue analyses many factors may contribute to failure. Among these factors are choice of glassware and apparatus; reaction conditions; stoichiometry; reagent choice, purity, and quality; use of catalysts; chemical stability of the derivative, and, direct manipulative losses. Often the reaction conditions may be too harsh, causing hydrolysis of the carbamate or the derivative. Theoretically, a procedure that derivatizes, for example, one *N*-methylcarbamate should be applicable to all *N*-methylcarbamates; however, for the above reasons it may not be unless modified and/or optimized. Alternatively, certain derivatives may be unacceptable because of sensitivity, specificity, volatility, separation on GLC columns, or excess reagent interfering with detectors. With ECD, care must be taken to remove excess electron capturing reagents and solvents which may produce unacceptable chromatographic results. Furthermore, insufficient details are provided on some methods, or optimization of the procedure has not been accomplished. For these reasons, many derivatization procedures, based on sound theoretical principles, either fail entirely or are used with several changes in methodology between laboratories.

A derivatization reaction may be designed to protect or reduce the polarity of a specific functional group, to introduce an electron capturing moiety, or to increase the volatility of a specific compound. However, few reagents or procedures are specific enough to derivatize only one type of functional group; attempts to derivatize the carbamate NH may result also in derivatization of a hydroxyl group in the molecule. For example, during acetylation of ring hydroxyl carbaryl or 3-hydroxy carbofuran a *N*-acyl *O*-acyl derivative may be obtained; dinitrophenylation (DNP) of a hydrolysis mixture of *N*-methylcarbamates may produce DNP-methylamine and/or the DNP ether of the phenol. Many procedures have been developed for one specific compound or metabolite and little discussion is provided on what other derivative might be formed. The following discussions describe those methods primarily developed for one specific derivative formation but where applicable secondary derivatives will be mentioned; additional details on these other derivatives are provided in the pertinent sections.

For the analysis of *N*-methylcarbamates by derivatization, there are two general approaches, namely, a derivatization of the intact pesticide and derivatization of a hydrolysis product, one of which is always the volatile methylamine. Excellent reviews on derivatization (primarily of pesticides) have previously been published.[238,242,243,250,554-557] Confirmation of pesticides via chemical derivatization is discussed by Chau in Volume I, Chapter 4.

A. Derivatives of Intact *N*-Methylcarbamates

Since the *N,N*-dimethylcarbamates are thermally stable, efforts were made to prepare the dimethyl analogs of the corresponding *N*-methylcarbamates prior to GC. The proton bonded to the carbamate nitrogen atom is slightly acidic, and methylation using methyl iodide or dimethyl sulfate must be carried out under basic conditions. Under these conditions, Mobam was found to first hydrolyze to 4-hydroxybenzothiophene (4-HBT) prior to methylation:[558]

Greenhalgh and Kovacicova[559] showed that the carbamate NH moiety is very reactive to base catalyzed methylation using NaH-CH₃I-DMSO under mild conditions:

$$R-\overset{\overset{H}{|}}{N}-CH_3 \xrightarrow{\text{base}} R-\overset{\ominus}{N}-CH_3 \xrightarrow{CH_3I} R-\overset{\overset{CH_3}{|}}{N}-CH_3$$

This procedure was investigated for the confirmation of two *N*-methylcarbamates (methiocarb and turbucarb) and a few other NH-containing pesticides.[560] The order of reactivity of the methylation reaction appears to be carbamates > phosphorami-dates-amides > phosphoroamidothioates > triazines > ureas. An alternative approach to methylation of intact *N*-methylcarbamates and other carbamates and ureas containing NH moieties is by on-column methylation which is discussed in Section X.D.

Silylation of many diverse chemical functional groups such as hydroxyl, amine, and amide are well known and the use of silylating reagents to derivatize many biological substances abound in the literature. Pierce[561] has published an excellent reference book on silylation.

Fishbein and Zielinski[562] chromatographed intact *N*-methylcarbamates as their trimethylsilyl (TMS) derivatives. Standards were reacted in pyridine with HMDS and TMS:

$$R-O-\overset{\overset{O}{\|}}{C}-\overset{\overset{H}{|}}{N}-CH_3 + \underset{\text{HMDS}}{(CH_3)_3SiNHSi(CH_3)_3} \xrightarrow{\text{TMS}} R-O-\overset{\overset{O}{\|}}{C}-\overset{\overset{Si(CH_3)_3}{|}}{N}-CH + NH_3$$

TMS derivatives are less polar than their corresponding parent molecule and as such shorter GLC retention times are found on nonpolar columns. Good peak symmetry of the derivatives were obtained, but unfortunately the retention time of TMS derivatives were very close to the parent compounds. Mexacarbate, when chromatographed as its TMS derivative showed two peaks, presumably the parent and phenol derivative. Ciba-10573 has been chromatographed as the TMS derivative,[363] but attempts to silylate Mobam were unsuccessful.[558]

If ECD is to be used, the presence of halogens in the silylating reagents, such as chloromethyldimethylsilyl or bromomethyldimethylsilyl (Figure 31) derivatives, would enhance detector response and increase retention time. However, Bache et al.[563] could not prepare these silyl derivatives of several intact carbamates; *O*-CMDMS derivatives have been prepared.[564]

Mathur et al.[565] showed that *N*-thiomethyl and *N*-thio-*p*-tolyl derivatives of carbaryl could be prepared; *N*-sulfenylated derivatives of carbamates had previously been investigated toxicologically.[566] The presence of the sulfur atom in the derivative allowed quantitation by FPD (sulfur mode) which showed some selectivity to background interferences. Unfortunately, derivatization of fortified crop with *p*-tolylsulfenyl chloride resulted in a background that was too severe to allow quantitation even after treatment of the final extract with HgCl₂-CaCO₃. *N*-thiomethylcarbaryl could be quantitated after the mercury treatment at levels above 5 ppm in crops. Extraction and cleanup was by Holden's method[379] and supplementary cleanup might allow these derivatives to be used at a lower residue concentration.

Since most *N*-methylcarbamates do not respond well to ECD, formation of a halogenated derivative is an excellent method of increasing the detectability of nonresponding compounds. Ralls and Cortes[356] brominated carbaryl with Br₂-CCl₄ in a sealed tube but a mixture of brominated derivatives was produced:

$(CH_2Cl)(CH_3)_2SiCl$ $(CH_2Br)(CH_3)_2SiCl$ $\left[(CH_2Cl)(CH_3)_2Si\right]_2NH$

A B C

FIGURE 31. Some halogenated silylating reagents.[515] (A) chloromethyldimethylchlorosilane, (B) bromomethyldimethylchlorosilane, and (C) 1,3-bis(chloromethyl)tetramethyldisilazane. (Courtesy of Pierce Chemical Company, Rockford, Ill., 1979 to 1980.)

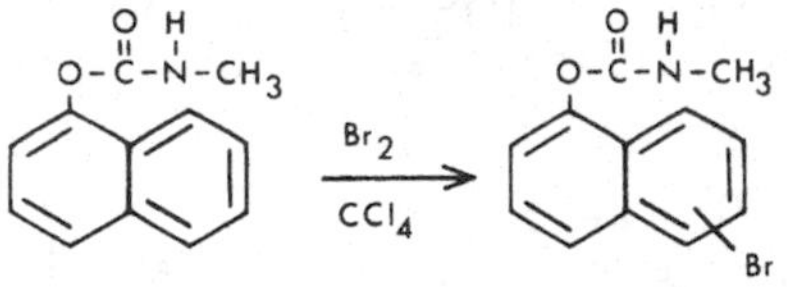

mixed products

Although good sensitivity was obtained, the formation of the mixed bromination products in varying yields and also the large proportion of derivatized sample extract results in this procedure being poor as a quantitative method although it may be used as screening method.

Acylation involves the introduction of an acyl group

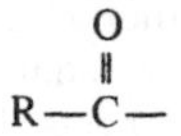

$$\underset{R-C-}{\overset{\overset{\textstyle O}{\|}}{}}$$

into a molecule by substitution of one of the replaceable (active) hydrogen atoms in alcohols, phenols, amines, and amides. A typical reaction is shown in Figure 32. *N*-acetyl-*N*-methyl and *N*-acyl-*N*-methylcarbamates have been synthesized and tested for their pesticidal properties.[42,567-571] For analytical purposes, acetyl derivatives have been prepared[558,572] for use with FID, but most analysts prefer to introduce electron capturing moieties to be used with the more selective and sensitive ECD. In general, halogens are preferred for increasing ECD response and sensitivity in the order $F < NO_2 \approx Cl < Br < I$.[573] Some increase in sensitivity also results as the number of electron capturing groups increase particularly with some halogen-specific detectors. Fluoro-compounds tend to be more volatile than the other halogen derivatives, but the GLC retention time of perfluoro derivatives tends to increase as the size of the acyl group increases. Typical acylating reagents are shown in Figure 33.

Pyridine catalyzed reaction with chloroacetic anhydride did not acylate the amide nitrogen of carbaryl,[564] although the trichloroacetyl derivative could be prepared from the anhydride.[574] *N*-chloroacetyl (Cl = 0,1,2,3) carbaryl was prepared using the appropriate acid chloride and pyridine catalyst.[574] In a GC study of these derivatives,[575] it was found that with the possible exception of *N*-acetyl carbaryl, these derivatives underwent thermal decomposition; the *N*-nitroso (C-NO) derivative also was thermally unstable.

Lau and Marxmiller[361] were able to chromatograph both isomers of Landrin® as their trifluoroacetyl (TFA) derivatives. Ethyl acetate was found to be the best reaction solvent and reaction at rt was preferred since high-temperature reaction increased the background. This procedure was also applied to carbofuran and its metabolites.[328]

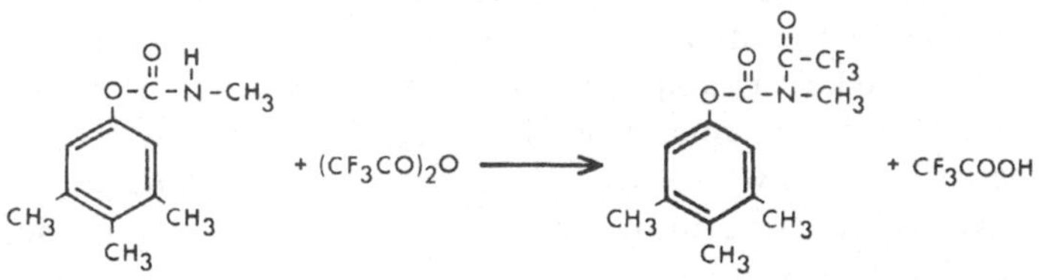

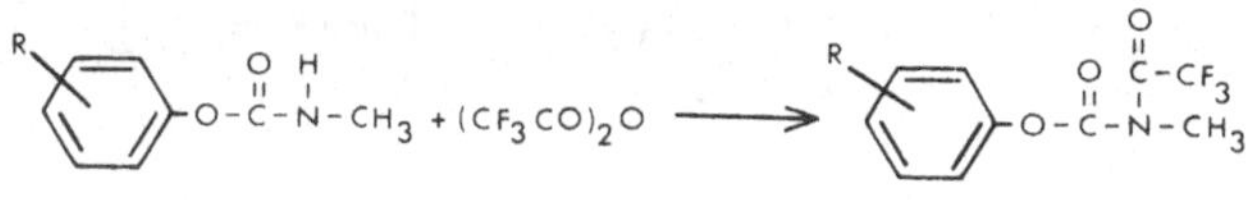

FIGURE 32.   *N*-acylation of a typical *N*-methylcarbamate.

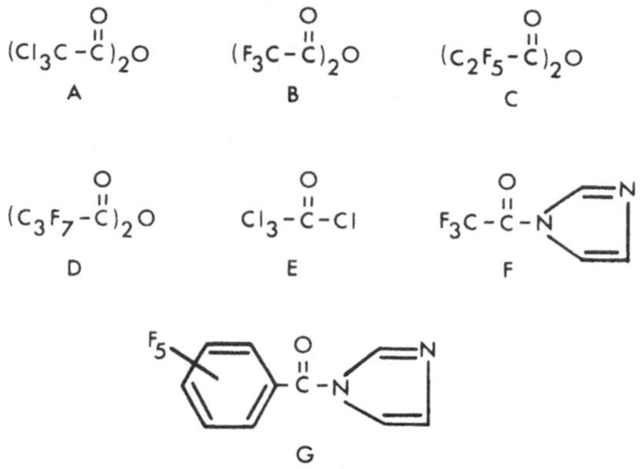

FIGURE 33.   Some acylation reagents. (A) trichloroacetic anhydride, (B) trifluoroacetic anhydride, (C) pentafluoropropionic anhydride, (D) heptafluorobutyric anhydride, (E) trichloroacetyl chloride, (F) *N*-trifluoroacetylimidazole, and (G) *N*-pentafluorobenzoylimidazole.

Greenhalgh et al.[365,514] demonstrated that methiocarb and its oxidation and hydrolysis products could be trifluoroacetylated (Figure 34). Using a selective sulfur detector, these (S-containing) products could then be analyzed at the 0.1 ppm level in virtually crude extracts of blueberries; for EC detection, additional cleanup would be required. Methiocarb sulfoxide and its phenol showed anomalous reaction with a di-TFA derivative being formed. A subsequent study[576] confirmed that compounds with a sulfoxide moiety (methiocarb, aldicarb) undergo a Pummerer reaction which involves rearrangement of the acylated sulfoxide to an α-acyloxymethyl sulfide (Figure 35). Under mild reaction conditions (30°C; 15 min) the trifluoroacetoxymethyl sulfide analogs were obtained, whereas more vigorous reaction conditions (100°C; 30 min) resulted in the di-TFA derivatives. This procedure may also be used for confirmatory purposes.[560,576]

Seiber[577] examined such *N*-perfluoro derivatives as TFA, pentafluoropropionic (PFP), and heptafluorobutyric (HFB) of several intact *N*-methylcarbamates. These acylating reagents, as their corresponding anhydrides, produced derivatives that had similar rates of formation, stability, spectral properties, and GLC retention times to the TFA analog. Only a two- to five-fold increase in electron capture response was found between the TFA and the HFB derivative. Reaction solvent polarity and temperature were found to govern the rate of formation of the derivatives. Trifluoroacetylation of eight *N*-methylcarbamates was demonstrated; Figures 27 and 36 show electron capture composite chromatogram of some TFA derivatives.

Lawrence et al.[375,477,478,578] also examined several perfluoro derivatives of intact *N*-methylcarbamates and compared their responses on electrolytic conductivity (ElCD) and ECDs. Perfluoro compounds were 10 to 100 times more sensitive to ECD than to the ElCD; however, since larger sample injections are possible with the latter, sensitivities in samples are about the same. The ElCD response increases as the number of fluorines increases (Table 23) whereas ECD response is dependent on the compound investigated.[477] 3-Hydroxy carbofuran which forms the di-perfluoro derivative had

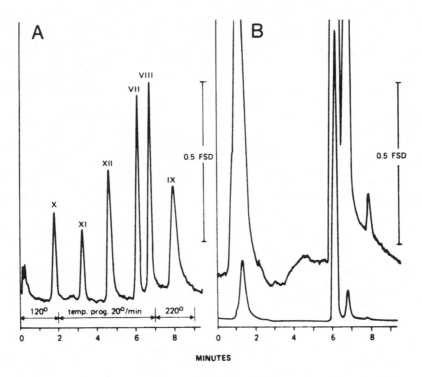

FIGURE 34. Gas chromatogram of the TFA derivatives of methiocarb and its metabolites, and a derivatized extract of field-treated blueberries. (A) Methiocarb phenol TFA (X) 3 ng, methiocarb sulfoxide phenol TFA (XI) 8 ng, methiocarb sulfone phenol TFA (XII) 6 ng, methiocarb TFA (VII) 5 ng, methiocarb sulfoxide di-TFA (VIII) 10 ng, and methiocarb sulfone TFA (IX) 12 ng; (B) acetone extract of field-treated blueberries after partitioning and derivatization. Operating parameters: column, 2 ft x ¼ in. o.d. packed with 5% DC-200; temperatures as shown; gas flow, nitrogen carrier, 50 m*l*/min; FPD (sulfur mode). (Reprinted with permission from Greenhalgh, R., Marshall, W. D., and King, R. R., *J. Agric. Food Chem.*, 24, 266, 1976. Copyright 1976, American Chemical Society.)

twice the response with ElCD and 2 to 3 times the response with ECD compared to the monoderivatized carbofuran or 3-keto carbofuran. With carbofuran, response was much greater to the derivative using ElCD in the halogen mode than to the parent compound during direct analysis (*N*-mode) which was less reproducible and also required considerable column conditioning. HFB derivatives were found to be 2 to 4 times more sensitive than their equivalent TFA derivatives, and the HFB and PFO (perfluorooctanoic) derivatives were found to be more stable and could be stored for several weeks prior to analysis. The choice of derivative may depend on available detectors, sensitivity required, and the volatility and peak shape obtained with a specific GLC system. Furthermore, crop matrices may produce varying backgrounds and windows may be found for only some compounds under certain conditions (Figure 37). Reaction of perfluoroacyl anhydrides proceeded much faster with the phenols than with the carbamates, but trimethylamine catalyzed the direct acylation. HFB derivatives were found applicable for 14 of 16 carbamates investigated (Figure 38); only aldicarb and benomyl did not react. Analysis of carbofuran and its two metabolites was demonstrated in several crops.

Shafik et al.[579] described a derivatization procedure using pentafluoropropionic anhydride to obtain stable and sensitive derivatives of intact *N*-methylcarbamates. This

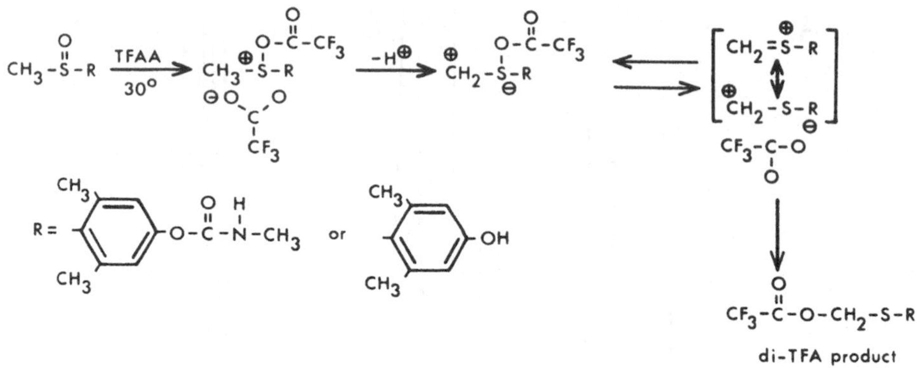

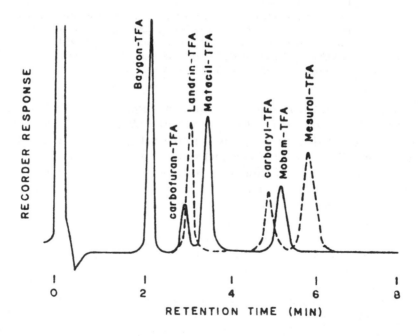

A

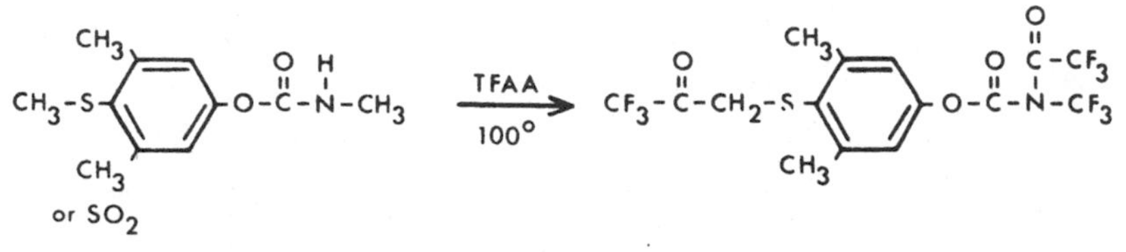

B

FIGURE 35.   Trifluoroacetylation reactions of methiocarb and its sulfoxide and sulfone metabolites. (A) Mechanism of the trifluoroacetylation of the methyl/aryl sulfoxide moiety. (B) Formation of the di-TFA derivative of methiocarb (or its sulfone) at elevated temperatures.[514,557,576]

FIGURE 36.   Composite gas chromatogram of seven methylcarbamate TFA derivatives, 2 ng each injected on 6% SE-30 column at 230°C and tritium ECD. (Reprinted with permission from Seiber, J. N., *J. Agric. Food Chem.*, 20, 443, 1972. Copyright 1972, American Chemical Society.)

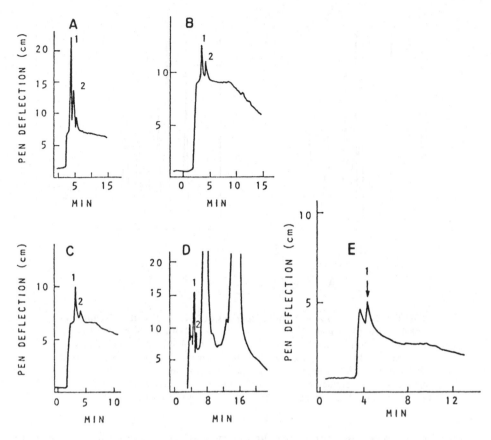

FIGURE 37. Chromatograms of carbofuran (1) and 3-keto carbofuran (2) as their N-HFB derivative at the 0.1 ppm level. (A) potato, 30-mg sample injected, 4 ×. (B) wheat, 14-mg sample injected, 8 ×. (C) corn, 20-mg sample injected, 8×. (D) turnip, 40-mg sample injected, 8 ×. Chromatogram of 3-hydroxy carbofuran (1, E) as N-HFB derivative in turnip at 0.1 ppm level. (E) turnip, 4-mg sample injected after cleanup. Column: 3% OV-1; temperature, 160°C; Coulson electrolytic conductivity detector in the halogen mode. (Adapted from Lawrence, J. F., Lewis, D. A., and McLeod, H. A., *J. Chromatogr.*, 138, 143, 1977. With permission.)

Table 23
COMPARISON OF RETENTION TIMES AND SENSITIVITIES OF SOME PERFLUORO DERIVATIVES OF CARBOFURAN

| | | Electrolytic conductivity detection | | | | ECD | | | |
| | | | Sensitivity | | | | Sensitivity | | |
Compound	Derivative	Retention time (min)	PH[a]	PA[b]	Temp. (°C)	Retention time (min)	PH[a]	PA[b]	Temp. (°C)
Lindane	—	2.4	1.3	0.29	160	2.2	83	16.7	115
Carbofuran	TFA	1.9	0.55	0.09	160	3.6	6	0.9	120
	PFP	1.8	1.0	0.17	160	3.4	15	2.3	120
	HFB	1.9	1.4	0.21	160	4.0	24	4.4	120
	PFO	3.9	2.0	0.51	160	3.7	71	11.0	140

[a] PH, peak height (in.) per nanogram (of parent compound), corrected to 32 X (ECD) and 4 X (electrolytic conductivity).

[b] PA, relative peak area per nanogram (of parent compound), corrected to 32 X (ECD) and 4 X (electrolytic conductivity).

Adapted from Ryan, J. J. and Lawrence, J. F., *J. Chromatogr.*, 135, 117, 1977. With permission.

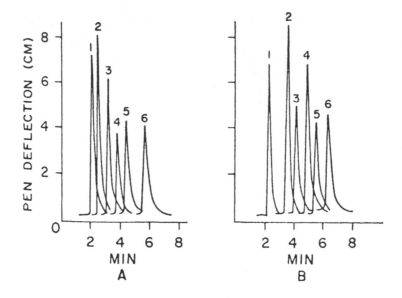

FIGURE 38.   Composite gas chromatograms of N-HFB *N*-methylcarbamates. Peak identity: (A) 1 methomyl, 2, propoxur, 3, carbofuran, 4, aminocarb, 5, mexacarbate, 6, Mobam. (B) 1, Landrin® and Meobal®, 2, carbanolate and bufencarb, 3, 3-hydroxy carbofuran, 4, butacarb, 5, carbaryl, 6, methiocarb. (Only aldicarb and benomyl did not give a response). Column, 3% OV-1; Coulson electrolytic conductivity detector in the halogen mode. (Reprinted from Lawrence, J. F., *J. Chromatogr.*, 123, 287, 1976. With permission.)

procedure has been recommended for the determination of these compounds in air[580] and in blood serum and fat samples.[240]

Chapman and Robinson[581] described a "simplified" method for the determination of residues of carbofuran and its metabolites using GC-mass fragmentography; unfortunately, few laboratories have a GC-mass spectrometer with chemical ionization[582] (MS-CI) available for routine analysis of samples (see Section XII.F). HFB derivatives were prepared using the Shafik et al.[579] method because that of Seiber[577] failed. During MS-CI several fragments were obtained, but the HFB-methylamine moiety (m/e 228) was found to be common to all three compounds and the intensity of the ion was directly proportional to the amount of material. Therefore, by single-ion monitoring, quantitation was possible at levels above 0.05 ppm in crops. The use of HFB derivatives has the advantage of producing high-molecular-weight fragments.

Carbaryl readily formed the TFA and HFB derivative at rt using either the anhydride or imidazole reagent (Figure 33), and pyridine catalyst.[564] The imidazole acylating reagents may be considered a cleaner reagent since it does not release free acid into the reaction mixture and thus requires no further cleanup; the reaction by-product is inert imidazole.[515] Most anhydride reactions require a large excess of reagent for quantitative derivatization, and hence excess anhydride and any free acid by-product must be removed. Even trace amounts of pyridine, the anhydride or acid in the sample extract may result in large tailing solvent fronts on the chromatogram. Partitioning of the extract-reaction mixture with water to hydrolyze the anhydride, or aqueous ammonia to extract the acid is often necessary prior to GLC. Delicate compounds may also be hydrolyzed.

As mentioned, pyridine acts as a catalyst in many of the reactions;[564,573] as in most acylation reactions, pyridine neutralizes the released acid and also shifts the equilibrium of the reaction to favor formation of the product.[583] Other compounds that have shown catalytic ability[584] for the acylation of *N*-methylcarbamates include trimethylamine,[478,515] methanesulfonic acid,[572,573] sulfuric acid,[567,568] and glass beads.[558] At rt, preparative scale trifluoroacetylation of carbofuran using stoichiometric portions of the anhydride yielded only 50% of the derivative after 18 hr.[584] Under these conditions, the dissociation of the acid conjugate of carbofuran was assumed to be rate limiting.

carbofuran – TFA   acid conjugate

The addition of a threefold excess of pyridine to carbofuran substantially affected both the rate and extent of the reaction. It appears therefore that the pyridine acts as both the catalyst and as a competitive base with respect to the formation of the trifluoroacetic acid conjugates. Using acetic anhydride and pyridine to acetylate carbaryl, 30 to 35% 1-naphthyl acetate was found in addition to the *N*-acetyl derivative; using methanesulfonic acid as the catalyst, quantitative yields of *N*-acetyl *N*-methyl-1-naphthyl carbamate were obtained.[572]

Perfluoro derivatives have been used to determine carbamates in crops,[365,372,373] water,[87,312] soil,[111] and fish and animal tissues.[87,328] Mass spectral fragmentation patterns of some perfluoro derivatives have been elucidated.[577,581,584-586]

1. Typical Acylation of Intact N-Methylcarbamates (Lawrence et al.[375,578])

To the intact *N*-methylcarbamates or an extract that has been evaporated just to dryness in a test tube add 80 $\mu\ell$ of heptafluorobutyric anhydride and 2.0 mℓ trimethylamine (0.025 *M* in benzene). Stopper and shake gently. Allow to react for 30 to 60 min at rt or at 60°C until derivatization is complete. Add 3 × 10 mℓ water, shaking each time and discarding the aqueous phase which contains the excess reagent. Use an aliquot for GLC examination. If extract is to be stored, dry benzene through Na$_2$SO$_4$ and keep in a tightly stoppered tube.

B. Derivatives of Hydrolysis Products of N-Methylcarbamates

The one reaction that is most characteristic of the *N*-methylcarbamates is that of hydrolysis.

As opposed to derivatizing the intact carbamate molecule (Section X.A), many methods have been reported that involve chemical modification of the hydrolysis products of carbamates; both the phenolic and methylamine moieties have been derivatized. In general, the procedure involves chemical hydrolysis (Section X.C) of the parent molecule followed by the derivatization step.

Hydrolysis of *N*-methylcarbamates can occur either chemically or biologically and as discussed previously (Section V), the corresponding phenol may be one of the major degradation products of these carbamates in environmental samples. Ring hydroxylation is another metabolic process whereby a phenolic compound may be formed and these metabolites may also be derivatized by some of these methods. Phenols are ubiquitous in nature and sample extracts after hydrolysis often contain many phenolic compounds in addition to the carbamate phenols. One major problem encountered with derivatization of carbamate phenols is that much of the co-extractive material is also derivatized unless suitable cleanup procedures are incorporated in the methodology. Some metabolites, such as *N*-hydroxymethylcarbamates, also produce the same phenol as the parent carbamate after hydrolysis:

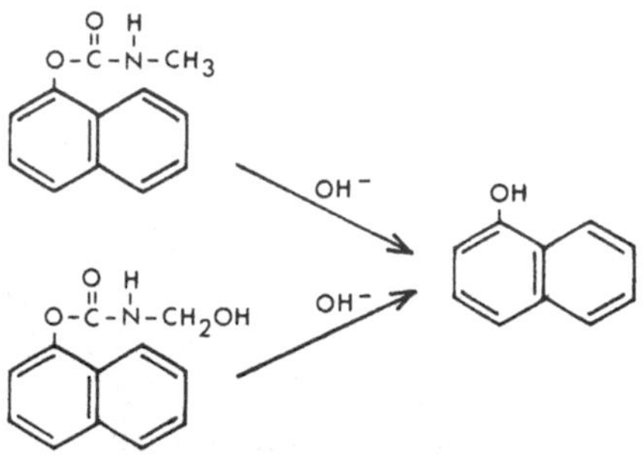

These considerations attest to the need for good methodologies, not only for the derivatization step itself, but also for the pre-GLC separation or cleanup of sample extracts. Nevertheless, many derivatization methods for carbamate phenols pervade the literature and it is not surprising that many of these schemes parallel those for the intact *N*-methylcarbamates. Many of the procedures and reagents are similar, and one must bear in mind that the following discussion focuses on *O*-derivatization although both *O*- and *N*-derivatives may be formed by the same reaction (see Section X.A).

1. Derivatization of Carbamate Phenols

Introduction of bromine to insensitive electron capturing compounds having some degree of unsaturation has been applied to carbamate phenols. Carbaryl was hydrolyzed, brominated, and esterified in a single step to yield brominated 1-naphthyl acetate,[316] which eluted as a single peak unlike the products obtained with bromination of the intact carbaryl.[356]

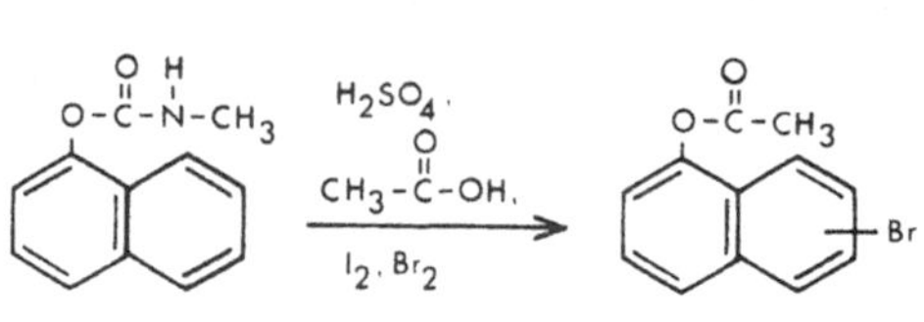

Although this method did not work for carbofuran and Mobam,[378] residues of carbaryl in cotton have been determined.[355] Several carbamates were hydrolyzed to their corresponding phenols and then brominated (but not acetylated) to derivatives which were highly sensitive to ECD.[357] For carbaryl, the derivative was identified as 2,4-dibromo-1-naphthol.

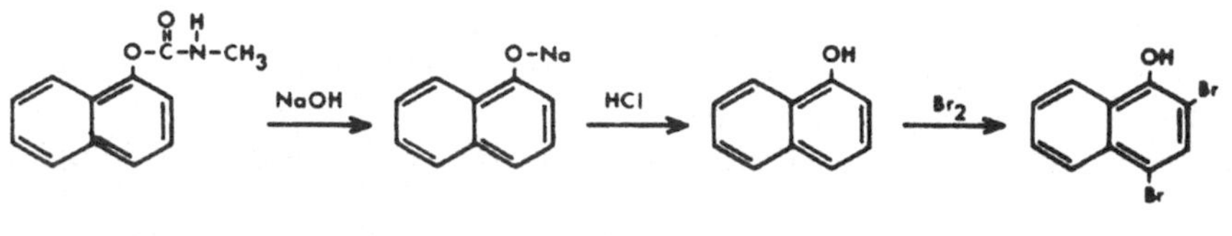

Although they could not form the *N*-halosilyl derivatives with the intact *N*-methyl-carbamates, Bache et al.[563] found they could prepare the *O*-chloro and the *O*-bromo-methyl dimethylsilyl derivatives of the phenols representing the metabolites of several carbamates.

mexacarbate phenol   CMDMS

1-Naphthol and several naphthalenediols were well separated by GLC and could be quantitated as their CMDMS derivatives.[564]

Bowman and Beroza[339] coupled dimethyl chlorothiophosphate in the presence of pyridine with the phenol of carbofuran. The dimethylthiophosphoryl derivative was determined by FPD.

Eight other phenols, derived from *N*-methylcarbamates, were also derivatized and were found to chromatograph well (Figure 39). These derivatives were felt to be useful since they could be determined by FPD in the phosphorus or sulfur mode and also by ECD. High sensitivities at the nanogram level were obtained and the choice of several detectors allows the analyst the option of using the detector that requires the least cleanup of extracts. Poor yields, however, were found with less acidic phenols and excess reagent produced large background currents with a thermionic detector.[313]

Moye[587] prepared esters of sulfonyl chlorides (sulfonates) by the reaction of 1-naphthol with halogenated benzenesulfonyl chlorides in a Schotten-Baumann type reaction.

a sulfonate

2,5-Dichloro-, 3,4-dichloro-, *p*-bromo-, and pentafluorobenzenesulfonates of 1-naphthol were prepared. Intact carbamates could also be reacted with a similar product being formed indicating that either hydrolysis followed by coupling or a displacement type reaction occurred. Aqueous buffer-acetone speeded up the reaction and also provided solubility for the excess sulfonyl chloride. The aqueous buffer provides a choice of optimum reaction conditions; at pH 12 conversion was complete in 5 min, but a rapid hydrolysis of the sulfonate then occurred, whereas at pH 8 a longer reaction time was required, but hydrolysis was not as pronounced. The sulfonates were thermally stable and sensitive to FPD (sulfur) or EC detection. Cleanup of sample interferences with dilute NaOH caused hydrolysis and low recoveries. Eight intact *N*-meth-

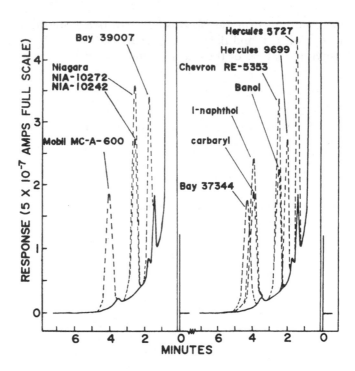

FIGURE 39.   Chromatogram of dimethylthiophosphoryl derivatives of nine carbamates and two phenols [NIA-10272 (carbofuran phenol) and 1-naphthol] obtained after FPD in the phosphorus mode. Operating parameters: column, 120 cm × 4 mm i.d. glass column packed with 10% DC-200; temperature, 190°C; gas flow, nitrogen carrier, 160 mℓ/min. Dotted lines are chromatograms of the compounds; solid lines are chromatograms of the reagent blank. Compound identities, NIA-10242 (carbofuran), BAY-37344 (methiocarb), and Chevron RE-5353 (bufencarb). (Reprinted from Bowman, M. C. and Beroza, M., *J. Assoc. Off. Anal. Chem.*, 50, 926, 1967. With permission.)

ylcarbamates reacted and responded well, however, aldicarb had a low response and no response was observed with methomyl (Figure 40).

As with the intact *N*-methylcarbamates, acylation reagents (Figure 33) are often employed to derivatize carbamate phenols. Many of the same considerations of reagent, catalyst, reaction conditions, cleanup, and so on, are also applicable.

Butler and McDonough[378] trichloroacetylated (TCA) the phenols of carbaryl, carbofuran, and Mobam to produce derivatives that could be successfully used to determine residues in crops at detection levels of 0.01 to 0.1 ppm using EC detection.

$$Ar-OH \ + \ Cl_3C-\overset{O}{\overset{\|}{C}}-Cl \ \xrightarrow{\ \Delta\ } \ Ar-O-\overset{O}{\overset{\|}{C}}-CCl_3$$

Pyridine was used as a catalyst and the reaction mixture in methylene chloride was heated at 100°C until the solvent evaporated. The method was evaluated against the bromination-acetylation scheme of Gutenmann and Lisk[316] with good agreement between the methods; however, the EC response of the TCA derivative was four times as great as that of the brominated derivative.[336] Since 3-hydroxy and 3-keto carbofuran metabolites show significant toxicity, it is important that they be included in any carbofuran analysis; trichloroacetylation was also extended to these compounds by Butler and McDonough.[320] It appeared that the phenolic and not the alcoholic hydroxyl of

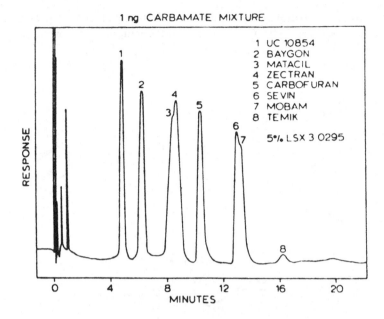

FIGURE 40. Chromatogram of 1 ng each of seven carbamates and two carbamoyl oximes after derivatization with 2,5-dichlorobenzenesulfonyl chloride. Methomyl (not shown) gave no response. Operating parameters: column, 6 ft × ¼ in. o.d. × 2 mm i.d. packed with 5% LSX-3-0295; temperature, 220°C; gas flow, nitrogen carrier, 60 ml/min; tritium ECD. (Reprinted with permission from Moye, H. A., *J. Agric. Food Chem.*, 23, 415, 1975. Copyright 1975, American Chemical Society.)

3-hydroxy carbofuran is derivatized since no product was obtained when derivatization of the intact carbofuran metabolite was attempted.

Lawrence and Ryan[478] formed a di-perfluoro derivative with 3-hydroxy carbofuran using HFBA.

One problem that is often attributed to this procedure is the need to use freshly distilled trichloroacetic chloride. Today, however, many chromatographic suppliers sell high-purity derivatization reagents and solvents, often in small quantities, that alleviate many of these types of problems.

Propoxur and its metabolites, *o*-hydroxy and *N*-hydroxymethyl propoxur have been determined in plant tissue,[219] and animal tissue and milk[220] as their trichloroacetyl derivatives. Propoxur and *N*-hydroxymethyl propoxur were separated by solvent partitioning, and then were base hydrolyzed to their phenol (*o*-isopropoxy phenol) and derivatized with trichloroacetyl chloride at rt (Figures 25 and 26). The ring hydroxyl of the intact *o*-hydroxypropoxur metabolite was derivatized in a similar manner but without the hydrolysis step. The *O*-glycosides and *O*-glucuronides were first released by acid hydrolysis.

Landowne and Lipsky[588] observed that of the mono-, di-, and tri-haloacetate derivatives of cholesterol, the chloroacetyl derivative was the most sensitive. Argauer[338,589] developed a method for the determination of carbanolate in crops based on hydrolysis and chloroacetylation of the phenol; nine other carbamates were also derivatized. The phenol was reacted with chloroacetic anhydride in dilute NaOH and benzene at rt for 3 min. Although standards showed good sensitivities but variable response characteristics and carbanolate was satisfactorily recovered from crops, the results with carbaryl suggested that possible modifications should be incorporated to clean up sample extracts that cause interferences and reduce the limit of detection.

Under these reaction conditions, the procedure was not suitable for the formation of the trifluoro- and trichloroacetyl derivatives. The perfluorobutyl derivative of 1-naphthol prepared in the pyridine-benzene solution in microgram amounts was unstable when water was introduced.

Shafik et al.[337] derivatized 1-naphthol in urine samples using chloroacetic anhydride, pyridine as a catalyst, and reaction at rt for 10 min. Cleanup of derivatized urine extracts by silica gel adsorption chromatography was used to determine 1-naphthol at less than 1 ppm using ECD. This method is also detailed in the EPA pesticide residue manual.[257] Propoxur was determined in air samples as the chloroacetyl derivative of the hydrolyzed carbamate;[590] mass spectral characteristics of the derivative were given.

In a similar manner, Khalifa and Mumma[564] showed that the TFA and HFB anhydrides could be used to derivatize both the intact carbaryl molecule as well as its aglycone metabolites (Section V):

N-hydroxymethyl carbaryl was hydrolyzed under the derivatization conditions and formed the perfluoro 1-naphthol derivative. TFA and HFB derivatives of 1-naphthol and naphthalenediols gave poor GLC separation, but their chloromethyldimethylsilyl derivatives gave good separation of these metabolites. Trifluoroacetylation of carbofuran-7-phenol showed poor reproducibility and sensitivity.[372]

Acetic anhydride was used to acetylate hydrolyzed carbanolate in crops prior to microcoulometric detection.[362] Paulson and Portnoy[591] showed that acetic anhydride and methanesulfonic acid could be used to remove the sulfate ester conjugate from a variety of compounds and convert them to their acetyl derivative in one step; recoveries were about 50% and detection was by FID.

This procedure was used to identify the metabolites of carbaryl in chicken urine.[592] Mass spectral characterization of some metabolites was shown. Monochloroacetylation has been used to detect N-methylcarbamates in grains and legumes.[593]

While perhaloacyl derivatives have been used to determine phenolic and hydroxy metabolites of N-methylcarbamates, this derivatization procedure is more widely accepted for the intact N-methylcarbamates (Section X.A). Trace phenols have been determined as their HFB derivatives,[594] which have been shown to possess the best combination of detector response and volatility (Table 23).[477,595] The fact that under the same reaction conditions both the N- and O-derivatives are formed and they may both be determined complements this technique. However, as shown by Archer,[372] these derivatives may not be applicable to all metabolites found in real samples. This also points out that no one method may be a panacea for the analysis of carbamates or even only one compound and its metabolites. One derivative may allow determination of some metabolites while a second derivative may be required for determination of some other residues.[564]

Reinheimer et al.[596] showed that phenols could be reacted with 2,4-dinitrofluorobenzene (FDNB) to form dinitrophenyl ethers (DNPE) and this principle has been employed to determine phenols in water using GLC.[597] Cohen et al.[313] demonstrated that some carbamates could be determined as the 2,4-dinitrophenyl derivative of their phenol moiety using EC detection:

The carbamate phenol was reacted with FDNB in acetone and a buffer to produce the sensitive derivative. Yields varied markedly with changes in the volume of reagent and pH of the buffer. Phosphate buffer at pH 11.0 gave good and consistent results; at this pH amines do not react with FDNB (see Section X.B.2). Carbamates could be determined in water at the 0.005 ppm level and in plant material at 0.1 ppm concentrations. Free phenols were removed by selective oxidation (Section VIII.A).

Holden[379] described a method for the determination of 13 *N*-methylcarbamates as their 2,4-dinitrophenyl ethers (Figure 41) in crops. Using EC detection, residues could be determined at levels as low as 0.05 ppm. This method is generally suitable for the determination of aromatic *N*-methylcarbamates that do not contain an aminophenyl group; aminocarb and mexacarbate were not recovered in the cleanup. Three dimethylcarbamates and aldicarb also did not form satisfactory derivatives. Carbamates were shaken with KOH and FDNB, than 5% borax was added and the mixture was heated on a steam bath. This method has been studied collaboratively[380] and has been adopted for propoxur, Landrin®, carbanolate, and carbaryl as the AOAC standard method.[255]

FDNB was also recommended for determining carbamates in a multi-class, multi-residue method for pesticides in water;[278] the intact carbamates were reacted at a carefully controlled pH of 9.4 with FDNB. A dinitrophenylation method using FDNB has been discussed for the determination of carbofuran in soil.[321] Residues of carbofuran,[110,598] Landrin®,[114] and carbaryl[83] have been determined in soil using this derivative. The carbamate acaricide Promacyl and two of its metabolites in tissue and milk of cattle have also been determined as their DNPE derivatives of the hydrolysis products.[599]

Dinitrophenylation with FDNB requires simple reaction conditions, results in stable derivatives, may be used in multi-residue applications, and the derivatives demonstrate good chromatographic behavior. The presence of the two nitro groups make these derivatives detectable by either nitrogen-specific or ECDs.[329,380] Another advantage of this method is that the dinitrophenyl ether derivatives produce GLC peaks which elute significantly later than those resulting from reagent impurities or other contaminants (Figure 41)[278,379] and also many of the other derivatives of carbamate phenols (Table 24).

In a study of the phenolic metabolites of carbofuran,[329] quantitative conversion of carbofuran-7-phenol and the 3-keto-7-phenol metabolite to their DNPE was achieved but derivatization of the 3-hydroxy-7-phenol metabolite was not adequate. Satisfactory derivatization of the 3-hydroxy-7-phenol to a 7-mono-substituted DNPE could only be achieved after "protecting" the 3-position. This was accomplished by reacting the 3-hydroxy-7-phenol with ethanol to selectively form the 3-ethoxy-7-phenol compound prior to dinitrophenylation; 3-hydroxy carbofuran was also ethoxylated as an aid in separation of the carbamate from the phenols resulting from loss of the carbamate group.

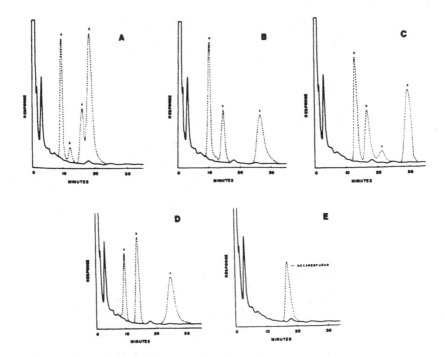

FIGURE 41.   Response (ECD) to carbamates recovered from kale at 0.05 ppm (broken lines) superimposed on chromatogram from unfortified kale sample (solid line) after derivatization with fluoro 2,4-dinitrobenzene. Operating conditions: column, 18 in. × ¼ in. o.d. glass column packed with 10% DC-200; temperatures, column 212°C, detector 218°C; gas flow, nitrogen carrier, 60 mℓ/min; Tritium ECD. Peak identity; (A) (a) propoxur, (b) Landrin® (minor isomer), (c) Landrin® (major isomer), and (d) carbanolate, (B) (a) Hercules 5727, (b) BAY 78537, and (c) Mobam, (C) (a) promecarb, (b) bufencarb (major isomer), (c) bufencarb (minor isomer), and (d) methiocarb, (D) (a) Hercules 9007, (b) carbofuran, and (c) carbaryl, (E) decarbofuran (2,3-dihydro-2-methyl-7-benzofuranyl methylcarbamate). (Adapted from Holden, E. R., *J. Assoc. Off. Anal. Chem.*, 56, 713, 1973. With permission.)

Table 24
COMPARISON OF RETENTION TIMES AND ELECTRON CAPTURE RESPONSE FOR THE PFB, DNT, AND DNP DERIVATIVES OF CARBAMATE PHENOLS

	PFB (195°C)		DNT (230°C)		DNP (250°C)	
Phenol	Retention time	Response	Retention time	Response	Retention time	Response
Phenol	0.14	0.18	0.48	0.14	1.5	0.23
p-Chlorophenol	0.30	0.29	0.89	0.19	2.5	0.31
3,4,5-Trimethyl-phenol	0.49	0.36	1.19	0.16	3.8	0.20
Carbofuran phenol	0.51	0.30	1.17	0.13	3.7	0.25
p-Nitrophenol	0.92	0.29	2.14	0.10	5.5	0.06
1-Naphthol	0.94	0.39	2.25	0.20	6.3	0.19

Note: Column, 6 ft, packed with 5% SE-30 on 60 to 80 Chromosorb W, DMCS treated. Temperatures as shown. All values relative to aldrin (1.00).

Reprinted from Seiber, J. N., Crosby, D. G., Fouda, H., and Soderquist, C. J., *J. Chromatogr.*, 73, 89, 1972. With permission.

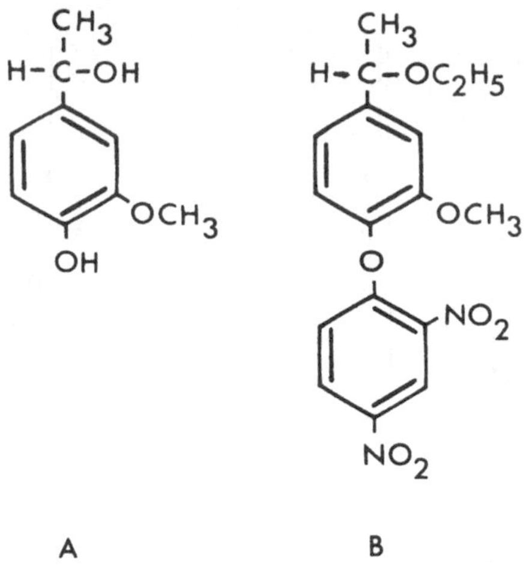

FIGURE 42. Structure of α-(4-hydroxy-3-methoxyphenyl)-ethanol (A), a plant constituent that interferes with the GLC determination of 3-hydroxy-7-phenol carbofuran using the ethoxylation-dinitrophenylation procedure.[273,329] (B) Structure of derivatized plant constituent. (Adapted from Holmstead, R. L., Allsup, T. L., and Fullmer, O. H., *J. Assoc. Off. Anal. Chem.*, 62, 89, 1979. With permission.)

With crop substrates, a plant component or breakdown product interfered with the determination of the 3-ethoxy-7-DNPE of carbofuran. The interference was of plant origin, although not present in all substrates, and was characterized as a vanillin derivative; the compound was identified as α-(4-hydroxy-3-methoxyphenyl)-ethanol [methyl-guaiacyl-carbinol] (Figure 42).[273] This interference could be resolved by de-ethoxylation of the derivatization mixture (after ethoxylation-dinitrophenylation) with 0.25 N HCl and extracting the DNPE derivatives; the resultant 3-hydroxy-7-DNPE of carbofuran had a different GLC retention time than the de-ethoxylated interference.

This study points out the importance of confirming residues and also of selecting a method or procedure that is essentially free of interferences. In some cases, sacrifice of time and recovery must be accepted in order to produce accurate residue data.

Kawahara described a novel reagent for determining phenols and mercaptans,[600] and organic acids[601] as their pentafluorobenzyl (PFB) ethers or esters using GLC and EC detection; these compounds were also determined in surface water.[602] Johnson[457] described a method for the preparation of PFB derivatives of two acids and three phenols from pesticides and a column cleanup procedure which separates the excess reagent from the derivatives. This fractionation separates the pesticide derivatives into groups of greater and lesser polarity and is thus an aid to identification. 1-Naphthol and *sec*-amyl phenol (bufencarb phenol) were derivatized using PFB bromide and K_2CO_3 at 50°C for 15 min.

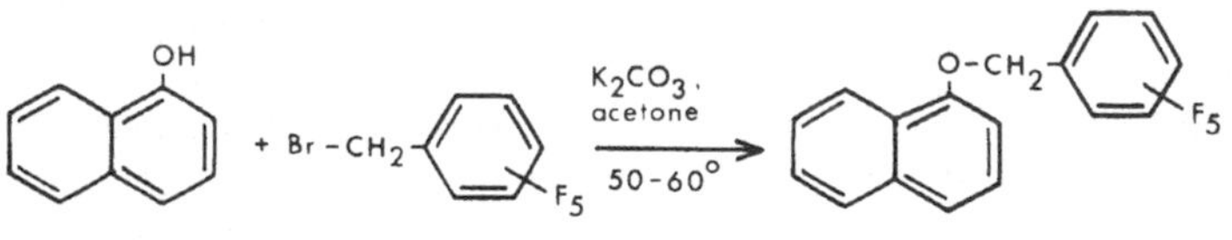

Derivatives were cleaned up on 1% deactivated silica gel using hexane-benzene mix-

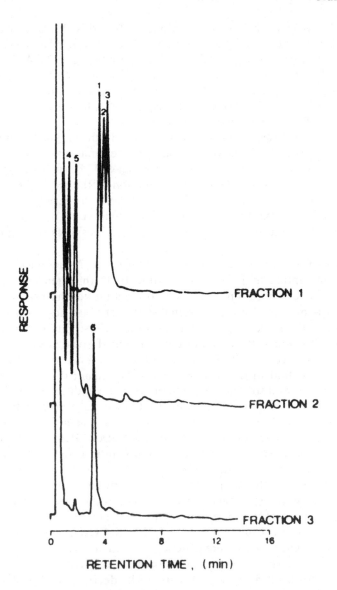

FIGURE 43. Gas chromatograms of the PFB derivatives of *N*-methylcarbamates on 3% OV-225 after fractionation on silica gel column. PFB derivatives of (1) methiocarb, (2) carbaryl, (3) Mobam, (4) propoxur, (5) carbofuran, and (6) 3-keto carbofuran. (Reprinted from Coburn, J. A., Ripley, B. D. and Chau, A. S. Y., *J. Assoc. Off. Anal. Chem.*, 59, 188, 1976. With permission.)

tures.[457,603] Sherma[240] mentioned that ten carbamates or metabolites could be analyzed as their PFB ethers by GLC after silica gel cleanup; no other details were given.

Coburn et al.[277] described a method for the analysis of 6 *N*-methylcarbamates (Figure 43) in water and soil using hydrolysis and formation of PFB derivatives; the detection limit was less than 1 μg of parent compound. Cleaner reactions and enhanced derivative stability resulted when K_2CO_3 was used instead of KOH as the base in the derivatization step. In the absence of K_2CO_3 anilines react with the PFB bromide to produce substituted amines.[240] Considerable improvement in the reaction background was also obtained when the hydrolysis step was accomplished at rt (Section X.C). Semiqualitative identification of residues is obtained by micro-column cleanup; best recoveries of each derivative in a single fraction was found with 1.5% deactivated silica gel.[240,277] No interferences were found with the several o.p. pesticides and phthalate esters studied that also hydrolyze and form PFB derivatives.[277,604] As in most cases of derivatization, detection limits are not limited by the sensitivity of the derivative per se, but rather by the reagent blanks, and in this case much of the reagent blank could be attributed to old organic solvents and reagents. Blanks were significantly reduced through the use of freshly prepared reagents.

This derivatization scheme was extended to several other aryl *N*-methylcarbamates, including aminocarb and mexacarbate, and the fractionation-cleanup was scaled up five times for convenience with no loss of compound.[315] Derivatization of 3-hydroxy carbofuran after hydrolysis resulted in the formation of the mono-PFB ether at the 7-position. Residues of carbamates have been determined in agricultural run-off water using this method.[100,101] Archer found the PFB derivative to be superior to the TFA derivative in determining carbofuran-7-phenol residues in alfalfa[372] and strawberries.[373]

The pentafluorobenzyl ether derivatives of hydrolyzed *N*-methylcarbamates exhibit excellent electron capture and electrolytic conductivity (halogen mode) response and are amenable to GLC separation on commonly used liquid phases.[277,315] Pentafluorobenzyl bromide is more specific than some of the stronger acylating reagents and it has been found to be stable in solution for several weeks;[277] however, some decrease in potency was observed with time, but concomitant derivatization of standards and samples reduces this problem.[315] The cleanup and fractionation step adds a twofold dimension to the method in that it removes excess reagent that might produce large tailing solvent fronts, and the fractionation provides an additional confirmation to the GLC retention time of peak identity. Some laboratories object to the cost of some derivatization reagents, however, the use of the same reagent for different purposes within the laboratory alleviates part of this consideration. Pentafluorobenzyl bromide has also been used for derivatization of phenoxyalkanoic acids[605-609] and phenols,[277,610] and the confirmation of o.p. pesticides.[604]

Seiber et al.[314] showed that 29 phenols and 4 phenol-generating pesticides could be determined by EC detection as their 2,6-dinitro-4-trifluoromethylphenyl ether derivatives. The phenols in acetone were reacted in the dark for 2 hr with 4-chloro-α,α,α-trifluoro-3,5-dinitrotoluene and a trace of KOH or K_2CO_3. Conversions exceeded 50% with either base. Higher yields were generally obtained with the more acidic phenols and with K_2CO_3; electron-attracting substituents (X, NR_3^+) increase the acidity of phenols while electron-releasing substituents (CH_3) decrease acidity. Use of KOH catalyst may cause hydrolysis of the derivative.

Greater than 80% recovery of carbaryl and carbofuran, and their respective phenols, in water fortified with 25 ppb of compound was observed, although lower concentrations may be determined in the absence of background interferences. The injection of acetone solutions containing the derivatives into an ECD is not recommended for continued good performance and sensitivity of the detector.

In general, ester derivatives of carbamate phenols are hydrolytically unstable whereas the ether derivatives show good stability and excellent GLC responses.[314,611] Seiber et al.[314] compared the formation and GLC characteristics of the pentafluorobenzyl (PFB), 2,4-dinitrophenyl (DNP), and 2,6-dinitro-4-trifluoromethylphenyl

(DNT) ethers of carbamate phenols. The reactivity of the three reagents and GLC characteristics of the corresponding derivatives appeared to be quite similar. The relative GLC retention times for the three derivatives varied regularly for the test compounds in the order PFB > DNT > DNP. Relative detector responses did not vary greatly between the derivatives although the PFB responses were consistently higher (Table 24). The presence of the trifluoromethyl group (in DNT) apparently confers greater volatility compared to the DNP ether. The differences in retention times suggests that the use of the three derivatives may be helpful in confirmatory qualitative analysis of the phenols (from carbamates) in question and hence they offer an alternative where interferences complicate the analysis of any one of the other derivatives. In general, a GLC liquid phase of intermediate polarity, such as OV-225, offers a means of increasing retention times (relative to SE-30) of the less polar phenol derivatives. Picryl ethers greatly increase retention times and may be used where volatility of the PFB, DNT, or DNP derivatives fail to remove the compound from solvent fronts or other early eluting co-extractives. Of course, co-extractives that are concomitantly derivatized will also show an increase in retention time.

a. Typical Methods for Derivatization of Carbamate Phenols

When conducting any derivatization procedure samples and standards should be reacted concurrently to account for any variations in reagents and recovery.

i. Pentafluorobenzylation (Coburn et al.[277] Ripley[315])

Evaporate the solution containing the hydrolyzed *N*-methylcarbamates (Section X.C.1) or the isolated phenols (Section VIII.A.1) with 2 m*l* *iso*-octane to 1 to 2 m*l* using rotary vacuum evaporation. Add 10 m*l* acetone, 25 μ*l* 5% aqueous K_2CO_3, 100 μ*l* pentafluorobenzyl bromide (1% in acetone) and allow reaction to proceed at 60°C for 20 min.

Rotavap derivatization mixture to 1 to 2 m*l* two times after adding 5 m*l* *iso*-octane each time to solution. Cleanup and fractionate derivatization solution on a 5 g silica gel column (Grade 950, previously equilibrated with 1.5% water). Pre-wash column with 25 m*l* hexane and add concentrated *iso*-octane solution to column. Rinse reaction flask with 5 m*l* hexane and add to column; wash column with 25 m*l* 5% benzene in hexane. Elute compounds as shown below. A 1-g micro-column may also be employed and the elution scheme scaled down five times. Concentrate all eluates to <5 m*l* after addition of 2 m*l* *iso*-octane using rotary vacuum evaporation. Dilute with *iso*-octane to 5 m*l* and examine by GLC.

Elution order of column and retention time of derivative[a]

Compound	Fr I	Fr II	Fr III	Fr IV
	(30 m*l* 25% B-H)[b]	(40 m*l* 75% B-H)	(100 m*l* 0.1% E-B)	(50 m*l* acetone)
Bendiocarb	2.1			
Bufencarb I	2.8			
Bufencarb II	3.1			
Carbaryl	7.50			
Carbofuran		4.4		
3-K carbofuran			6.7	
3-OH carbofuran				7.1
Methiocarb		4.0		
Propoxur		3.2		

[a]   Retention time in min on 3.6% OV-101/5.5% OV-210 column at 200°C.

[b]   B = benzene, H = hexane, E = ether.

ii. Dinitrophenylation (Holden[379])

Evaporate extract containing *N*-methylcarbamates (Sections VII.A.1a. or VIII.A.1) and 2 drops acetophenone to 1 to 2 m*l* using a rotavap, then remove flask from water bath and continue evaporation until dryness. Add 100 m*l* water, 2 m*l* 0.5 *N* KOH, and 1 m*l* 1-fluoro-2,4-dinitrobenzene (6% in acetone). Stopper and mix 20 min at high speed on a mechanical shaker. Add 10 m*l* 5% aqueous borax, swirl to mix, and heat on a steam bath for 10 min. Add 5 m*l* *iso*-octane, stopper and shake at high speed for 3 min, and then pour into 250-m*l* separatory funnel. Drain aqueous phase, and rinse *iso*-octane twice with water. Drain *iso*-octane into test tube and examine by GLC.

iii. DNT, PFB, or DNP ethers of carbamate phenols (Seiber et al.[314])

To a solution containing the carbamate phenols (Section X.C.1) in 0.2 m*l* acetone in a 10-m*l* volumetric flask, add 20 μ*l* 5% aqueous K_2CO_3 or KOH. Agitate briefly, and add 9 m*l* acetone and 0.25 m*l* of (5 mg/m*l*) DNT chloride, PFB bromide, or DNP fluoride in acetone. Adjust volume to 10 m*l* with acetone, stopper and shake vigorously for 30 sec, and allow to stand in the dark for 2 hr. Inject solution directly into GLC.

2. Derivatization of Amine Hydrolysis Products of N-Methylcarbamates

The GC behavior of simple aliphatic amines has been reviewed,[612,613] but neither the sensitivity nor the resolution has been adequate for residue analysis, and as such, amines require derivatization prior to GLC analysis. The carbamates and some other classes of pesticides produce amines as one of the products of hydrolysis. In the laboratory, chemical hydrolysis under either acidic or basic conditions is the most common method of isolating the amines and many of the procedures outlined below (Section X.C) may be employed. From *N*-methyl and oxime carbamates the product is methylamine, whereas the *N,N*-dimethyl and dithiocarbamates yield dimethylamine; thiocarbamates yield various dialkyl amines, and ureas and *N*-phenylcarbamate herbicides yield anilines (Section XI.B). Although many of the derivatization procedures discussed below show applicability to amines in general, consideration should only be given to those procedures forming chemical products indicative of the parent carbamate(s). Care must also be exercised during the hydrolysis step since methylamine is very volatile (bp −6.3°C) and it may be lost from the reaction mixture.[614]

As mentioned the amine hydrolysis products from all *N*-methylcarbamates and oxime carbamates, and the *N,N*-dimethylcarbamates are the same being methyl- and dimethylamine, respectively. In these cases, derivative formation may indicate the type of parent compound but does not differentiate between the parent carbamate(s) producing the various amines. Although the phenolic derivatives (Section X.B.1) produce greater specificity in the analysis, determination of the methyl- or dimethylamine portion of the carbamate molecule allows one to check if any insecticidal carbamate may be present in the sample.[389] Therefore these derivatives may be useful in determining a "total carbamate" value or ascertaining if any carbamate is present in a sample prior to subsequent qualitative analysis. Alternatively the carbamates may be separated prior to the derivatization step, however, this is time consuming and not always feasible.

Day et al.[615] used 1-fluoro-2,4-dinitrobenzene (FDNB) to prepare derivatives of C_1-C_4 primary and secondary amines. Heating with NaOH effected reaction and also hydrolyzed excess reagent. Holden et al.[389] found the procedure of Day et al.[615] unsatisfactory for the low levels of pesticides found in crops. They found the coupling reaction required a fixed pH and recommended the use of a strong buffer (borax) to maintain the slightly alkaline pH:

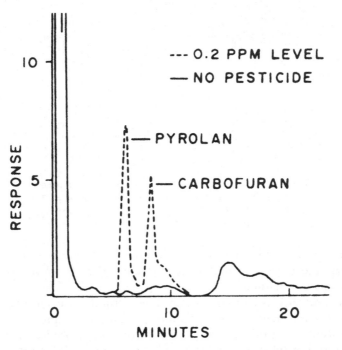

FIGURE 44. Typical chromatograms obtained when Pyrolan and carbofuran were added to spinach. Peaks represent the 2,4-dinitroaniline derivatives from the fortified sample (broken line) superimposed on the chromatogram of the unfortified sample (solid line). Chromatographic conditions: column, 4 ft × ¼ in. o.d. packed with 2% XE-60; temperature, 190°C; gas flow, nitrogen carrier, 100 mℓ/min; ECD. (Reprinted with permission from Holden, E. R., Jones, W. M., and Beroza, M., *J. Agric. Food Chem.*, 17, 56, 1969. Copyright 1969, American Chemical Society.)

After alkaline hydrolysis, acidification and removal of the phenols and other extraneous material, and neutralization of the reaction mixture, the liberated methyl- or dimethylamine was reacted with FDNB in 5% borax solution on a steam bath for 30 min. Whereas Day et al.[615] found heating the reaction mixture with NaOH hydrolyzed excess reagent, Holden et al.[389] found this step tended to destroy the *N*,*N*-dimethyl-2,4-dinitroaniline and furthermore the destruction was almost complete in the absence of an organic solvent (benzene); excess reagent can be removed with the aqueous phase while the derivatives remain in the benzene. The procedure was shown applicable for at least two dimethylcarbamates and eight methylcarbamates. Interferences (erroneous high results) may result from other pesticides that contain an amide or urea group that could also release methyl- or dimethylamine (Section XI.B). Likewise, *N*-methylcarbamates with a dimethylamino group (aminocarb, mexacarbate) can split off the dimethylamino group to produce both the methyl and dimethyl derivative. Rigorous control of hydrolysis conditions may minimize these interferences. Figure 44 shows a

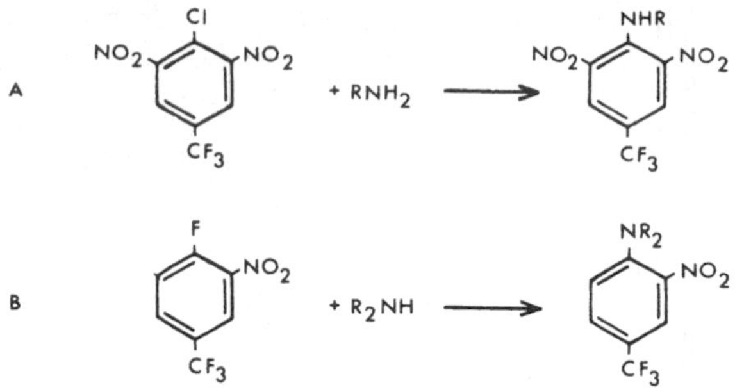

FIGURE 45.   Reaction of DNT (A) and MNT (B) reagents
with amines. (Reprinted with permission from Crosby, D. G.
and Bowers, J. B., *J. Agric. Food Chem.*, 16, 839, 1968. Copyright 1968, American Chemical Society.)

typical chromatogram of the methylamine and dimethylamine derivatives resulting
from carbofuran and Pyrolan, respectively. Residues of carbofuran have been determined in corn using this method.[616]

Sumida et al.[614] described a rapid and sensitive method for determining microgram
amounts of 5 *N*-methylcarbamates as dinitrophenyl-methylamine (DNP-MA). To
avoid the hydrolysis-partitioning step whereby losses could occur, simultaneous hydrolysis and dinitrophenylation was employed. Maximum yield of DNP-MA was obtained at pH 9.0. Glycine was added to remove excess reagent which overshadowed
the DNP-MA peak; DNP-glycine was thus formed but it remained in the aqueous
solution. This method was then employed to determine Meobal® residues in rice
grain.[617]

Holden et al.[389] showed that the dinitrophenylation procedure was applicable to aldicarb. Mendoza and Shields[618] modified the dinitrophenylation and applied it to the
determination of methomyl in rapeseed oil. Hydrolysis of methomyl was found unsatisfactory using the Sumida et al.[614] method; however, the carbamate could be hydrolyzed with NaOH immediately before the coupling step. Low derivatization results
were also found to occur if the FDNB was not thoroughly dissolved in the aqueous
reaction mixture, a problem that could be solved by dissolving the reagent in dioxane
prior to addition of the borate solution. Maximum interaction between methylamine
and FDNB was found to occur at 82°C. An interference was found just after the DNP-
MA peak and although glycine reduced this interference, low-level quantitation was
difficult unless the DNP-MA peak was significantly higher than the interference. A
micro-column of silica gel was found satisfactory for the cleanup of extracts to permit
quantitation of the methomyl derivative.[456] FDNB was also observed to be very reactive with methanol or ethanol resulting in a major peak before the derivative.[615] This
procedure worked for carbaryl and mexacarbate, although dimethylamine was readily
hydrolyzed from the phenolic moiety of mexacarbate as observed previously.[389,619]

Crosby and Bowers[430] demonstrated that primary and secondary amines undergo
facile reaction in alkaline solution with 4-chloro-α,α,α-trifluoro-3-nitrotoluene or
α,α,α-4-tetrafluoro-3-nitrotoluene to produce substituted anilines; products are 2,6-
dinitro-4-trifluoromethyl (DNT) and 2-nitro-4-trifluoromethyl (MNT) anilines, respectively (Figure 45). These derivatives undergo GC and may be detected down to 50 pg
by EC detection. Amines were reacted in acetone with reagent solution, sodium tetra-
borate and KOH for 2 hr at 55°C. These derivatives showed greater response, better
resolution, and shorter retention times than the DNP homologs. Alcohols present in
samples and solvents were found to interfere. This procedure, although applicable to
methylamine formed from the hydrolysis of *N*-methylcarbamates, may serve better as

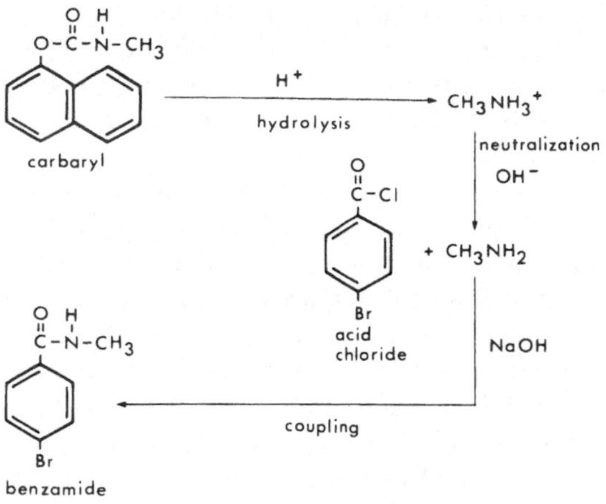

FIGURE 46. Acid hydrolysis of carbaryl to methylamine, and subsequent coupling with acid chloride to derive 4-bromo-N-methylbenzamide. (Adapted with permission from Tilden, R. L. and Van Middelem, C. H., *J. Agric. Food Chem.*, 18, 154, 1970. Copyright 1970, American Chemical Society.)

a confirmatory test for other compounds such as the thiocarbamate herbicides. (Section XI.B).

Although halogenated acyl derivatives of amines have been prepared,[620,621] they were not specifically applied to methylamine or other carbamate amine hydrolysis products.

Tilden and Van Middelem[622] examined aniline, amide, sulfonamide, and benzamide derivatives of methyl- and dimethylamine. The benzamide derivative was found to be superior from the point of view of sensitivity to EC detection, ease of preparation, and thermal and chemical stability during GLC analysis. A reaction sequence for carbaryl is shown in Figure 46. Carbaryl was determined in field crops,[622] and the method and results compared well with the colorimetric procedure.[345]

C. Analytical Hydrolysis of N-Methylcarbamates

Since many of the methods for determining N-methylcarbamates rely on derivatization of the hydrolysis products, a quantitative conversion to the phenol or methylamine is essential. N-methylcarbamates are esters of carbamic acid and as such they can be hydrolyzed using concentrated acid or aqueous base to form the phenol and acid; the acid is not stable and further decomposes into carbon dioxide and methylamine.

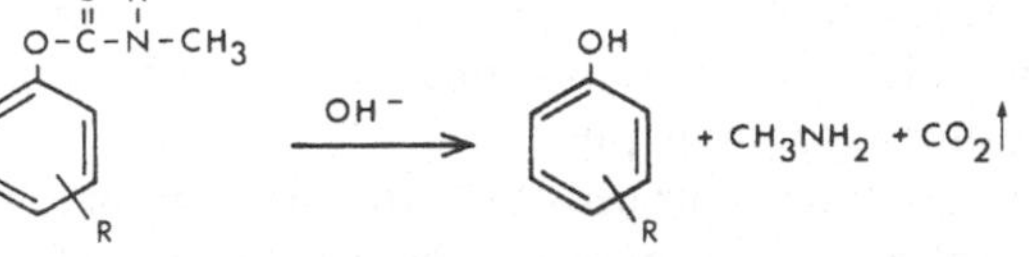

$$\text{(5)}$$

Acid hydrolysis is used, generally, when one wishes to isolate methylamine for derivatization.[85,424,622] In acid, the methylamine exists as the salt which is water soluble whereas in alkali it exists as the free methylamine and is organically soluble. Benzene is generally used to extract methylamine from basic solution. Methylamine may also be obtained after base hydrolysis of the carbamates.[389] Unlike the acid-catalyzed hydrolysis, alkaline hydrolysis of carbamate esters is rapid and irreversible and is the method of choice to obtain the carbamate phenols. An additional benefit of base hydrolysis is the formation of the alkaline phenoxide which is water soluble. The alkaline solution may be extracted to remove neutral extraneous co-extractive material and after acidification, organic extraction isolates the phenol for derivatization. The carbamate phenols produced by hydrolysis may also be steam distilled after acidification of the hydrolysis solution.[339,362]

Base hydrolysis follows second order kinetics with the rate dependent on the concentration of both the carbamate and the base; the rate is also influenced by the structure of the carbamate with electronegative groups on the phenol ring accelerating the hydrolysis (see also Section II.B and References 1,28,29,31,32,73,89 and 90). Examination of the analytical conditions used for base hydrolysis of N-methylcarbamates shows a wide range in concentration of base, temperature of reaction, and the time required for hydrolysis. Usually KOH or NaOH in methanol is used to effect hydrolysis with the concentration ranging from 0.02 N to 2.5 N. Butler and McDonough found 0.1 N methanolic NaOH suitable to hydrolyze most carbamates,[336,378] however, the addition of 15% water was found to be optimum for the hydrolysis of 3-hydroxy carbofuran.[320] Hydrolysis may be carried out at rt or at elevated temperatures with the time required being a function of the temperature. Coburn et al.[277] observed that propoxur, carbofuran, 3-keto carbofuran, methiocarb, carbaryl, and Mobam required 4 hr at rt whereas at 60°C only 40 to 45 min was required for maximum hydrolysis. Reproducibility decreased with the more volatile compounds and the "background" was more severe at the higher temperature. Often it is convenient to use overnight hydrolysis at rt. In a multi-residue method, it is imperative that all the carbamates be hydrolyzed quantitatively.

The derivatization of the phenol or methylamine may be carried out in alkaline solution[314,338,389,614] or after evaporation of the hydrolysis solution.[320,336,362,378] Isolation of the phenol may be achieved by acidification of the hydrolysate and extraction with an organic solvent. One should be aware that most carbamate phenols are extractable under these conditions; however, the phenols from the aminophenyl N-methylcarbamates (aminocarb, mexacarbate) are extracted from a neutral bicarbonate solution. The choice of solvent is dependent upon the partition coefficient of the phenol; some selection may be based on the requirement of the subsequent derivatization step. No one solvent has shown universal applicability for all phenols. Typical solvents for the extraction of the phenols include benzene,[219,220,339,424,590] pentane,[362] ethyl ether,[357] methylene chloride,[340] and chloroform.[364] Suffett[283] examined p-values of carbamates and their phenols with different solvents and at various pHs under conditions of constant temperature and ionic strength. Coburn et al.[277] found benzene to give generally good results with a minimum of co-extractives from natural samples; methiocarb was only 50 to 60% recovered. Subsequently, methylene chloride was found to give better recoveries[100] but like ethyl ether and ethyl acetate showed increased backgrounds from samples; ethyl ether demonstrated depressed recoveries. Similar results were obtained by Ripley[315] during the extraction of 3-hydroxy carbofuran-7-phenol; depressed recoveries were found with most solvents except anhydrous ethyl ether which was also found to be satisfactory for the other carbamate phenols.

A prime concern in any analytical procedure is to account for all the parent compound as well as those metabolites that may also be present in a sample. Conjugation of carbamate metabolites by plant or animal systems is one method of detoxifying xenobiotics. Since many of these conjugates are esters, they are prone to hydrolysis by esterases to produce a phenol (the aglycone) from the conjugate. Hence esterases may be employed to hydrolyze such carbamate conjugates. For example, 3,4-dichlorobenzyl glucuronide can be hydrolyzed by β-glucuronidase to yield 3,4-dichlorobenzyl alcohol;[623] 1-naphthyl sulfate can be hydrolyzed by aryl sulfatase;[592] β-glucosidase can be used to hydrolyze propoxur-O-glucosides.[219] Acid hydrolysis is also often used to release conjugated glucosides in plant material;[160,220,364] one of the most common applications is the digestion of crop material in hot 0.25 *N* HCl to release 3-hydroxy carbofuran (Section VII.A.4). Refluxing with concentrated HCl has been used to hydrolyze 1-naphthol conjugates in urine.[337] Once released, these phenols are available for determination usually via derivatization.

1. Typical Hydrolysis of N-Methylcarbamates to Isolate the Phenols (Coburn et al.,[277] Ripley[315])

To a concentrated solution containing the *N*-methylcarbamates in a round-bottom flask add 2 mℓ 10% methanolic KOH. Allow to hydrolyze at 60°C for 20 min (or overnight at rt). A concentration series of carbamate standards should be hydrolyzed concurrently with samples. Transfer hydrolysis solution with 50 to 60 mℓ distilled water rinsings to a 500-mℓ separatory funnel and extract with 50 mℓ CH_2Cl_2 which is discarded. Acidify hydrolysis solution to pH <2 with about 0.3 to 0.5 mℓ 50% H_2SO_4 and extract with two 50-mℓ aliquots of anhydrous ethyl ether (benzene and CH_2Cl_2 have also been successfully used for most carbamate phenols). If analysis is to include aminomethylphenyl *N*-methylcarbamates, adjust hydrolysis solution to pH 6.5 to 7.0 with phosphate buffer or $NaHCO_3$ and re-extract with solvent which is added to previous extracting solvent. Add 2 mℓ *iso*-octane (or other suitable keeper solution) and rotavap solvent to 1 to 2 mℓ. Proceed to derivatization step (Section X.B.1.a).

D. On-Column Reactions

Reaction GC[527,624] has evolved to include any structural change of a compound occurring in a GC.[366] One such on-column change has already been discussed under direct GLC of carbamate phenols. *N*-methylcarbamates are thermally unstable and as such they often decompose on a GLC column to methylisocyanate and the respective phenol (Section IX.B). A reaction zone of phosphoric acid at the front of a GLC column may be employed to facilitate this decomposition thus allowing quantitation of the eluted phenol;[340] alternatively, the carbamate may be injected with base.[509]

Another example of on-column reaction may involve mixing the compound of interest with a derivatization reagent and injection of this solution into the GLC. The heat of the injection block promotes the derivatization and the eluted derivative may be determined. An example is on-column methylation of carbamates (and ureas) containing a NH moiety using trimethylanilinium hydroxide (TMAH; Methelute[515]).[625-627] Reaction products are shown in Figure 47; ureas and *N*-acyl carbamates produce the expected *N*-methyl carbamate derivatives; however, oxime and *N*-methylcarbamates produce methoximes and anisoles, respectively. The latter two groups of compounds appear to react in two steps: first there is the loss of the carbamate group to the hydroxy moiety, which is then methylated. Carbamates are combined in a syringe with a two molar excess of TMAH and injected into the GLC; with a FID, increases in sensitivity were 30% for IPC and 400% for carbaryl. Single products are obtained from compounds with only one group available for methylation, but this procedure does not distinguish between parent compounds and hydrolysis products unless they are previously separated. Furthermore, aldoximecarbamates were not readily derivatized since during the thermal decomposition the nitrile or other smaller fragments are formed rather than the oxime.

Another application of reaction GC is transesterification (ester interchange). Gaylord and Sroog[628] showed that certain carbamates could undergo such a base-catalyzed reaction. The reaction was postulated to follow carbonyl addition, and the reaction was favored when R′ is aromatic rather than aliphatic:

$$RN\overset{H}{\underset{}{-}}\overset{O}{\overset{\parallel}{C}}-O-R' + R''OH \rightleftharpoons \underset{\underset{R''OH}{}}{RN\overset{H}{-}\overset{O}{\overset{\parallel}{C}}-O-R'} \rightleftharpoons RN\overset{H}{-}\overset{O}{\overset{\parallel}{C}}-O-R'' + R'OH$$

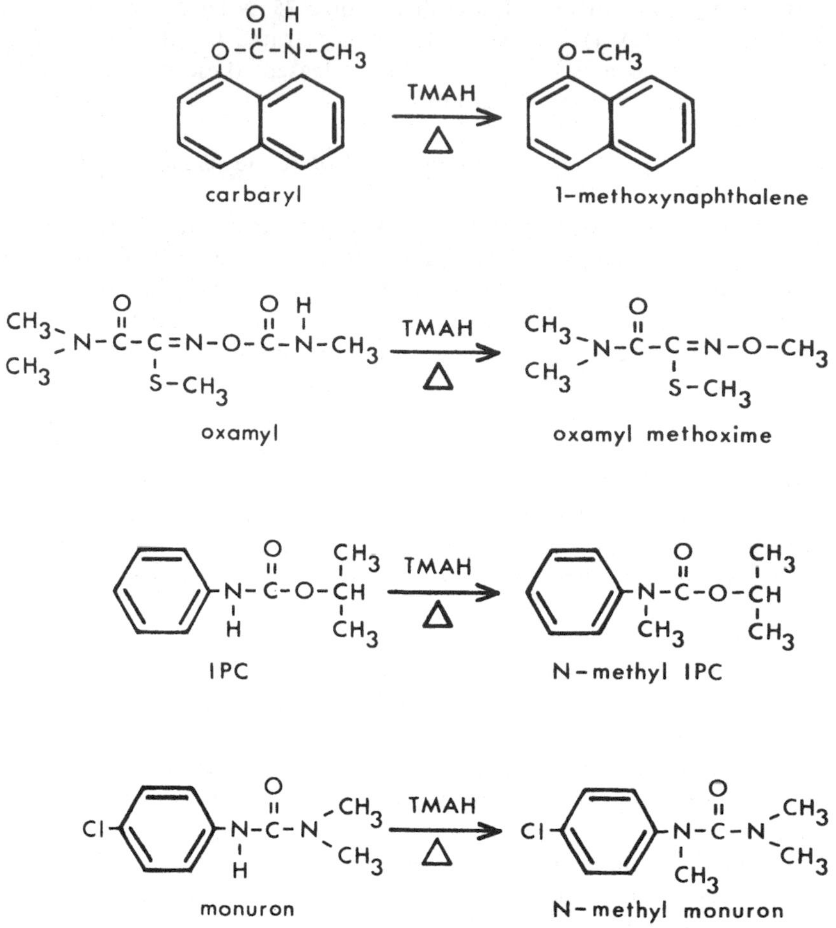

FIGURE 47.   On-column derivatization of carbamates and ureas with trimethylanilinium hydroxide (TMAH).[625-627]

Moye[366] showed that seven *N*-methylcarbamates and two oxime carbamates could be made to undergo transesterification with methanol in the injection port of a GC. The first 6 in. of the GLC column (injection port) was packed with untreated glass micro beads, and the carbamate in 5×10^{-3} *M* methanolic-KOH was injected on a Porapak P® column at 180°C; the optimum injection port temperature was found to be 215°C and detection was by rubidium sulfate AFID. Conversions were nearly quantitative even at the 1-ng level, and the product was confirmed to be methyl *N*-methyl-carbamate.

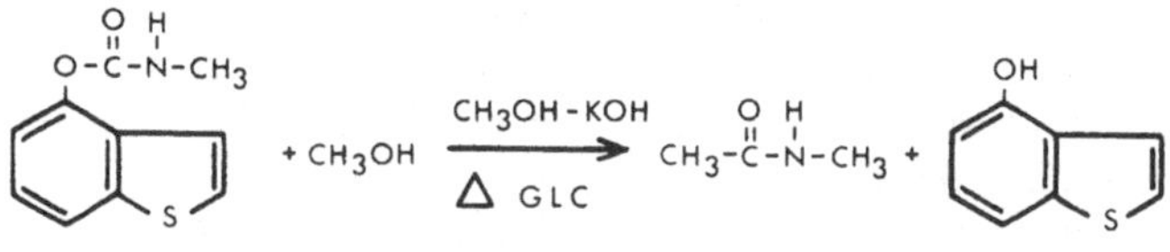

This reaction was shown to be instantaneous and could also occur in the presence of other short-chain alcohols (Figure 48).

Crop extracts did not appear to interfere with the GLC analysis or impede the reaction. Mobam in lettuce and soil,[366] carbofuran in lettuce,[369] soil[322,629] and grass samples,[629] and methomyl in tobacco[402] have been determined using this method. The transesterification procedure can also be extended to include o.p. pesticides.[630]

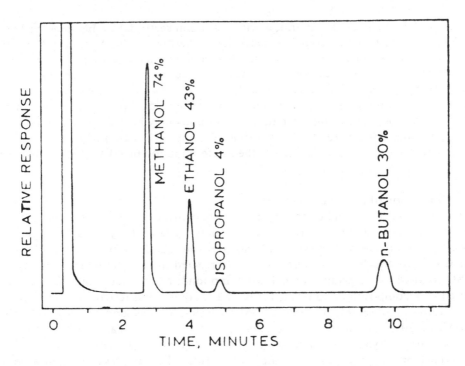

FIGURE 48. On-column transesterification. Peaks obtained upon individual injections of Mobam in various alcohols. Operating parameters: column, 5 ft × 1/8 in. packed with 80/100 mesh Porapak P® (first 6 in. packed with untreated glass microbeads; temperatures, column 180°C, injector 220°C; gas flow, helium carrier, 27 mℓ/min; AFID. All injections in 5 × 10^{-3} M NaOH in the alcohol. (Reprinted with permission from Moye, H. A., *J. Agric. Food Chem.*, 19, 452, 1971. Copyright 1971, American Chemical Society.)

Like the derivatization of the amine portion of N-methylcarbamates, the transesterification procedure using methanol only produces one peak for all the N-methylcarbamates, namely methyl N-methylcarbamate. However, transesterification should be useful when a survey of suspected samples is desired to confirm or deny the presence of any N-methylcarbamate. It may also be useful in disappearance studies of a single carbamate after known application to a field crop.[366]

XI. DERIVATIZATION OF OTHER CARBAMATE PESTICIDES

Many of the noninsecticidal carbamates (Table 1) may be determined by direct GLC, or quickly and easily by other techniques, without resorting to chemical derivatization-GLC (Section IX.D and References 213 and 250). In many cases, chemical derivatization of these carbamates is used primarily as a confirmatory technique although in a multi-residue application or in natural samples these derivatives could be formed concomitant with the insecticidal carbamates. Therefore, one must be aware that any derivatization procedure may not be specific per se for the compound investigated and in fact many other pesticides and natural product co-extractives may also be derivatized thus complicating chromatograms and making identification difficult. However, the pre-derivatization steps or GLC conditions may be selective in removing many possible interferences. Furthermore, it is not surprising that many of the methods developed primarily for the noninsecticidal carbamates follow closely to or are analogous to those methods for the insecticidal carbamates.

As with the insecticidal carbamates, derivatization procedures for the noninsecticidal carbamates may focus on the intact molecule or one of hydrolysis products. Derivatization-GLC determination of noninsecticidal carbamates have been reviewed previously.[250,631-634] Much of the attention with the noninsecticidal carbamates has been directed to the carbamate herbicides, particularly the *N*-phenylcarbamates; *N*-phenylcarbamates are similar in structure to the *N*-phenylurea herbicides and thus many of the methods apply equally to both these types of herbicides. All thiocarbamate herbicides have been successfully determined directly by GLC and derivatization is used mainly for confirmatory purposes. The carbamate fungicides are not easily determined by GLC or via derivatization; since these compounds may be determined quickly and easily, although not separately, by colorimetry (Section XII.E.1) it is not surprising that little attention has been devoted to their determination by GLC or derivatization-GLC.

A. Derivatization of the Intact Compound

Alkylation has been applied to the intact carbamates to produce either more stable or more volatile compounds prior to GLC analysis. Greenhalgh and Kovacicova[559] showed that pesticides with NH or NH$_2$ moieties could undergo base-catalyzed alkylation. IPC, pentanochlor, and terbucarb were reacted at rt with NaH-methyl iodide-DMSO to yield the respective methylated compound; reaction at higher temperatures resulted in decomposition; however, for the other compounds tested, elevated reaction temperatures are desirable or necessary. Methylation in this manner may also be used to confirm other carbamates and pesticides.[559,560] Wien and Tanaka[626] showed that IPC could be methylated on-column with TMAH (Section X.D, Figure 47).

Lawrence et al.[426,427,432] demonstrated that CIPC, IPC, and swep as well as seven *N*-phenylureas could be determined or confirmed in crops as their methylated derivative. Detection was by Coulson electrolytic conductivity in the *N*-mode and cleanup was not required. Derivatives containing a chlorine atom could also be determined with a chlorine-specific detector.[427] Subsequently, it was shown that the methylated derivative could be confirmed following alkaline hydrolysis to the anilines.[426] Less than 1 hr at 100°C was required for hydrolysis of the carbamates; however, overnight was needed for some of the ureas. The same GLC conditions were employed for the determination of the anilines; retention times of the anilines were less than one-half of the methylated product.

Fishbein and Zielinski[562] prepared trimethylsilyl (TMS) derivatives of CIPC and IPC for FID-GLC. As with the *N*-methylcarbamates there was little change in retention time from the parent compounds to the derivative; however, there was a general improvement in peak appearance. Halomethyldimethylsilyl derivatives of the *N*-phenylcarbamates could not be prepared.[564]

B. Derivatization of Hydrolysis Products

Many procedures call for hydrolysis of carbamate pesticides prior to derivatization

Table 25
AMINES FROM HYDROLYSIS OF DIFFERENT CLASSES OF PESTICIDES

Class	Example	Amine
N-methylcarbamates	Carbaryl	CH_3NH_2
	Mexacarbate	CH_3NH_2 and $(CH_3)_2NH$
Oxime carbamates	Aldicarb	CH_3NH_2
N,N-dimethylcarbamates	Pyrolan	$(CH_3)_2NH$
Thiocarbamates	EPTC	$(C_3H_7)_2NH$
Dithiocarbamate salt	Metham-sodium	CH_3NH_2
Dithiocarbamate ester	Sulfallate	$(C_2H_5)_2NH$
Amide	Dicrotophos	$(CH_3)_2NH$
Phosphoramidate	DMPA	$(CH_3)_2CHNH_2$
N-phenylcarbamate	CIPC	$ClC_6H_4NH_2$
Ureas	Linuron	$ClC_6H_4NH_2$
Anilide	Propanil	$Cl_2C_6H_3NH_2$

Adapted with permission from Crosby, D. G. and Bowers, J. B., *J. Agric. Food Chem.*, 16, 839, 1968. Copyright 1968, American Chemical Society.

or for confirmatory purposes. Original methods for CIPC were based on hydrolysis[52,53,420] and GLC detection of the liberated chloroaniline or derivatization for GLC or colorimetric analysis. A few cautionary notes must be mentioned with regards to the hydrolysis of the carbamates (Section X.C). Either acidic or basic hydrolysis has been employed to liberate the amine or aniline from the carbamate pesticides. Under these conditions many different pesticides and natural products are similarly hydrolyzed and thus the determination of the hydrolysis products may be nonspecific. Several different pesticides yield the same or similar hydrolysis products (Table 25). Furthermore some of the amines, particularly dimethyl- and methylamine, are very volatile and care must be exercised during weighing of standards and the hydrolysis step to prevent loss of these compounds. Volatile amines are usually handled as their chloride salts; standards are prepared in 0.1 N HCl and distillates are trapped in dilute HCl solutions.

Ripley et al.[613] found that direct base hydrolysis of substrates and distillation resulted in appreciable amounts of methyl- and dimethylamine, as well as several other amines. It appears that methyl- and dimethylamine are ubiquitous in nature and are produced from some natural products. Substrates extracted with methylene chloride prior to hydrolysis showed only minimal sample (or reagent) interference (Figure 44).[389,622] Acid hydrolysis and distillation of sugar beets resulted in no background interference with the MNT derivative from EPTC.[430]

Spengler and Hamroll[635] investigated the uncatalyzed thermal breakdown and reaction GC using a KOH impregnated column for CIPC, IPC, and several ureas. Using the KOH column, reproducible conversions were obtained and the anilines were determined by FID-ECD. Separation and detection of anilines by FID-GLC has also been reported.[636]

Lawrence[432] showed that thiocarbamates could be confirmed after methoxylation (Figure 49). The thiocarbamate linkage is broken with sodium methoxide and the corresponding methylcarbamate was formed; these products eluted from the GLC faster than the parent compounds. Since the sulfur atom was lost, detection was based on nitrogen.

CIPC and three ureas were determined by Gutenmann and Lisk[423] following a one-step hydrolysis of the pesticide to the chlorinated aniline and bromination to the bromochloroaniline derivative:

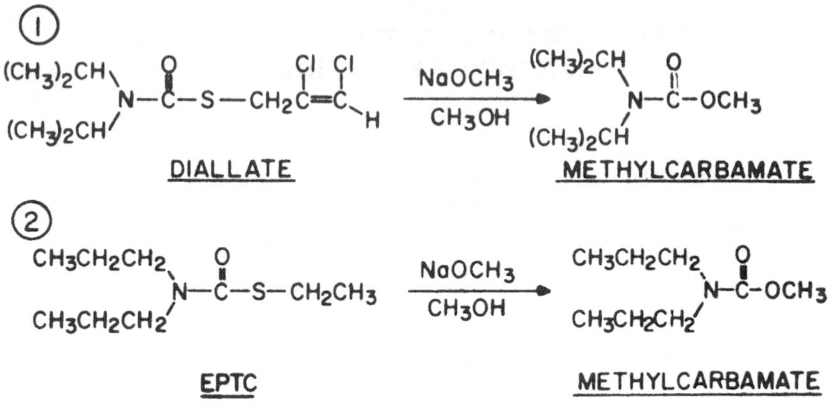

FIGURE 49.   Methoxylation reaction for diallate and EPTC with sodium methoxide-methanol. (Reprinted with permission from Lawrence, J. F., *J. Agric. Food Chem.*, 24, 1236, 1976. Copyright 1976, American Chemical Society.)

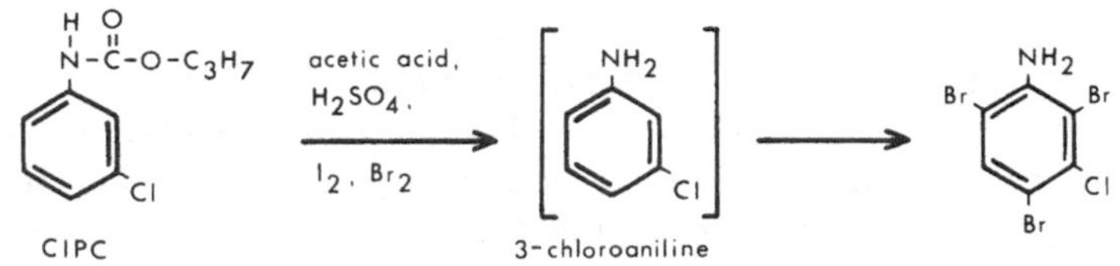

CIPC was mixed with glacial acetic acid, sulfuric acid, and a solution of acetic acid saturated with iodine crystals and 5% (by volume) bromine. This mixture was reacted at 130°C for 1 hr and then water and base were added and the resultant 2,4,6-tribromo-3-chloroaniline was extracted and analyzed by ECD-GLC. This procedure was also used to determine carbaryl.[357]

Several reagents have been used to derivatize amines (Section X.B.2). Day et al.[615] showed that primary and secondary amines could be reacted with 1-fluoro-2,4-dinitrobenzene (FDNB). Cohen and Wheals[424] showed that urea and *N*-phenylcarbamate herbicides could be hydrolyzed and derivatized with FDNB on a silica gel chromatoplate. The DNB derivative was then eluted from the adsorbent and determined by ECD-GLC.

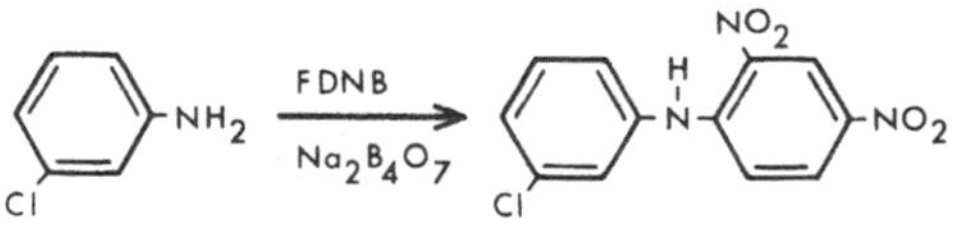

This procedure was used to determine *N*-phenylcarbamates and ureas in river water down to 1 ppb and in soil and plant material to 0.02 to 0.05 ppm. A similar reaction was employed to determine the amine or phenols of *N*-methylcarbamate insecticides.[379,380,389] FDNB was preferred over HFBA which was prone to atmospheric hydrolysis.[424]

Crosby and Bowers[430] used halodinitrotoluene reagents in alkaline solution to prepare substituted aniline derivatives. Greater detector response, better resolution, and shorter retention times were observed compared with the DNB derivatives. These derivatives have been used to determine amines and anilines resulting from the hydrolysis of carbamate pesticides (Figures 45, 50, 51).

Various acyl and benzyl derivatives of amines and phenols have been prepared. McCurdy and Reiser[621] used TFA to determine fatty amines in water. Clarke et al.[620]

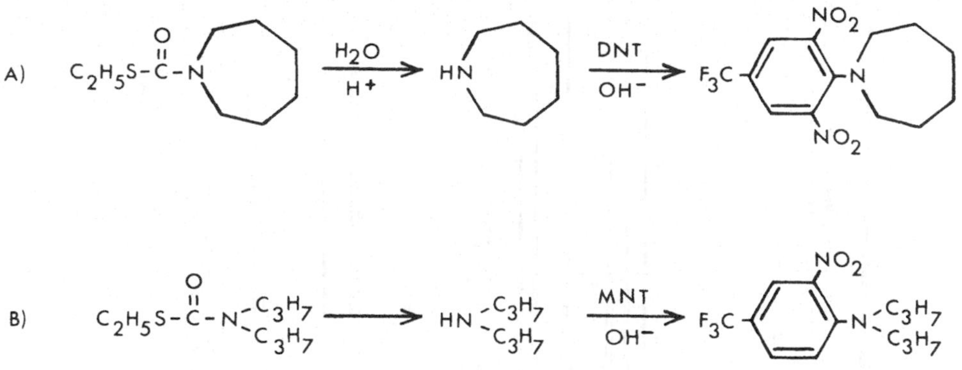

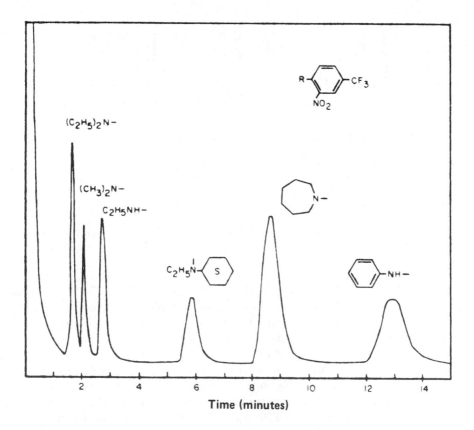

FIGURE 50.   Formation of (A) DNT derivative from molinate and (B) MNT derivative from EPTC. (Adapted with permission from Crosby, D. G. and Bowers, J. B., *J. Agric. Food Chem.*, 16, 839, 1968. Copyright 1968, American Chemical Society.)

FIGURE 51.   Gas chromatogram of a mixture of MNT derivatives of various amines after hydrolysis of carbamates (Table 25). Operating parameters: column, 10 ft × 1/8 in. o.d. stainless steel packed with 3% SE-30; temperature, 180°C; gas flow, nitrogen carrier, 30 ml/min; EC (tritium) detection. (Reprinted with permission from Crosby, D. G. and Bowers, J. B., *J. Agric. Food Chem.*, 16, 839, 1968. Copyright 1968, American Chemical Society.)

compared ECD response and GLC characteristics of several haloacyl derivatives. Moffet et al.[637] examined perfluorobenzene derivatives of primary and secondary amines using ECD-GLC and Hartvig et al.[638] studied the ECD-GC of tertiary amines as pentafluorobenzyl carbamates. While none of these procedures were specifically applied to pesticide residues, the principles indicate that these reagents could be employed.

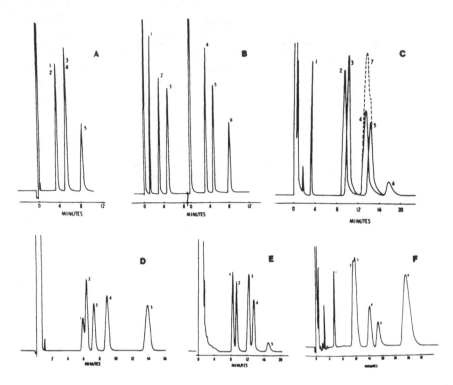

FIGURE 52.   Chromatograms of derivatives of substituted anilines. (A) heptafluorobu-
trylamides, (B) pentafluoropropylamides, (C) 2,4-dinitrophenyl, (D) pentafluorobenzyl,
(E) 2,6-dinitro-4-trifluoromethylphenyl, and (F) *p* -bromobenzamide derivatives. Peak
identities: aniline (1-B, C, F), *m*-chloroaniline (1-A, E, F; 2-B, C, D), *p*-chloroaniline (2-
A, D, E; 3-C, F; 4-B), 3-chloro-4-methylaniline (3-A, B, D, E; 4-C, F), *p*-bromoaniline
(4-A, D, E; 5-B, C, F), 3,4-dichloroaniline (5-A, D, E; 6-B, C, F). Columns: (A) 5% OV-
210, 160°C, (B) 5% OV-210, 155°C, (C) 3% OV-1, 200°C, (D) 4% SE-30/6% OV-210,
190°C, (E) 3% OV-1, 200°C, and (F) 1% E-301/0.1% Epon® 1001, 180°C. (Reproduced
from Bradway, D. E. and Shafik, T., *J. Chromatogr. Sci.*, 15, 322, 1977. By permission
of Preston Publications, Inc.)

Bradway and Shafik[639] studied several of these derivatives for the ECD-GLC analysis of anilines resulting from hydrolysis of pesticides. Stability of reagents and derivatives, ease of handling, reaction conditions, and GC characteristics were all studied
for the preparation of amides using haloacyl anhydrides or imidazoles, *p*-bromoben-
zoyl bromide, pentafluorobenzoyl chloride, pentafluorobenzyl bromide, and nitro
and/or trifluoromethylnitro substituted phenyl halides. At a high concentration of aniline (500 ng) all reagents were satisfactory but at a lower concentration (50 ng) broad
solvent fronts from most of the reagents obscured the aniline derivative peak. Clean
chromatograms were obtained with PFPA and HFBA (Figure 33) and the PFPA derivative was better than the HFBA. Unfortunately, the GLC separation characteristics
were less than ideal (Figure 52), but this may have been due to choice of column;
sensitivities were less than with some of the other derivatives. Nevertheless, the use of
PFPA was recommended for use and the applicability to urine and water samples
demonstrated. Qualitative confirmation of the anilines may be obtained with any of
the other derivatives. Mitten[640] also found that HFB derivatives of CIPC to be the
most useful.

Ripley et al.[613] prepared pentafluorobenzamide derivatives of C_1-C_5 mono- and di-
alkylamines. These derivatives were easily separated on an Ultra-Bond® column and
could be determined with a N-P detector. Excess reagent and by-products produced

dirty ECD chromatograms. Following alkaline hydrolysis and distillation of the volatile amines from crop substrates, these nitrosoamine precursors could be quantitated.

Chapman and Harris[534] noted that direct GLC of oxime carbamates (methomyl, oxamyl) resulted in broad tailing peaks with retention times corresponding to the respective oxime compound.[210] In contrast to the behavior of the parent compounds and underivatized oximes, the oxime-TMS ethers readily produced symmetrical peaks of consistent size. *N,O-Bis* (trimethylsilyl)acetamide (BSA) was used to form the TMS ethers of the free oximes or from the parent compounds after hydrolysis.

C. Ethylenebisdithiocarbamate (EBDC) Fungicides and Ethylenethiourea (ETU)

The EBDC and dithiocarbamate fungicides represent an agriculturally important class of compounds; unfortunately, the nonvolatility of these compounds precludes direct determination by GLC (Section IX.D). Furthermore, they are heat labile and during GLC analysis the compounds pyrolyze to carbon disulfide or ETU.[545] The most convenient analytical procedure for these compounds is by carbon disulfide evolution and colorimetric determination (Section XII.E.1). As yet there are no derivatization methods for these compounds, although Onuska and Boos[549] derivatized similar compounds (*N,N*-dialkyldithiocarbamates) to their *S*-alkyl esters using iodoalkanes or diazomethane.

The CS_2-colorimetric method for dithiocarbamates is nonspecific since thiram, the EBDCs, and dithiocarbamates (Table 1) all react similarly. Concern has been raised about the EBDC fungicides because of their breakdown to ETU.[212,641] Newsome[550] described a method to measure a representative compound (ethylenediamine) which is produced on hydrolysis of EBDCs.

The EBDC is hydrolyzed by refluxing with 1 *N* HCl containing stannous chloride. The released ethylenediamine was isolated by ion-exchange chromatography and then trifluoroacetylated prior to GLC analysis. Other EBDC metabolites such as ethylenebis(isothiocyanate) and ethylenethiuram monosulfide, may be determined by direct GLC using an ECD,[642,643] while ethyleneurea and 2-imidazoline may be determined after derivatization.[644,645]

In any discussion of EBDC fungicides it would be remiss to omit ETU. It has been shown that ETU is present in EBDC formulations,[646,647] is a metabolic breakdown product[648,649] and is present after field application of EBDCs,[213-215,650-652] and it may be formed during cooking of EBDC-treated crops.[653,654] ETU has also been shown to be carcinogenic,[655,656] tumorigenic,[657] and teratogenic.[658] Furthermore, ETU is water soluble and has been shown to move systemically upwards from soil into plants;[659,660] ETU is mobile in moist soils,[661] although only for short periods of time.[39,661,662]

Direct GLC analysis of ETU residues is not practical since the peak tails because of its high polarity. Onley and Yip,[663] and Haines and Adler[664] reported difficulties in chromatographing ETU directly by GLC. However, Otto et al.[665] were successful in determining ETU directly using a 90-cm × 2.5-mm column packed with 3% Versamid 900 and using a column temperature of 260°C and a sulfur detector. With formulations, Bontoyan et al.[646,647] were able to chromatograph ETU directly on a 2% Carbowax® column and quantitate with a thermal conductivity detector.

The more common method for the determination of ETU is via chemical derivatization using the following general scheme:[557]

Onley and Yip[663] reacted ETU in NaOH with methanolic 1-bromobutane to form the *S*-butylated derivative that was determined using a thermionic detector; the method was later modified to reduce interferences from EBDCs[666] and subsequently was studied collaboratively.[667] Haines and Adler[664] used 1-bromobutane in the presence of DMF and sodium borohydride to form the same derivative with determination by FPD in the sulfur mode; the derivative was characterized as 2-*n*-butylmercapto-2-imidazoline. Newsome[668] used benzyl chloride in aqueous methanol to prepare the *S*-benzyl derivative, and after cleanup by solvent partitioning, it was trifluoroacetylated for quantitation by ECD-GLC. Several modifications to this basic reaction have been tried. Nash substituted pentafluorobenzoylation for trifluoroacetylation,[669] and *o*-chlorobenzyl chloride for benzyl chloride.[670] King[671] prepared a *m*-trifluoromethylbenzyl derivative; either the first derivative (*m*-TFM benzyl ETU) or the subsequent *N*-trifluoroacetylated derivative may be quantitated by GLC using an ECD or FPD (S mode). A nitroso derivative has been used for polarographic determination.[672,673] Recently, Singh et al.[674] described an extractive acylation technique for determining ETU from water using dichloroacetic anhydride.

The variety of derivatives employed to determine ETU attests to the utility of forming a derivative to assist in an assay for a troublesome compound in plant material.[669] Unfortunately, no concensus among analysts exists and residues of ETU have been reported using most of these procedures. While finding a suitable, sensitive method for the determination a compound may be pleasurable, that in itself does not contribute to an analytical method. Several complications arise in all these procedures. Since ETU is water soluble, difficulties may arise in extracting the compound; water-miscible solvents have been successfully employed. However, ETU is unstable in alcoholic extracts and they have to be analyzed immediately.[669] Furthermore, decomposition of EBDCs to ETU has been shown to occur during maceration and storage of samples or extracts.[270] Moreover, refluxing is often employed for derivative formation, and it is possible to cause conversion of EBDC to ETU when EBDC-treated crops are analyzed.[652] Some of these problems have been discussed.[212,215,652,675]

XII. OTHER DETERMINATIVE METHODS

A. HPLC

HPLC is a technique that is rapidly gaining prominence in the field of separation analysis of small amounts of materials. Analogous to TLC or GLC, HPLC involves a differential migration process wherein sample components are selectively retained by a stationary phase.[676] Since the *N*-methylcarbamates are thermally labile and often break down during direct GLC analysis, HPLC is rapidly becoming an alternative method for their analysis. Multi-residue direct analysis of carbamates by HPLC is relatively simple and has efficiencies comparable to GLC although sensitivities may be somewhat poorer. Much of the column theory is analogous to GLC. Several good texts on liquid chromatography are available[676-680] and review articles on HPLC of pesticides have been published.[681-684] Theory of HPLC columns are available in several journals such as *Journal of Chromatographic Science* and technical information on HPLC columns is available from most commercial chromatographic supply companies.

In liquid chromatography (LC) the mobile phase is a solvent and the stationary phase may be a solid such as microparticular or pellicular silica gel (adsorption chromatography), a liquid coated on a solid support (liquid-liquid, partition, or bonded-phase chromatography), ionic resins (ion-exchange chromatography), or gels (exclusion chromatography). The two main types of LC used in (non-ionic) pesticide analysis are the adsorption and liquid-liquid modes.

Adsorption, or normal phase, LC is recommended for the separation of moderately polar water-insoluble compounds. Normal phase entails the use of a polar stationary phase and a nonpolar mobile phase. In the adsorption mode, HPLC of carbamates on silica is preferred since alumina packing produces low resolution and dissymmetric peak shape; the mobile phase is usually isopropanol in hexane.[685] Physically coated liquid stationary phases (such as β,β-oxydipropionitrile) are being used less often since chemically bonded phases have been introduced. Modern technology has produced chemically bonded liquid phases on siliceous supports that allow these columns to be operated in either the "normal" or "reverse" phase mode. Several HPLC liquid stationary phases with polar functional groups (cyano, amino) have been chemically bonded to microparticulates or glass beads; these phases are actually adsorbents but they can function as both partition or adsorption media. These columns are used for nonpolar and polar compounds and selectivity is achieved through changes in solvent polarity.

Today, one of the most versatile and popular modes in HPLC is the "reverse phase". In reverse-phase HPLC, both the stationary and mobile phase polarities are reversed relative to "normal" phase HPLC; thus, the stationary chemically bonded phase is nonpolar (typically C_{18}, octadecylsilane) and the mobile phase is polar. In general, the mobile phase is a combination of water-miscible solvents (methanol, acetonitrile) and water.

TLC (Section XII.B) is a useful technique for selecting a solvent system for elution of compounds. Direct extrapolation from TLC to HPLC can frequently be achieved with the added benefit of easier quantitation, greater speed and efficiency, and if desired, convenient sample collection; sometimes a slight change in mobile phase composition may be required to adjust for differences in adsorbent activity.[676] Recently, some of the column packing materials for HPLC have become available on TLC plates. Thus, reversed-phase TLC using silica gel plates coated with octadecylsilane (C_{18}) can be used to work out HPLC separations, solvent systems, and retention times before HPLC analysis (or vice versa).

Unlike the myriad of liquid phases used in GC, there are a limited number of common columns, manufactured under different trade names, available for HPLC. However, in HPLC the co-partner to the stationary phase is the mobile phase and it is manipulation of this parameter that produces much of the selectivity and separations available with HPLC today.[686] The mobile phase may be isocratic (constant composition) or varied in concentration (gradient elution, solvent programming) throughout the analysis. The elutropic series (Volume I, Chapter 2) is a useful means of selecting solvent polarities. Special consideration in the choice of solvent should be given to purity, detector compatibility, and UV cut-off. For example, benzene is a strong UV absorber and therefore is not used as a mobile phase with a UV detector; sample extracts should not be in benzene or aliphatic hydrocarbons that contain aromatics since a peak is produced that is often mistaken for a carbamate pesticide.

Many of the carbamates can be analyzed in both the normal and reverse-phase modes (Figure 53). Choice of LC mode is often dependent on compound polarity (note reversal of elution order between operating modes) but also the separation of the compound peak(s) from sample background. Most crop extracts contain interferences (i.e., UV absorbers) that mask part of the chromatogram and the compound must be resolved from these co-extractives. Modification of the mobile phase will allow optimi-

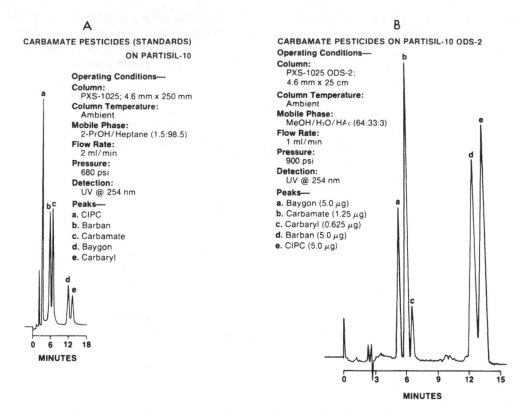

FIGURE 53.   Elution of some carbamates by normal (A) and reverse phase (B) HPLC. (Courtesy Whatman Chemical Separation Division, Clifton, N.J., 1978.)

Table 26
CHANGES IN MOBILE PHASE TO ELUTE COMPOUNDS FROM HPLC COLUMNS

Media	Compounds elute with solvent front. Change solvent to	Compounds elute at long retention time. Change solvent to
Adsorption	Less polar	More polar
Polar bonded phases	Less polar	More polar
Reverse phase, C_{18}	More polar	Less polar

Courtesy Whatman Chemical Separation Division, Clifton, N.J.

zation of the separation. Typical changes in solvent polarity that effect eluting compounds are shown in Table 26. In general, when using reverse-phase LC, the more nonpolar the compound the higher the proportion of organic solvent; as the polarity of the compound increases, the water content of the mobile phase is increased to effect separation.

Lawrence and Turton[687] tabulated HPLC data on analysis conditions for 166 pesticides. While the carbamates may be determined individually by HPLC,[84,358,414,683,688-689] the real advantage of HPLC is its applicability to multi-residue analysis.[690] Thurston[691] achieved partial separation of 23 carbamates isocratically on normal or reverse-phase columns. Unfortunately, the wide range of polarities of the carbamates precludes effective, fast separation of these compounds isocratically; however, solvent programming can be used successfully to effect their separation. Moye[681] resolved ten carbamate pesticides in 20 min on a μC_{18} reverse-phase column by solvent

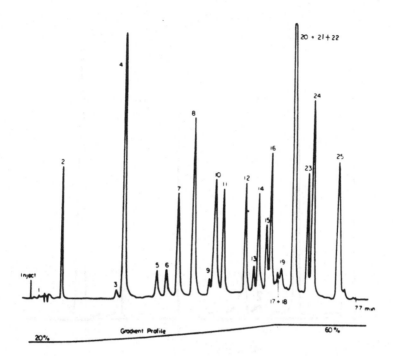

FIGURE 54. Separation of carbamates by reverse phase HPLC. Column, μ-Bondapak C_{18}; mobile phase 20 to 60% acetonitrile in water, concave gradient (60 min); flow rate, 1.0 mℓ/min; UV detection at 220 nm (1.0 AUFS); injection volume, 1.0 μℓ. Peaks: (1) solvent front, (2) methomyl, (3) aldicarb, (4) isolan, (5) propoxur, (6) carbofuran, (7) Mobam, (8) carbaryl, (9) Landrin®, (10) IPC, (11) carbanolate, (12) methiocarb, (13) mexacarbate, (14) phenmedipham, (15) CIPC, (16) EPTC, (17) bufencarb, (18) captafol, (19) barban, (20) cycloate, (21) vernolate, (22) pebulate, (23) butylate, (24) diallate, and (25) triallate. (Reproduced from Sparacino, C. M. and Hines, J. W., *J. Chromatogr. Sci.*, 14, 549, 1976. By permission of Preston Publications, Inc.)

programming from 30 to 50% acetonitrile in water. If partial resolution is achieved for some compounds under one set of conditions, they are easily separated with another mobile-phase concentration.

The development of more efficient columns has led to better separations.[692,693] Sparacino and Hines[685] studied the separation of 30 carbamate pesticides on small particle, high efficiency columns using both the normal- and reverse-phase modes. In the normal phase, the overall selectivity of Si-10, CN-10, and NH-10 were generally similar although each showed selectivity to certain compounds. Reverse-phase chromatography gave generally superior results to the normal phase; acetonitrile-water gave the best overall results on a μC_{18} column. Relative elution orders are provided for both LC modes as well as chromatograms of separations of various carbamate mixtures under a variety of chromatographic conditions; an example of the separation of 25 carbamates by reverse phase HPLC is shown in Figure 54. This work is invaluable to anyone pursuing HPLC of carbamate pesticides.

Not only can the parent carbamates but also their metabolites can be determined by HPLC, for example, aldicarb and its sulfone and sulfoxide,[685] carbaryl and naphthols,[513,688] carbofuran, 3-hydroxy and 3-keto carbofuran,[374,375,694] mexacarbate and its metabolites,[84] and ETU.[666,697]

Most of the HPLC methods in the literature deal only with standards or with formulations.[513,688,689] Examples of analysis of carbamates in water[691,695] or foods[358,374,414,696] have been shown but few references on HPLC of environmental residues have been published.[156]

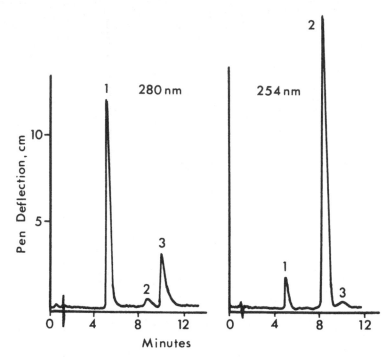

FIGURE 55.   Chromatograms of a mixture of 200 ng of carbofuran (1), 135 ng of 3-keto carbofuran (2), and 200 ng of 3-hydroxy carbofuran (3) at 254 and 280 nm detection. Mobile phase, 5% 2-propanol in *iso*-octane; column, 25 cm × 2.2 mm i.d. packed with LiChrosorb® Si 60; UV detector sensitivity 0.01 absorbance units full-scale. (Reprinted with permission from Lawrence, J. F. and Leduc, R., *J. Agric. Food Chem.*, 25, 1362, 1977. Copyright 1977, American Chemical Society.)

In the past, one of the major drawbacks to analyzing pesticide residues by HPLC has been the general lack of sensitivity relative to EC-GLC. Although many of the carbamates contain an aromatic ring and exhibit strong UV absorption, sensitivities have been 2 to 3 times less than that of EC-GLC; Thurston[691] observed that the sensitivity of carbamates to UV detection ranged from 20 to 1500 ng. Part of the problem lay with the use of a fixed wavelength UV detector (usually 254 or 280 nm). Variable wavelength UV detectors have allowed for an increase in sensitivity.[694] Sparacino and Hines[685] found that except for carbaryl, methomyl, and Mobam, the λ_{max} for most carbamates lay in the region 190 to 210 nm where they had an extinction coefficient of >9,000; further, extinction coefficients at λ_{max} could be 2 to 3 orders of magnitude larger than at 254 or 280 nm. Indeed, they found that low wavelength UV monitoring is a significant approach to a nearly universal LC detector. Effects of wavelength on sensitivity are illustrated in Figure 55 and Table 27. Special consideration must be paid to the solvent purity and UV cut-off when working at low UV wavelengths.

Development of new LC detectors such as fluorometric, EC, IR, and the variable wavelength scanning UV has improved detection limits. Reviews of LC detectors have been made.[479,681,698-700] Many of these detectors may be used separately or in combination to take advantage of the properties of the compounds.

While direct fluorescent detection of carbamates[332,354,701] may be limited, derivatization prior to HPLC may significantly increase sensitivity and selectivity.[555,702] Since many of the derivatives are less polar than their parents, easier separations may be possible. Fluorogenic labeling of *N*-methylcarbamates (see Section XII.D) has been applied prior to HPLC by Frei and co-workers[695,703] (Figure 56); detection limits of 1

Table 27

SENSITIVITIES OF SOME CARBAMATE PESTICIDES TO UV DETECTION AFTER HPLC. MINIMUM DETECTABLE QUANTITY (MDQ) AT WAVELENGTH OF MAXIMUM UV ABSORPTION AND AT 254 AND 280 NM

Compound	MDQ ng	(λ)	MDQ-254nm ng	MDQ-280nm ng	½FSD ng	(λ)
Barban	1.8	(206)	18.4	99.2	30.0	(206)
CIPC	1.0	(207)	13.8	—	7.0	(207)
Isolan	9.5	(205)	1620.0	—	38.0	(205)
IPC	3.3	(199)	18.1	132.8	33.2	(199)
EPTC	5.1	(205)	610.5	—	68.0	(205)
Carbaryl	3.6	(222)	8.7	10.9	45.2	(222)
Mexacarbate	11.0	(201)	10.2	—	88.0	(196)
Carbanolate	6.0	(198)	68.5	—	40.0	(198)
Aldicarb	5.3	(207)	14.3	—	60.0	(207)
Aldicarb sulfoxide	27.0	(210)	23.2	—	240.0	(210)
Aldicarb sulfone	61.3	(210)	2385.0	—	328.0	(210)
Landrin®	1.4	(198)	83.3	—	42.0	(198)
Cycloate	11.4	(206)	508.0	—	88.8	(206)
Methomyl	18.2	(233)	12.1	—	136.0	(233)
Mobam	11.7	(223)	2.0	73.9	92.0	(223)
Diallate	9.5	(202)	328.0	—	94.0	(202)
Carbofuran	1.1	(205)	59.2	4.3	5.7	(205)
Butylate	4.5	(205)	603.0	—	50.1	(205)
Triallate	22.5	(205)	28.8	—	39.2	(205)
Pebulate	59.0	(205)	970.0	—	188.0	(205)
Vernolate	10.9	(204)	808.5	—	94.0	(204)
3-Hydroxycarbofuran	1.2	(200)	55.5	11.7	38.7	(200)
Propoxur	13.6	(200)	39.7	136.0	54.5	(200)
Bufencarb	37.0	(200)	118.8	—	224.0	(200)
Methiocarb	1.2	(202)	18.3	35.0	6.5	(202)
1-Naphthol	0.5	(210)	22.6	0.7	4.4	(210)

Reproduced from Sparacino, C. M. and Hines, J. W., *J. Chromatogr. Sci.*, 14, 549, 1976. By permission of Preston Publications, Inc.

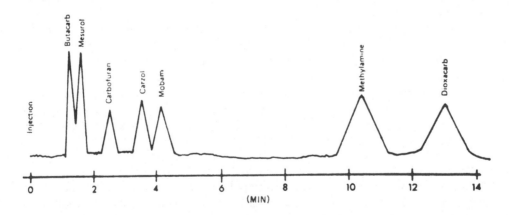

FIGURE 56. HPLC separation of dansylated carbamates. Column conditions: dimension 1m × 2.4 mm i.d., mobile phase 2% acetone in hexane, plain Corasil®, linear velocity 0.7 cm/sec; detector, Turner® fluorometer (filters, 810 primary, 827 secondary). (Reproduced from Frei, R. W., Lawrence, J. F., Hope, J., and Cassidy, R. M., *J. Chromatogr. Sci.*, 12, 40, 1974. By permission of Preston Publications, Inc.)

to 10 ng per 4-μl injections are reported. This procedure appears suitable for the determination of environmental residues such as in water, soil, and plants. Lawrence and Leduc[358,694] compared the fluorescent and direct UV detection of carbaryl and carbofuran; the response per nanogram of carbaryl as a dansyl derivative by fluorescence was about six- to eightfold greater than with direct UV detection of carbaryl and about 1.3 times more sensitive than UV detection of the derivative at 254 nm, while carbofuran was 10 times more sensitive by dansylation-fluorometric determination. Selectivity by fluorescence was also superior.

In the above examples, the derivatives were prepared prior to HPLC analysis. An alternative approach is to use post-column derivatization.[702,704-706] Hence, the HPLC column may be used to separate the compounds with their characteristic functional groups intact prior to on-line derivatization and detection.

Moye and co-workers[707,708] reported on the separation of seven carbamates by HPLC; as the carbamates eluted they were hydrolyzed in line under alkaline conditions to methylamine, which was subsequently reacted with o-phthalaldehyde and 2-mercaptoethanol. The resulting fluorophore was monitored with a fluorescence detector. Krause[709] evaluated and further refined this method.

Another method of increasing sensitivities is to inject a large sample volume; unlike GLC this is practical in HPLC. Depending on the size of the sample-injection loop, up to 2 ml of sample can be injected, but some detrimental effect on column efficiency may be obtained.[710] Usually it is preferred to inject the same solvent as the mobile phase. Extracts may also be concentrated prior to HPLC, but depending on sample type, background interferences and other problems may occur. Kikta and Stange[711] recommended using an internal standard, such as phenones, to compensate for daily variations in chromatographic separation and sensitivity; this application was shown for a carbofuran mixture.

Two further aspects of HPLC that may be overlooked are its potential as a screening device[696,712] and as a sample cleanup[713,714] or extraction-concentration technique (Section VII.A.1).[715] Sample extracts may be cleaned up by HPLC in lieu of or as an adjunct to column chromatography (Section VIII) and the judicious choice of mobile phase(s) makes it possible to elute and collect the compounds of interest free of interferences for subsequent quantitation by GLC or other HPLC techniques. One drawback to this procedure is the possible deterioration of the HPLC column due to contamination from the crude extract; special pre- or guard columns have been designed to alleviate part of this problem.

At the present time, HPLC may be somewhat limited in its applicability to routine determination of trace quantities of carbamates in environmental samples; however, future developments in the field of HPLC columns and detectors may make this technique the instrument of choice for analysis of the heat-labile carbamates as well as many other classes of pesticides. Nevertheless, the multi-residue applicability is well documented and more and more laboratories are using HPLC for the analysis of carbamates. HPLC equipment has also been interfaced with mass spectrometers (Section XII.F) for quantitation and identification of compounds.[441,716,717]

B. TLC

TLC refers to the liquid-solid chromatographic separation of compounds on a thin-film adsorbent. TLC has found wide application in the field of pesticide analysis due to its simplicity, selectivity, and availability in most laboratories. Prior to GC, TLC methods were developed for individual or different classes of pesticides, but recently, TLC applications have been limited to cleanup of samples, metabolic studies, and perhaps most importantly it is used as one of the confirmatory evidences for clinical and medico-legal cases in which positive unambiguous identification and quantitative determinations, particularly in the latter category, are essential.[718] Newer techniques, such as mass spectroscopy (Section XII.F) may replace this confirmatory aspect but to date few laboratories have this capability and may still rely heavily on TLC to complement other analytical techniques. Stahl[719] edited a monograph on TLC applications for the laboratory, and Aly and Faust[720] have reviewed the significance of TLC analysis to the field of water pollution.

As with many TLC methods, no one adsorbent or solvent system has been demonstrated as superior for the analysis of carbamate pesticides. Choice of adsorbent depends in part on the compound(s) to be separated and on the method of visualization; basic alumina may cause hydrolysis of some carbamates to their phenols during the separation. Many different solvent systems have been employed to achieve significantly different R_f values for confirmatory purposes as well as to achieve separation of compounds from co-extractives. In all cases, co-chromatography with authentic standards must be carried out in order to make qualitative and quantitative conclusions. Much effort has also been employed in the area of detection methods for carbamates on TLC plates; among these are spectrophotometric, chromogenic, fluorogenic, and enzyme-inhibition techniques. In general, sensitivities 10 to 100 times greater than colorimetric methods are achieved and these levels often approach those obtainable with GLC.

Chiba and Morley[721] showed that carbaryl and 1-naphthol could be determined in crops without cleanup using TLC. After separation on silica gel, the plates were sprayed with methanolic KOH followed by p-nitrobenzenediazonium fluoborate. Finocchiaro and Benson found that cleanup was in fact necessary for many crops prior to the TLC step, which was accomplished on alumina oxide G plates.[255,348,349]

Many of the chromogenic agents used in PC[27] or colorimetry (Section XII.E) may also be used in TLC. Some sprays, however, may not be specific only to carbamates and indeed some may react with most pesticide classes. Finocchiaro and Benson[722] studied the TLC behavior of 19 carbamates and 3 urea pesticides with regard to adsorbents, developing systems, chromogenic agents and their use for quantitation of the pesticides in crops. The compounds studied could be divided into four groups based on their R_f values and their response to chromogenic agents. Silica and alumina were used as adsorbents and different solvent systems were employed to separate the various classes of pesticides. Selectivity was achieved by using different sprays.

In a comparison of silica, alumina, and polyamide layers, two chromatographic systems on alumina or polyamide were found most suitable for separation of 12 N-methylcarbamates used in Japan; detection was by p-nitrobenzenediazonium fluoroborate.[723] Polyamide layers were previously used with various solvent systems to separate 22 carbamate pesticides.[724] These layers were found to be superior to silica gel layers for detection of UV-absorbing compounds; polyamide layers produce a bright background in transmitted UV light allowing a lower detection limit. Pinacryptol yellow which was used as the chromogenic spray produced dark grey spots against a light (fluorescent) background.

El-Dib[308] studied the detection, separation, and identification of N-methylcarbamates and related ureas in water samples on acidic silica gel TLC plates. Six solvent systems were found to give selective elution properties based on the relative basic characteristics of the compounds. In general, the R_f values for carbamates tended to increase in the following order: N,N-dimethylcarbamates > N-methylcarbamates > N-phenylcarbamates. After heating the plates, the resultant phenols or heterocyclic enols were visualized with p-dimethylaminobenzaldehyde while the anilines were further diazotized and reacted with 1-naphthol.

Carbamates may also be determined nondestructively *in situ* on silica gel TLC plates by reflectance spectroscopy after visualizing the plates with a reagent such as 9-dicyanomethylene-2,4,7-trinitrofluorene that forms visible pi-complexes with the pesticides.[725,726] Using this technique, the hydrolysis rates of some carbamate pesticides were studied.[30]

Ramasamy[718] showed that 11 o.p. and 7 carbamate pesticides as well as some of their phenols could be separated and identified by TLC. The combination of 2 adsorbents, 3 solvent systems and 3 chromogenic sprays provide a choice of 18 different procedures, and based on the aim of the analytical study, a system could be selected for routine screening of the compounds. Typical TLC results are shown in Table 28. Hence, TLC may be used for the identification and detection of carbamates in various substrates such as autopsy tissue.[727] High-performance TLC may be also used with comparable detection limits to those obtained with TLC or HPLC.[728]

The versatility of TLC to pesticide analysis is well documented. Selectivity may be achieved through the appropriate selection of adsorbent, solvent system, and chromogenic spray. Not all applications of TLC, however, are for quantitative analysis.[729] Many radiotracer studies involve the separation of metabolites using one- or two-dimensional TLC and detection of the metabolites by radioautography[79,185,243] (Figure 57). TLC may also be used as a cleanup technique either to separate pesticides from co-extractives or to isolate a specific compound.[353,424,437,730,731] The appropriate area of the TLC plate may be scraped and the compound (spot) subjected to subsequent analysis. In this manner distinct compounds may be isolated for confirmation by MS or other spectrographic technique.

C. Enzymatic Techniques

The carbamate insecticides, like the o.p. pesticides, are cholinesterase inhibitors and it is not surprising that analytical procedures were developed based on this fact. Most of the analytical procedures involve TLC separation of compounds followed by their determination by enzyme inhibition (TLC-EI). After TLC separation, the plates are sprayed with an enzyme-buffer solution and allowed to incubate. Subsequently the plates are sprayed with an enzyme substrate and because the enzyme is inhibited by the pesticide, the substrate at the TLC spot will not be hydrolyzed and shows up as a colored area on the background. Mendoza[732-734] has reviewed the TLC-EI technique. A brief discussion of some applications is presented below, and the reader is referred to other review articles for additional details.[1,242,251,735]

Initial attempts to determine carbaryl were based on existing EI methods for o.p. pesticides.[736] Mendoza et al.[737] showed that ten o.p. pesticides and carbaryl could be resolved by TLC on silica gel and determined with good sensitivity and reproducibility by EI. Steer-liver homogenate was used as the esterase and indoxyl or substituted indoxyl acetates were used as the substrate; at pHs greater than 8.0, pesticide inhibition spots were well defined and appeared as white spots on a highly colored background. Wales et al.[350] extended this procedure to five other carbamates with a sensitivity of 1 μg or less; carbaryl could be detected at 5 ng and the described screening procedure demonstrated that carbaryl could be determined at the tolerance level in several commodities.

Silica gel was shown to be the best adsorbent for TLC-EI work.[738] Bee-head esterase was found to be the most sensitive esterase for both o.p. and carbamate insecticides, but it requires more effort to prepare than other esterases. Pig-liver esterase is much more sensitive than beef-liver esterase for carbamates whereas the opposite was found for the o.p. pesticides. Detection levels for five carbamates were lower than those reported for GLC although this fact was determined prior to development of many of the derivatization procedures. Mendoza and Shields[739] indicated that TLC and colorimetry using indophenyl acetate are complimentary techniques.

Voss[740] described an automated procedure based on cholinesterase inhibition for the determination of o.p. and carbamate pesticides. The lower limit of detection depends

Table 28
$R_F \times 100$ AND COLOR OF TLC SPOTS WITH CARBAMATES AND SOME PARENT PHENOLS

Insecticide	Plate[a]	$R_F \times 100$ in solvent systems 1	2	3	Sprays A	B	C
Mobam	a	26	52	12	Bl-G	Pu	G
	b	34	81	21	Light Bl	Pu	W
Phenol (from Mobam)	a	36	49	21	Dark G appears without heating	Pu	Dark G without UV
	b	24	56	23	Pu-Bl turns dark Bl on heating	Pu	Br
Propoxur	a	26	52	14	G-Br	Pu	G
	b	38	76	32	Light G	Pu	Br
Phenol (from propoxur)	a	49	69	46	G-Br	Pu	G
	b	59	74	52	Pu-Bl turns Bl on heating	Pu	NR
Landrin®	a	26	58	17	Light Br	Light Y-Br	G
	b	53 67[b]	82	35	Light Y-Br	Light Y-Br	Light Br
Phenol (from landrin)	a	34	52	24	Light Br	Light Y-Br	G
	b	58 68[b]	60	30 13[b]	G-Br	Light Y-Br	Light Br
2,3,5-Isomer of landrin	a	25	55	17	G	NR	G
	b	58 68[b]	76	30 40[b]	Light G	Pu-Br	W
Carbamult	a	27	59	19	G	Pu-Br	G
	b	83	76	36 51[b]	Light G	Light Br	W
Phenol (from carbamult)	a	37	55	28	G	Pu-Br	G
	b	67	60	35	Sky Bl turns G-Bl on heating	Light Br	Light Br
OMS 15	a	25	64	19	G-Bl	Pu-Br	G
	b	59	75	35	Light G	Orange-Br	Light Br
Phenol (from OMS 15)	a	35	75	28	Dark G-Bl	Pu-Br	G
	b	64	57	32	Sky Bl turns G-Bl on heating	Orange-Br	Light Br
Carbaryl	a	18 34[b]	67	10	Intense G-Br	G	G
	b	44	76	24	Bl-G turns G-Br on heating	Sky blue to Bl-G	W
OMS 1028	a	32	71	20	Bl-G	Pu	NR
	b	65	76	33 45[b]	Light G	Pu	W

Note: Solvent systems: (1) hexane-acetone (3 + 1 v/v); (2) chloroform-acetone (9 + 1 v/v); (3) hexane-acetone (5 + 1 v/v). Sprays: (A) 2,6-dibromo-*p*-benzoquinone-4-chlorimine; (B) KOH followed by *p*-nitrobenzenediazonium fluoroborate; (C) bromine fumes followed by 4-methylumbelliferone in ammonia solution. Color code: Bl, blue, Br, brown, G, gray, Gr, green, Pu, purple, W, white, Y, yellow. NR, no reaction.

[a] a, Silica gel G plate; b, aluminum oxide G plate.
[b] Denotes R_f value with spray reagent C alone. All spots with spray reagent C appear as fluorescent or quenched areas, or both.

Adapted from Ramasamy, M., *Analyst (London)*, 94, 1075, 1969. With permission.

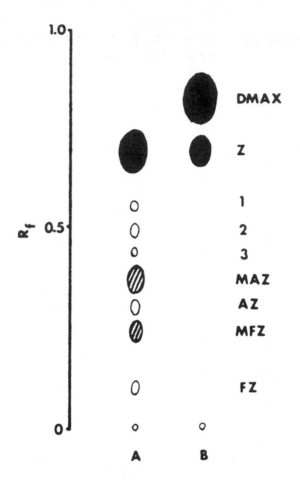

FIGURE 57.   Representation of the autoradiograms of the
two types of pathways of mexacarbate metabolism from
silica gel TLC plates developed by using ethyl ether-hexane-
ethanol (77:20:3). (A) metabolic pattern of bacteria HF-3;
(B) metabolic pattern of *Trichoderma viride*. Compound
identity: DMAX, 4-dimethylamino-3,5-xylenol; Z, 4-di-
methylamino-3,5-xylyl   methylcarbamate   (mexacarbate);
MAZ, 4-methylamino-3,5-xylyl methylcarbamate; AZ, 4-
amino-3,5-xylyl methylcarbamate; MFZ, 4-methylformam-
ido-3,5-xylylmethylcarbamate; FZ, 4-formamido-3,5-xylyl
methylcarbamate; 1,2,3, unidentified metabolites. (Reprinted with permission from Benezet, H. J. and Matsumura, F., *J. Agric. Food Chem.*, 22, 427, 1974. Copyright
1974, American Chemical Society.)

primarily on the inhibition potency of the particular compound under investigation
and upon whether the system was operated in a pre-inhibition (most sensitive) or simultaneous inhibition (least sensitive) mode. Although 40 samples per hour could be
determined, no differentiation between inhibitors may be made and this technique is
limited to studies on individual compounds.

The enzymatic approach to multi-residue analysis is somewhat limited especially
since it is specific only to esterase inhibitors and not to any one class of pesticides.
There are also inherent difficulties such as obtaining ultrapure compounds and enzymes, the short shelf-life of the enzyme, and the need for rigidly controlled experimental conditions.

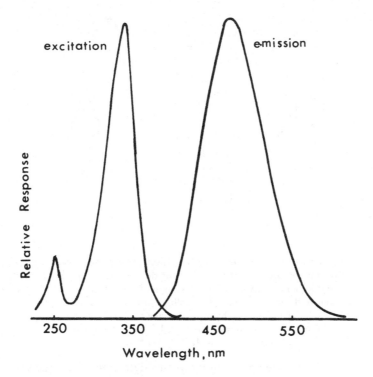

FIGURE 58.   Fluorescence excitation and emission spectra of carbaryl in 0.1 *N* NaOH (1 μg/m*l*). (Reprinted with permission from Argauer, R. J., Shimanuki, H., and Alvarez, C. C., *J. Agric. Food Chem.*, 18, 688, 1970. Copyright 1970, American Chemical Society.)

D. Fluorescence

Fluorescence describes the phenomenon whereby compounds immediately emit radiant energy after first having absorbed radient energy of some particular frequency.[741] In most cases the emitted radiation is of a longer wavelength than the absorbed radiation. Fluorescent methods are now appearing in the literature as this technique often affords the desired sensitivity for residue analysis. Development of specific instrumentation to complement basic residue analytical instruments has been slow, but many new fluorescent detectors are now appearing on the market.

There are several different applications of fluorescence to the detection of pesticide residues. Residues may be determined directly in a spectrophotofluorometer. For example, carbaryl exhibits a fluorescence (excitation 290 nm; emission 350 nm) in methanol and when hydrolyzed in basic solution the fluorescence shifts to that characteristic of the salt of 1-naphthol (excitation 340 nm; emission 480 nm) (Figure 58).[701] Many papers describe the determination of carbamates by *in situ* fluorometric measurement on TLC plates.[742] Development of detectors for GLC and HPLC has allowed spectrofluorometric determination of eluted compounds.[513,525] Fluorescent determination by TLC[743] or HPLC[703] may also be applied to nonfluorescent pesticides following fluorogenic labeling with a derivative such as dansyl chloride.

Eberle and Gunther[505] studied the fluorescence of carbaryl, mexacarbate, dimetilan, isolan, and Pyrolan and found that none of the parent carbamates fluoresced in the 300 to 700 nm region. On hydrolysis, both 1-naphthol and 4-dimethylamino-3,5-xylenol (from mexacarbate) fluoresced while the 2-pyrazolones did not. Subsequent studies indicated that some of these carbamates do in fact fluoresce in other solvents. The fluorescence of 17 insecticidal carbamates and 1-naphthol in ethanol was studied by

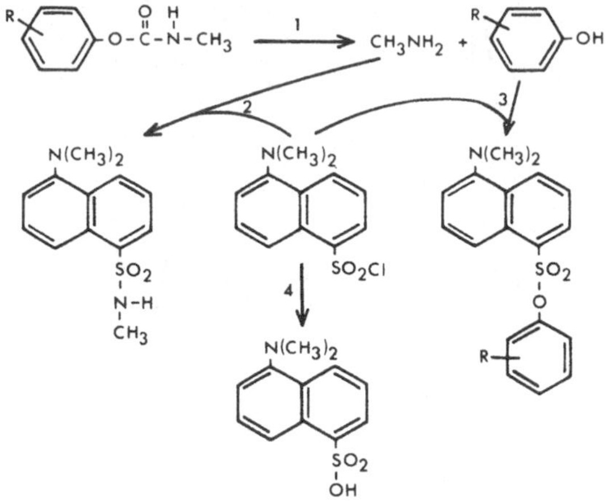

FIGURE 59. The reaction of dansyl chloride with *N*-methylcarbamates. (1) Hydrolysis of the carbamate to the phenol and methylamine, (2) coupling of the dansyl chloride to the methylamine, (3) coupling of the dansyl chloride to the phenol, and (4) hydrolysis of the dansyl chloride to sulfonic acid. (Adapted from Frei, R. W. and Lawrence, J. F., *J. Chromatogr.*, 61, 174, 1971. With permission.)

Bowman and Beroza[744] and the applicability of this detection technique was shown for residues of propoxur, aminocarb, carbofuran, Mobam, carbaryl and 1-naphthol in milk. Detection limits ranged from 0.06 ppm for 1-naphthol to 3 ppm for Mobam.

Spectrophotofluorometry is especially suited to the analysis of carbaryl due to its natural fluorescence and to 1-naphthol due to the sensitive fluorescent nature of the naphthoate anion. Residues have been reported in honey bees,[332] fish,[745] bean and tomato leaves,[354] and fruit and vegetables[358,746] using fluorescence. Carbaryl formulations have also been analyzed by this technique.[513,701] The value of this method over colorimetric analysis is that it eliminates the color reagent and increases sensitivities, but the problem still remains that one must remove interfering materials that absorb excitation radiation or the emitted fluorescence, or fluoresce strongly themselves.[354]

Several fluorogenic spray reagents have been developed for visualizing carbamates by fluorescence. Sulfur-containing pesticides, such as methiocarb, were resolved by TLC and exposed to bromine vapors and then sprayed with solutions of metal ions or pH-sensitive fluorescent indicators;[747-749] sensitivity was only about 10 times better than colorimetrically resulting in a 100 to 300 ng detection limit. Several flavones have been employed as fluorogenic sprays for polar carbamates separated on nonpolar cellulose TLC plates.[750,751] Detection limits of 0.01 to 0.06 μg have been reported. Since other polar pesticides such as the organophosphates and *s*-triazines were also visualized, this method may be limited to a screening procedure.

An alternative to fluorescent sprays is fluorogenic labeling which involves the formation of fluorescent derivatives that may be separated by TLC and quantitatively analyzed by *in situ* fluorescence. Most of this work has been carried out by Frei and co-workers[743,752-754] and the subject has been reviewed.[555,556,755,756] The carbamates are hydrolyzed and the liberated methylamine and/or phenolic hydrolysis products may be reacted with a fluorogenic reagent. The reagent most often used is dansyl chloride (5-dimethylaminonaphthalene-1-sulfonyl chloride);[757] a typical reaction scheme is shown in Figure 59. Linear calibration curves were found up to 300 to 400 ng per spot and good reproducibility was observed above 10 ng per spot; visual detection limits approached 5 ng or less per spot whereas 1 ng could be detected instrumentally. Natu-

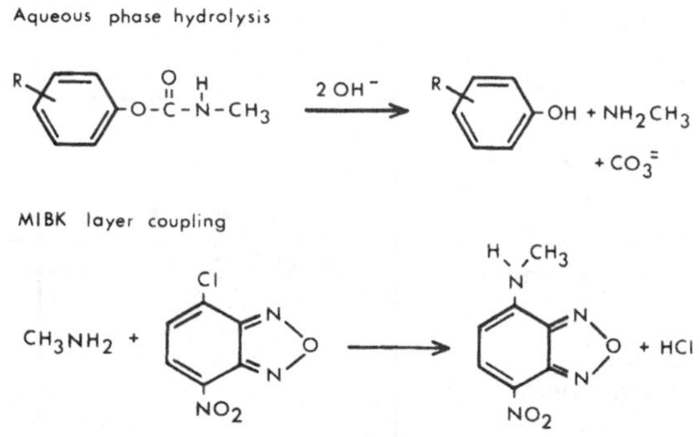

Aqueous phase hydrolysis

MIBK layer coupling

FIGURE 60. Reaction scheme for the formation of NBD-methylamine from *N*-methylcarbamates. (Reprinted with permission from Lawrence, J. F. and Frei, R. W., *Anal. Chem.*, 44, 2046, 1972. Copyright 1972, American Chemical Society.)

ral water samples were successfully analyzed at 2 ppb with no interferences from co-extractives. At this concentration in tap, lake, and sea water, recovery of methiocarb and Landrin® as their dansyl phenol was 93 to 106% while recovery of Landrin®, methiocarb, and propoxur as the dansyl methylamine (total carbamate) was 86 to 104%.[743]

Dansyl chloride forms fluorescent derivatives with both the amine and phenol hydrolysis products resulting in two derivatives suitable for quantitative fluorometric analysis. NBD-Cl reagent (4-chloro-7-nitrobenzo-2,1,3-oxadiazole) is superior to dansyl chloride in terms of selectivity. It does not form fluorescent derivatives with phenols, thiols, alcohols, or anilines but does react well with alkylamines to form highly fluorescent products.[619.758] A reaction scheme for the formation of NBD-methylamine derivatives of *N*-methylcarbamates is shown in Figure 60. Fluorogenic labeling with dansyl chloride or NBD-Cl offers two simple and sensitive methods for the determination of carbamates. Dansyl chloride is specific for individual carbamates, however, NBD-Cl offers more selectivity between *N*-methyl and *N,N*-dimethylcarbamates. Good recoveries have been shown for many carbamate insecticides in water and soil substrate at environmental concentrations (1 ng/spot) for mexacarbate, aminocarb, Landrin, and carbofuran. Comparative thin-layer chromatograms are shown in Figure 61.

Multi-residue determination of carbamates by *in situ* fluorometry has shown that TLC compares favorably with GLC methods in both sensitivity and reproducibility, and also has the advantage of not requiring a rigorous cleanup. Fluorogenic labeling may also be used in conjunction with HPLC methods either as a pre-[703] or post-[707.708] column reagent (Section XII.A). Unfortunately these techniques have not yet been widely used in the analysis of carbamate residues.

E. Colorimetric Methods

Prior to the development of GC, colorimetric methods were developed for carbaryl and some procedures were extended to other carbamates. The general approach is to couple a chromogenic reagent with the parent compound or a hydrolysis product, such as the phenol, and determination of the colored complex spectrophotometrically at the appropriate wavelength;[759] the sensitivity varies depending on the compound but sub-microgram amounts have been reported. In general, these levels are not as sensitive compared to many of the instrumental techniques and hence colorimetry is not generally used for residue analysis of environmental samples. This technique also suffers from a lack of selectivity. Several carbamates can be coupled with the same reagent;

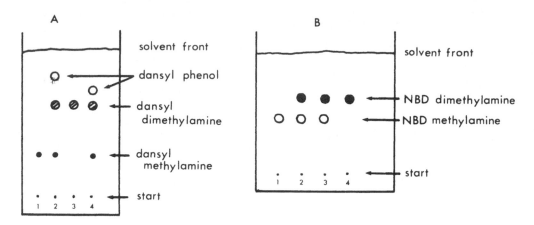

FIGURE 61.   Comparative thin-layer chromatograms of fluorescent derivatives. (A) Dansyl derivatives of aminocarb and mexacarbate eluted on silica gel G with benzene-acetone (98 + 2). Spot identity: 1, dansyl methylamine standard; 2, dansyl mexacarbate reaction mixture; 3, dansyl dimethylamine standard; 4, dansyl aminocarb reaction mixture. (B) NBD derivatives of aminocarb and mexacarbate eluted on silica gel G with tetrahydrofuran-chloroform (2 + 98). Peak identity: 1, NDB methylamine standard; 2, NBD mexacarbate reaction mixture; 3, NBD aminocarb reaction mixture; 4, NBD dimethylamine standard. (Adapted from Frei, R. W. and Lawrence, J. F., *J. Assoc. Off. Anal. Chem.*, 55, 1259, 1972. With permission.)

however, measurements are usually made at different wavelengths and due to the broad peak maxima quantitation of several compounds in a sample may be difficult. Therefore this technique is limited to screening or quantitative analysis of single compounds after exposure or after preseparation of mixtures. Also, as in any colorimetric method, problems can arise due to co-extractive interferences or lack of stability of the colored complex.

Miskus et al.[342] showed that carbaryl could be determined after hydrolysis to 1-naphthol and coupling with *p*-nitrobenzenediazonium fluoborate to form a purple complex in basic solution which could be determined at 590 nm. To achieve greater specificity and reduce cleanup requirements, Johnson[107,344,345] modified the procedure to permit color development in acetic acid wherein a yellow-orange complex with a maximum absorbance at 475 nm is obtained. Several changes[346] in the extraction and cleanup procedure were required to make the method suitable for many substrates, but the colorimetric basis remained the same and is one of the AOAC recommended methods for carbaryl.[255] The coupling reaction has been extended to several other phenolic carbamates such as carbofuran[158] and Mobam[341] and has been used as a rapid screening procedure for carbaryl.[760] Carbaryl formulations have also been determined using this method.[761]

Ramasamy[762] described a simple, quick, and reliable method for the colorimetric determination of eight carbamate insecticide residues. Carbamates which yielded phenols with free para-positions on hydrolysis were reacted with diazotized 3-nitroaniline-4-sulfonic acid to form an orange-red complex that could be measured from 416 to 540 nm depending on the carbamate. Aly[311] described an alternative method for the spectrophotometric determination of carbaryl in natural water. Carbaryl is hydrolyzed to 1-naphthol which is reacted with 4-aminoantipyrine and an oxidizing agent (potassium ferricyanide) in alkaline media to form a red-colored complex that could be determined at 500 nm. Subsequently, this method was applied to other carbamate insecticides that on hydrolysis formed a phenol or heterocyclic enol.[73] All compounds were determined at the same wavelength. Mexacarbate, as its phenol, may be determined after reaction with leutoarsenotungstic acid and measurement of the color intensity at 700 nm.[85] 2,6-Dibromo-*p*-benzoquinonechloroimide can be used to couple directly with 2-isopropoxyphenol (from propoxur) in a weakly alkaline medium to give a blue indophenol derivative which is determined at 595 nm.[763]

Other chromogenic reagents have been employed to estimate carbaryl directly rather than as a 1-naphthol-complex. Diazotized 2,5-dichloroaniline sulfate reacts with carbaryl to form a ruby-red color measurable at 510 nm while 1-naphthol produces a distinct and less sensitive yellow-red color; other carbamates do not react.[764] This procedure has been applied to the analysis of formulations[764] and several crops.[765] Carbaryl can also be estimated directly using diazotized o-toluidine hydrochloride, which produces a crimson-red color in basic solution with a maximum absorbance at 520 nm.[347] Carbofuran may be determined after electrophilic reaction with diazotized aniline to form a yellow compound with an absorption maximum at 460 nm.[766]

Total aldicarb has been determined in several substrates using a colorimetric procedure based on the specificity of the carbamoyloxime group.[390-392,767-769] Unfortunately it does not allow for the determination separately of the more toxic sulfoxide metabolite. Aldicarb and its metabolites are hydrolyzed with base to form their oximes, which are then hydrolyzed in acid to release hydroxylamine. The hydroxylamine is oxidized with iodine to nitrous acid, which stoichiometrically diazotizes sulfanilic acid, which in turn couples with 1-naphthylamine to form a related quantity of dye in solution that is measured spectrophotometrically at 530 nm.

Colorimetric methods have also been widely used for the determination of N-phenylcarbamates (Table 14). In general, the N-phenylcarbamates (IPC, CIPC) are hydrolyzed in acid or base, and the resultant aniline or chloroaniline is steam distilled into dilute HCl. The distillate is then diazotized with nitrous acid and coupled with N-(1-naphthyl) ethylenediamine. The aniline (IPC) complex is determined at 555 nm while the chloroaniline (CIPC) complex is determined at 540 nm.[420]

Thiocarbamates may be hydrolyzed in sulfuric acid to their respective amines, which are converted to the yellow complex of cupric dithiocarbamate that has an absorption peak at 440 nm.[428]

1. Colorimetric Method for Dithiocarbamate and Ethylenebisdithiocarbamate Fungicides

The most expedient way to analyze for these residues (Table 1) is via acid hydrolysis to evolve carbon disulfide which in turn is reacted to form a yellow cupric salt that is measured colorimetrically (435 nm). In 1951, Clarke[770] and Lowen[771] first showed the applicability of this technique, based on earlier work,[772,773] for the analysis of maneb residues in field crops. Since that time the method has become somewhat standardized although numerous minor and major modifications have been made.[40,213,551] Most noteworthy was the change in chromogenic solution from Viles reagent[772] to Cullen's reagent (cupric acetate monohydrate in diethanolamine-ethanol).[774] There are two main disadvantages to this procedure: first, it is nonspecific since alkyl monodithiocarbamates, bisdithiocarbamates, and thiuram disulfides all react similarly, and second, the method is relatively insensitive (about 1 μg detection limit). In most cases, however, it is sufficient to determine that a CS_2 evolving compound (hence these fungicides) have caused the contamination. If the history of the sample is known, quantitation may be based on the contaminant, otherwise it is usually expressed as the zineb equivalent. Semiqualitative confirmation of the compound identity may be obtained by PC.[775] Depending on the size of the digestion flask,[444] the sample size may be increased to obtain sufficient sensitivity, however, these compounds are relatively nontoxic (LD_{50} > 1000) and thus a high contamination level is required to warrant an environmental or health hazard. Such a case might occur after an accidental spill or misapplication. As discussed earlier (Section XI.C) a toxic carcinogen (ethylenethiourea) is associated with the EBDC fungicides.

F. Mass Spectrometry (MS)

MS is a valuable tool for the identification, confirmation, and elucidation of the structure of organic compounds including pesticides. A sample (compound) in the gaseous state is subjected to an ionization process in a mass spectrometer which produces a mass spectrum that is composed of the relative number of ions of different mass/charge ratios (m/e) that co-exist for a set of conditions defined by the apparatus.[776] A gas (or liquid) chromatograph can be interfaced with a mass spectrometer to form a GC/MS system which allows each eluting GC peak to be analyzed by the MS; in this case the MS may be the most definitive detector for a GC.[777] During GC/MS, eluting peaks may be observed and saved for analysis by using a total ion monitor or a single ion monitor. A complete discussion of MS instrumentation and analysis of carbamate pesticides by MS is beyond the scope of this work, and only the salient points in MS of carbamates are discussed below based primarily on electron impact (EI) mass spectroscopy. Those readers wishing more details are referred to the work of Safe and Hutzinger[776] or other reference material on mass spectroscopy[778-785] and MS of pesticides.[582,586,786-788]

Although an unique mass spectrum is not always obtained for a compound because of instrumental conditions, the fragmentation patterns are usually similar and comparison of samples with standard spectra or subjecting a spectrum to computer library searches helps the analyst in the identification of an unknown. As in all forms of chromatography or spectroscopic studies of unknowns, the analysis of pure standards under the same conditions is imperative. This is especially true in GC/MS studies of carbamates, as this class of compounds is particularly prone to GC effects and quite often the injected compound is not the one that is detected (Section IX). It may be for this reason that much of the controversy on direct GC of carbamates exists in the literature.

Large differences in fragmentation patterns and m/e intensities may be obtained because of the analytical conditions under which the spectra were obtained. For example, in the direct inlet mode the whole molecule is subjected to MS analysis, whereas in GC/MS it is the eluted material that is analyzed. Therefore, in GC/MS, the molecule is also subjected to various on-column effects such as the thermal breakdown of *N*-methylcarbamates to their respective phenol. The choice of ionization process can also have a significant effect upon the mass spectrum that is produced (Figure 62).

EI mass spectral studies of carbamates have been made[586,786,787,789-791] and the results show similar fragmentation patterns that allow identification of this class of compounds. In general, the *N*-methylcarbamates produce a very weak molecular ion ($M <$ 10%) and an intense peak (usually the base peak) at M-58 indicating loss of methylisocyanate, or at M-57 showing rearrangement (Figure 63). Subsequent breakdown is based on the resultant phenol since both the parent phenol and carbamate spectra are virtually identical below the m/e of the phenol (Figure 64).

N,N-dimethylcarbamates have a characteristic base peak at m/e 72 due to the *N,N*-dimethylisocyanate ion. (Phenyl ureas may also have a base peak at m/e 72).

N-phenylcarbamates exhibit an intense anilino ion (m/e 93) formed, presumably, by a McLafferty rearrangement with transfer of a hydrogen atom of the ethyl group. CIPC and IPC have m/e 43 ion [(CH₃)₂CH]⁺ as the base peak.

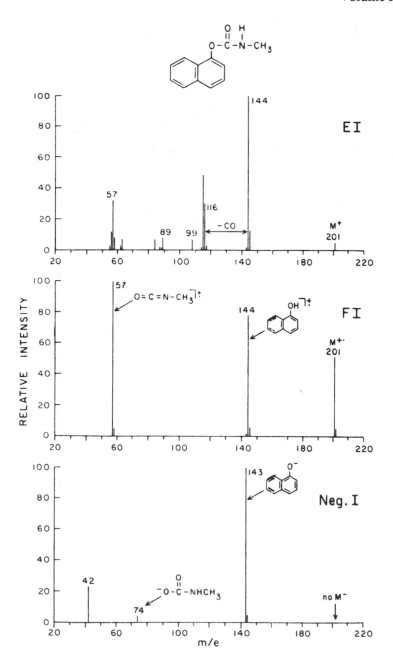

FIGURE 62. Mass spectra of carbaryl obtained under different ionization conditions. Top, 70 eV electron impact mass spectrum; middle, field ion mass spectrum; bottom, negative ion mass spectrum.[776]

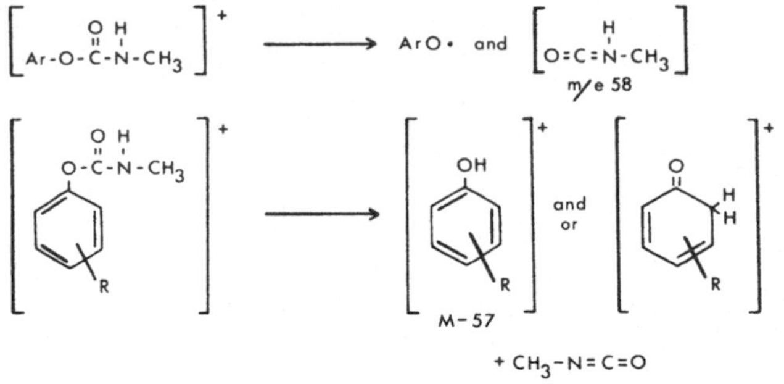

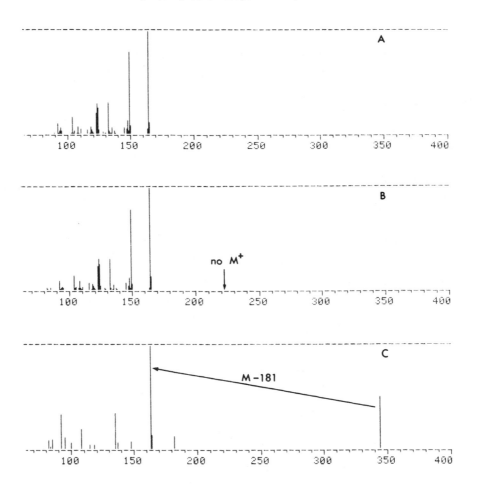

FIGURE 63. Fragmentation of a typical *N*-methylcarbamate. (Adapted from Damico, J. N. and Benson, W. R., *J. Assoc. Off. Anal. Chem.*, 51, 347, 1968. With permission.)

FIGURE 64. Mass spectra of (A) carbofuran phenol, (B) carbofuran, and (C) carbofuran-pentafluorobenzyl ether derivative obtained on a Hewlett-Packard® 5992 GC/MS.

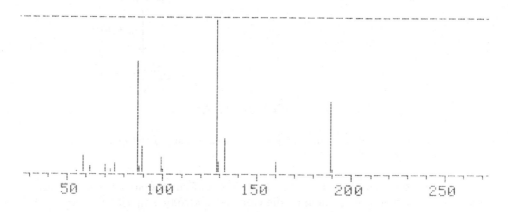

FIGURE 65. Fragmentation of EPTC (thiocarbamate).[776]

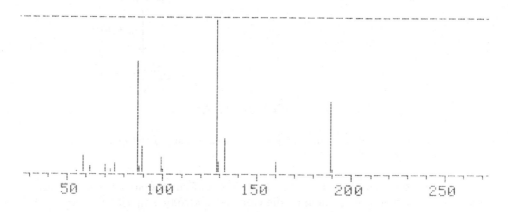

FIGURE 66. Mass spectrum of EPTC obtained with a Hewlett-Packard® 5992 GC/MS.

A typical fragmentation pattern for the thiocarbamates is shown for EPTC (Figures 65 and 66). Oxime carbamates and metabolites show facile cleavage to both the sulfur and oxime moieties (Figure 67) during direct inlet MS; however, during GC/MS a complete spectrum is rarely obtained and the eluted species is usually a smaller fragment containing the sulfur atom (Section IX.D). The dithiocarbamates do not chromatograph well and would not be expected to show a good mass spectrum on GC/MS examination.

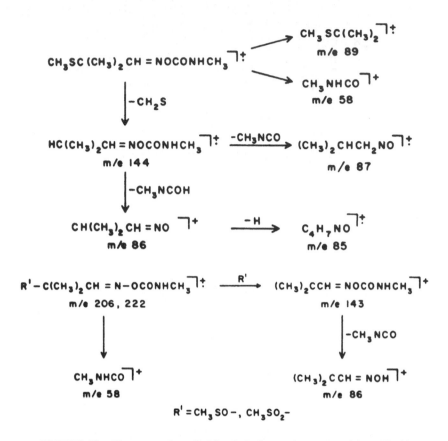

FIGURE 67.   Fragmentation of aldicarb (oxime carbamate) and its sulfoxide and sulfone metabolites.[776]

MS has also been employed to elucidate and confirm the structures of many synthetic carbamate metabolites such as the isomeric hydroxy-1-methylcarbamates[792] and hydroxymethylcarbamates,[226] and also biological products.[218,236,273,793]

Many of the derivatives mentioned previously that are employed for analytical quantitation have also been examined by MS.[277,585] Derivatization is used in GC to produce a more stable or suitable compound for detection, and the same advantages apply to GC/MS analysis of carbamates. Typical of the many derivatives is the presence of moderate to intense molecular ion for the derivative as well as the ion for the cleavage of the derivative fragment from the carbamate to produce the phenol or methylamine moiety (Figures 64 and 68).

MS may also be used for quantitative analysis. Often, the intensities of ions from fragmentation are erratic and make quantitation difficult; however, Chapman and Robinson[581] found that methane-chemical ionization of HFB derivatives of carbofuran and its carbamate metabolites produced an ion at m/e 228 corresponding to HFB-methylamine and which had an intensity directly proportional to injected amounts. Other indicative ions were also produced (Figure 68). Hence, GLC separation of the derivatives followed by chemical ionization and single (multi) ion monitoring will allow both quantitation and qualitation. Sensitivity of 0.05 ppm in crops was reported. This method should be applicable to any N-methylcarbamate-HFB derivative.

Using chemical ionization GC/MS, Hall and Harris[521] found that all the investigated carbamates (Table 22) were readily protonated with isobutane reactant gas to form the M + 1 ion which is most significant (Figure 69). Several carbamates also produced

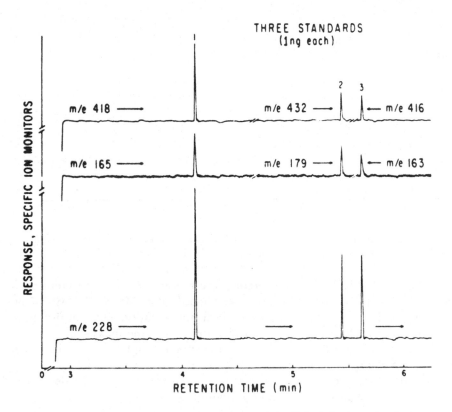

FIGURE 68. Reconstituted mass fragmentogram of the HFB derivatives of carbofuran (1), 3-keto carbofuran (2), and 3-hydroxy carbofuran (3). m/e 228 = methylamine-HFB; m/e 165 and 179 correspond to the phenols of carbofuran and 3-keto carbofuran, respectively, and m/e 163 is the phenol-H_2O from 3-hydroxy carbofuran; m/e 418 and 432 correspond to the HFB-parent molecule for carbofuran and 3-keto carbofuran, respectively, and m/e 416 is the 3-hydroxy carbofuran diHFB-HFB acid. (Adapted from Chapman, R. A. and Robinson, J. R., *J. Chromatogr.*, 140, 209, 1977. With permission.)

secondary ions (M-n) which were attributed to the corresponding phenols or isocyanates. Using this system, it was confirmed that the carbamates were chromatographed as the parent compound; carbaryl was the only compound to show extraneous peaks.

G. Other Spectroscopic Techniques

Other spectroscopic techniques, such as NMR, IR, and UV offer unique and selective means for the identification of compounds. However, these techniques suffer from a general lack of sensitivity for pesticide residue analysis; these limitations have been discussed.[777,794-796] Furthermore, these techniques require extensive cleanup of samples to obtain "pure" compounds for analysis. Compilations of NMR[797,798] and IR[799,800] spectra of carbamates have been made. Despite the limitations of sensitivity, IR analysis has been conducted for residues[317,801] but more often on formulations.[802-804] The main advantage of these techniques, however, is the identification, structural elucidation, and confirmation of pesticides, their metabolites, and their degradation products.

XIII. CONFIRMATION OF RESIDUES

In many cases, particularly where a high concentration of contaminant is apparent, residues should or must be confirmed. The most generally accepted method of confirmation of pesticide identity is on two or more GC columns of different polarity, and/

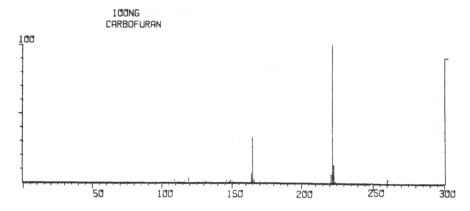

FIGURE 69.   Chemical-ionization mass spectrum using iso-butane reaction gas of carbofuran. (Reprinted from Hall, R. C. and Harris, D. E., *J. Chromatogr.*, 169, 245, 1979. With permission.)

or with different detectors if heteroatoms are present, or with different derivatives (Volume I, Chapter 4). Many laboratories concede that MS or GC/MS of pesticide residues is one of the best techniques for confirming identity. TLC confirmation, once used more extensively prior to the introduction of routine mass spectrometers, may also be used as may confirmation by p-values, polarography, spectrometry (IR, UV, NMR), or via chemical derivatization. In general, pesticide residues are confirmed by analysis under different analytical conditions (i.e., different columns, detectors, derivatives, or methodologies) or by ancillary techniques.

XIV. APPENDIX

Cross-Reference of Pesticides Mentioned in Text

Common or Trade Name	Chemical or Common Name
aldicarb	*
aminocarb	*
Banol	carbanolate
Baygon®	propoxur
BPMC	2-*sec*-butyl phenyl *N*-methylcarbamate
bufencarb	*
butacarb	3,5-di-*t*-butylphenyl *N*-methylcarbamate
butylate	*
Bux®	bufencarb
carbanolate	*
carbaryl	*
carbofuran	*
CDEC	sulfallate
chlorpropham	CIPC
chlorxylam	carbanolate
CIBA-10573	*O*-(1,3-dithiolan-2-yl)-phenyl *N*-methylcarbamate
CIPC	*
cycloate	*

* See Table 1 for chemical name and structure.

XIV. APPENDIX (continued)

Common or Trade Name	Chemical or Common Name
diallate	*
dimetilan	*
EPTC(Eptam)	*
ferbam	*
Furadan®	carbofuran
HRS-1422	3,5-diisopropylphenyl N-methylcarbamate
IPC	*
isolan	1-isopropyl-3-methylpyrazolyl-(5)-dimethylcarbamate
Lannate®	methomyl
Landrin®	no common name*
maneb	*
Matacil	aminocarb
Meobal®	3,4-dimethylphenyl N-methylcarbamate
Mesurol®	methiocarb
metalkamate	bufencarb
metacaptodimethur	methiocarb
methiocarb	*
methomyl	*
mexacarbate	*
Mobam	no common name *
molinate	*
nabam	*
oxamyl	*
pebulate	*
pentanochlor	N-(3-chloro-4-methylphenyl)-2-methylvaleramide
phemedipham	3-[(methoxycarbonyl)amino]phenyl (3-methyl-phenyl)carbamate
pirimicarb	*
promacyl	3-methyl-5-isopropylphenyl-N-(n-butanoyl)-N-methylcarbamate
promecarb	3-methyl-5-isopropylphenyl-N-methylcarbamate
propham	IPC
propoxur	*
Pyramat	2-n-propyl-4-pyrimidinyl-6-dimethylcarbamate
Pyrolan	3-methyl-1-phenyl-5-pyrazolyl dimethylcarbamate
Sevin®	carbaryl
sulfallate	2-chloroallyl diethyldithiocarbamate
Sutan	butylate
swep	*
Temik®	aldicarb
terbucarb(terbutol)	2,6-di-tert-butyl-p-tolyl N-methylcarbamate
thiofanox	*
thiram	*
triallate	*
UC-10854	3-isopropylphenyl methylcarbamate
vernolate	*
XMC	3,5-dimethylphenyl N-methylcarbamate
Zectran®	mexacarbate
zineb	*

REFERENCES

1. **Kuhr, R. J. and Dorough, H. W.**, *Carbamate Insecticides: Chemistry, Biochemistry, and Toxicology,* CRC Press, Boca Raton, Fla., 1976.
2. **Tisdale, W. H. and Flenver, A. L.**, Derivatives of dithiocarbamic acid as pesticides, *Ind. Eng. Chem.,* 34, 501, 1942.
3. **Gysin, H.**, Über einige neue Insektizide, *Chimia,* 8, 205, 1954.
4. **Lambrech, J. A.**, (Union Carbide Corp.), U.S. Patent 2,903,478, September 8, 1959.
5. **Kolbezen, M. J., Metcalf, R. L., and Fukuto, T. R.**, Insecticidal activity of carbamate cholinesterase inhibitors, *J. Agric. Food Chem.,* 2, 864, 1954.
6. **Friesen, G.**, The effect of phenylurethan on germination of seedling growth of the oat and wheat, *Planta,* 8, 666, 1929.
7. **Templeman, W. G. and Sexton, W. A.**, Effect of some arylcarbamic esters and related compounds upon cereals and other plant species, *Nature (London),* 156, 630, 1945.
8. **Harman, M. W. and D'Amico, J. J.**, (Monsanto Chemical Co.), U.S. Patent 2,854,467, September 30, 1958; U.S. Patent 2,919,182, December 29, 1959.
9. **Tilles, H. and Antognini, J.**, (Stauffer Chemical Co.), U.S. Patent 2,919,327, November 17, 1959.
10. **Payne, L. K., Jr., Stansbury, H. A., Jr., and Weiden, M. H. J.**, The synthesis and insecticidal properties of some cholinergic trisubstituted acetaldehyde O-(methylcarbamoyl)oximes, *J. Agric. Food Chem.,* 14, 356, 1966.
11. **Johnson, O,**, CW report on pesticides '72. I, *Chem. Week,* 34, June 21, 1972. II, *Chem. Week,* 18, July 26, 1972.
12. **Smart, N. A.**, Collaborative studies of methods for pesticide residue analysis, *Residue Rev.,* 64, 1, 1976.
13. **Sherma, J.**, Manual of Analytical Quality Control for Pesticides and Related Compounds in Human and Environmental Samples, EPA-600/1-76-017, Thompson, J. F., Ed., Health Effects Research Laboratory, U.S. Environmental Protection Agency, Research Triangle Park, N.C., February 1976.
14. **Conant, J. B.**, *The Chemistry of Organic Compounds,* Rev. ed., Macmillan, New York, 1939, chap. 14.
15. **Melnikov, N. N.**, Chemistry of pesticides, *Residue Rev.,* 36, 1971.
16. **Schlagbauer, B. G. L. and Schlagbauer, A. W. J.**, The metabolism of carbamate pesticides - A literature analysis. I, *Residue Rev.,* 42, 1, 1972.
17. **Packer, K.**, Ed., *Nanogen Index, A Dictionary of Pesticides and Chemical Pollutants,* Nanogen International, Freedom, Calif., 1975.
18. **Martin, H. and Worthing, C. R.**, Eds., *Pesticide Manual,* 4th ed., British Crop Protection Council, Croyden, England, 1974.
19. Guide to the Chemicals Used in Crop Protection, 6th ed., Agriculture Canada Publication No. 1093, Department of Supply and Service, Ottawa, 1973.
20. **Zweig, G.**, Ed., *Analytical Methods for Pesticides and Plant Growth Regulators,* Vol. I to IX, Academic Press, New York, 1963 to 1977.
21. *Herbicide Handbook of the Weed Science Society of America,* 3rd ed., Weed Science Society of America, Champaign, Ill., 1974.
22. **Sharvelle, E. G.**, *The Nature and Use of Modern Fungicides,* Burgess Publishing, Minneapolis, Minn., 1961.
23. **Beroza, M., Inscoe, M. N., and Bowman, M. C.**, Distribution of pesticides in immiscible binary solvent systems for cleanup and identification and its application in extraction of pesticides from milk, *Residue Rev.,* 30, 1, 1969.
24. **Gunther, F. A., Westlake, W. E., and Jaglan, P. S.**, Reported solubilities of 738 pesticide chemicals in water, *Residue Rev.,* 20, 1, 1968.
25. **Bailey, G. W. and White, J. L.**, Herbicides: a compilation of their physical, chemical, and biological properties, *Residue Rev.,* 10, 97, 1965.
26. **Abdel-Wahab, A. M. and Cassida, J. C.**, Photooxidation of two 4-dimethylaminoaryl methylcarbamate insecticides (Zectran and Matacil) in bean foliage and of alkylaminophenyl methylcarbamates on silica gel chromatoplates, *J. Agric. Food Chem.,* 15, 479, 1967.
27. **Krishna, J. G., Dorough, H. W., and Casida, J. E.**, Synthesis of N-methylcarbamates via methyl isocyanate-C[14] and chromatographic purification, *J. Agric. Food Chem.,* 10, 462, 1962.
28. **Dittert, L. W. and Higuchi, T.**, Rates of hydrolysis of carbamate and carbonate esters in alkaline solution, *J. Pharm. Sci.,* 52, 852, 1963.
29. **Casida, J. E., Augustinsson, K. -B., and Jonsson, G.**, Stability, toxicity, and reaction mechanism with esterases of certain carbamate insecticides, *J. Econ. Entomol.,* 53, 205, 1960.

29a. **Fukuto, T. R., Fahmy, M. A. H., and Metcalf, R. L.,** Alkaline hydrolysis, anticholinesterase, and insecticidal properties of some nitro-substituted phenyl carbamates, *J. Agric. Food Chem.,* 15, 273, 1967.

30. **MacNeil, J. D., Frei, R. W., and Frei-Häusler, M.,** Electron donor-acceptor reagents in the analysis of pesticides, *Int. J. Environ. Anal. Chem.,* 2, 323, 1973.

31. **Wolfe, N. L., Zepp, R. G., and Paris, D. F.,** Carbaryl, propham and chlorpropham: a comparison of the rates of hydrolysis and photolysis with the rate of biolysis, *Water Res.,* 12, 565, 1978.

32. **Mabey, W. and Mill, T.,** Critical review of hydrolysis of organic compounds in water under environmental conditions, *J. Phys. Chem. Ref. Data,* 7, 383, 1978.

33. **Stedman, E.,** Studies on the relationship between chemical constitution and physiological action. I. Position isomerism in relation to miotic activity of some synthetic urethanes, *Biochem. J.,* 20, 719, 1926.

34. **Daly, N. J. and Ziolkowski, F.,** The thermal decomposition of carbamates. II. Methyl *N*-methylcarbamate, *Aust. J. Chem.,* 25, 1453, 1972.

35. **Freed, V. H., Hague, R., and Vernetti, J.,** Thermodynamic properties of some carbamates and thiolcarbamates in aqueous solution, *J. Agric. Food Chem.,* 15, 1121, 1967.

36. **Smith, A. E. and Fitzpatrick, A.,** The loss of five thiolcarbamate herbicides in nonsterile soils and their stability in acidic and basic solution, *J. Agric. Food Chem.,* 18, 720, 1970.

37. **Patchett, G. G., Batchelder, G. H., and Menn, J. J.,** Eptam, in *Analytical Methods for Pesticides, Plant Growth Regulators, and Food Additives,* Vol. 4, Zweig, G., Ed., Academic Press, New York, 1964, chap. 12.

38. **Patchett, G. G., Batchelder, G. H., and Menn, J. J.,** Tillam, in *Analytical Methods for Pesticides, Plant Growth Regulators, and Food Additives,* Vol. 4, Zweig, G., Ed., Academic Press, New York, 1964, chap. 25.

39. **Lyman, W. R. and Lacoste, R. J.,** New developments in the chemistry and fate of ethylenebisdithiocarbamate fungicides, in *Pesticides, Lectures held at the IUPAC 3rd Int. Congr. Pesticide Chemistry, Helsinki, Environ. Qual. Saf., Suppl. Vol. 3),* Coulston, F. and Korte, F., Eds., Georg Thieme, Stuttgart, 1975, 67.

40. **Ripley, B. D.,** Residues of ethylenebisdithiocarbamates on field-treated fruits and vegetables, *Bull. Environ. Contam. Toxicol.,* 22, 182, 1979.

41. **Metcalf, R. L. and Fukuto, T. R.,** Some effects of molecular structure upon anticholinesterase and insecticidal activity of substituted phenyl *N*-methylcarbamates, *J. Agric. Food Chem.,* 15, 1022, 1967.

42. **Miskus, R. P., Look, M., Andrews, T. L., and Lyon, R. L.,** Biological activity as an effect of structural changes in aryl *N*-methylcarbamates, *J. Agric. Food Chem.,* 16, 605, 1968.

43. **Gilbert, E. E., Peterson, J. O., and Walker, G. L.,** Insecticidal evaluation of isomeric methylthioisopropylphenyl *N*-methylcarbamates, *J. Agric. Food Chem.,* 16, 787, 1968.

44. **Mahfouz, A. M. M., Metcalf, R. L., and Fukuto, T. R.,** Influence of the sulfur atom on the anticholinesterase and insecticidal properties of thioether *N*-methylcarbamates, *J. Agric. Food Chem.,* 17, 917, 1969.

45. **Scharpt, W. G.,** (FMC Corporation), U.S. Patent 3,474,170, October 21, 1969; U.S. Patent 3,474,171, October 21, 1969.

46. **Metcalf, R. L., Fukuto, T. R., Collins, C., Borck, K., El-Aziz, S. A., Munoz, R., and Cassil, C. C.,** Metabolism of 2,2-dimethyl-2,3-dihydrobenzofuranyl-7-*N*-methylcarbamate (Furadan) in plants, insects, and mammals, *J. Agric. Food Chem.,* 16, 300, 1968.

47. **Skraba, W. J. and Young, F. G.,** Radioactive Sevin (1-naphthyl-1-carbon-14 *N*-methylcarbamate), a convenient synthesis, *J. Agric. Food Chem.,* 7, 612, 1959.

48. **Dorough, H. W. and Casida, J. E.,** Nature of certain carbamate metabolites of the insecticide Sevin, *J. Agric. Food Chem.,* 12, 294, 1964.

49. **Knaak, J. B., Tallant, M. J., Bartley, W. J., and Sullivan, L. J.,** The metabolism of carbaryl in the rat, guinea pig, and man, *J. Agric. Food Chem.,* 13, 537, 1965.

50. **Stansbury, H. A., Jr. and Miscus, R.,** Sevin, in *Analytical Methods for Pesticides, Plant Growth Regulators, and Food Additives,* Vol. 2, Zweig, G., Ed., Academic Press, New York, 1964, chap. 39.

51. **Metcalf, R. L., Osman, M. F., and Fukuto, T. R.,** Metabolism of ^{14}C-labeled carbamate to $^{14}CO_2$ in the housefly, *J. Econ. Entomol.,* 60, 445, 1967.

52. **Gard, L. N. and Ferguson, C. E., Jr.,** IPC, in *Analytical Methods for Pesticides, Plant Growth Regulators, and Food Additives,* Vol. 4, Zweig, G., Ed., Academic Press, New York, 1964, chap. 14.

53. **Gard, L. N. and Ferguson, C. E., Jr.,** CIPC, in *Analytical Methods for Pesticides, Plant Growth Regulators, and Food Additives,* Vol. 4, Zweig, G., Ed., Academic Press, New York, 1964, chap. 7.

54. Bartley, W. J., Heywood, D. L., Steele, T. E. N., and Skraba, W. J., Synthesis of C¹⁴-labeled 2-methyl-2-(methylthio)propionaldehyde O-(methylcarbamoyl)oxime, *J. Agric. Food Chem.*, 14, 604, 1966.

55. Tilles, H., Thiolcarbamates. Preparation and molar refractions, *J. Am. Chem. Soc.*, 81, 714, 1959.

56. O'Brien, R. D., *Insecticides: Action and Metabolism,* Academic Press, New York, 1967.

57. Metcalf, R. L. and Fukuto, T. R., Effects of chemical structure on intoxication and detoxication of phenyl N-methylcarbamates in insects, *J. Agric. Food Chem.*, 13, 220, 1965.

58. Ashton, F. M. and Crafts, A. S., *Mode of Action of Herbicides,* John Wiley & Sons, New York, 1973.

59. Kearney, P. C. and Kaufman, D. D., *Herbicides: Chemistry, Degradation and Mode of Action,* Vol. 1 and 2, 2nd ed. Marcel Dekker, New York, 1975.

60. Corbett, J. R., *The Biochemical Mode of Action of Pesticides,* Academic Press, New York, 1974.

61. Klingman, G. C., Ashton, F. M., and Noordhoff, L. J., *Weed Science: Principles and Practices,* John Wiley & Sons, New York, 1975.

62. Kaars Sijpesteijn, A., Dekhuijzen, H. M., and Vonk, J. W., Biological conversion of fungicides in plants and microorganisms, in *Antifungal Compounds,* Vol. 2, Siegel, M. R. and Sisler, H. D., Eds., Marcel Dekker, New York, 1977, chap. 3.

63. Engst, R. and Schnaak, W., Residues of dithiocarbamate fungicides and their metabolites on plant foods, *Residue Rev.*, 52, 45, 1974.

64. Ludwig, R. A., Thorn, G. D., and Unwin, C. H., Studies on the mechanism of fungicidal action of metallic ethylene bis dithiocarbamates, *Can. J. Bot.*, 33, 42, 1955.

65. Owens, R. G., Organic sulfur compounds, in *Fungicides, an Advanced Treatise,* Vol. 2, Torgeson, D. C., Ed., Academic Press, New York, 1969, chap. 5.

66. Ludwig, R. A. and Thorn, G. D., Chemistry and mode of action of dithiocarbamate fungicides, *Adv. Pest Control Res.*, 3, 219, 1960.

66a. Thorn, G. D. and Ludwig, R. A., *The Dithiocarbamates and Related Compounds,* Elsevier, Amsterdam, 1962.

67. Lukens, R. J., *Chemistry of Fungicidal Action,* Springer-Verlag, New York, 1971.

68. Edwards, C. A., Ed., *Environmental Pollution by Pesticides,* Plenum Press, London, 1973.

69. Edwards, C. A., *Persistent Pesticides in the Environment,* 2nd ed., CRC Press, Boca Raton, Fla., 1973.

70. Gerakis, P. A. and Sficas, A. G., The presence and cycling of pesticides in the ecosphere, *Residue Rev.*, 52, 69, 1974.

71. Haque, R. and Freed, V. H., Behaviour of pesticides in the environment: "Environmental chemodynamics", *Residue Rev.*, 52, 89, 1974.

72. Aly, O. M. and El-Dib, M. A., Studies of the persistence of some carbamate insecticides in the aquatic environment, in *Fate of Organic Pesticides in the Aquatic Environment,* Advances in Chemistry Series 111, Gould, R. F., Ed., American Chemical Society, Washington, D.C., 1972, chap. 11.

73. Aly, O. M. and El-Dib, M. A., Studies of the persistence of some carbamate insecticides in the aquatic environment. I. Hydrolysis of Sevin, Baygon, Pyrolan and dimetilan in waters, *Water Res.*, 5, 1191, 1971.

74. Karinen, J. F., Lamberton, J. G., Stewart, N. E., and Terriere, L. C., Persistence of carbaryl in the marine estuarine environment. Chemical and biological stability in aquarium systems, *J. Agric. Food Chem.*, 15, 148, 1967.

75. Lamberton, J. G. and Claeys, R. R., Degradation of 1-naphthol in sea water, *J. Agric. Food Chem.*, 18, 92, 1970.

76. Aly, O. M. and El-Dib, M. A., Photodecomposition of some carbamate insecticides in aquatic environments, in *Organic Compounds in Aquatic Environments,* Faust, S. J. and Hunter, J. V., Eds., Marcel Dekker, New York, 1971, chap. 20.

77. Sikka, H. C., Miyazaki, S., and Lynch, R. S., Degradation of carbaryl and 1-naphthol by marine microorganisms, *Bull. Environ. Contam. Toxicol.*, 13, 666, 1975.

78. Bollag, J. -M., Czaplicki, E. J., and Minard, R. D., Bacterial metabolism of 1-naphthol, *J. Agric. Food Chem.*, 23, 85, 1975.

79. Benezet, H. J. and Matsumura, F., Factors influencing the metabolism of mexacarbate by microorganisms, *J. Agric. Food Chem.*, 22, 427, 1974.

80. Ahlrichs, J. L., Chandler, L., Monke, E. J., and Reuszer, H. W., Effect of Pesticide Residues and Other Organo-Toxicants on the Quality of Surface and Ground Water Resources, PB Report No. 211080, U.S. National Technical Information Service, Springfield, Va., 1970.

81. Sheridan, R. P. and Simms, M. A., Effect of the insecticide Zectran (mexacarbate) on several algae, *Bull. Environ. Contam. Toxicol.*, 13, 565, 1975.

82. Paris, D. F. and Lewis, D. L., Chemical and microbial degradation of ten selected pesticides in aquatic systems, *Residue Rev.*, 45, 95, 1973.

83. **Caro, J. H., Freeman, H. P., and Turner, B. C.,** Persistence in soil and losses in run-off of soil-incorporated carbaryl in a small watershed, *J. Agric. Food Chem.,* 22, 860, 1974.

84. **Hosler, C. F., Jr.,** Degradation of Zectran in alkaline water, *Bull. Environ. Contam. Toxicol.,* 12, 599, 1974.

85. **Mathews, E. W. and Faust, S. D.,** The hydrolysis of Zectran in buffered and natural waters, *J. Environ. Sci. Health,* B12, 129, 1977.

86. **Wauchope, R. D. and Haque, R.,** Effects of pH, light and temperature on carbaryl in aqueous media, *Bull. Environ. Contam. Toxicol.,* 9, 257, 1973.

87. **Kanazawa, J.,** Uptake and excretion of organophosphorus and carbamate insecticides by fresh water fish, Motsugo, *Pseudorasbora parva, Bull. Environ. Contam. Toxicol.,* 14, 346, 1975.

88. **Eichelberger, J. W. and Lichtenberg, J. J.,** Persistence of pesticides in river water, *Environ. Sci. Technol.,* 5, 541, 1971.

89. **Faust, S. D. and Gomma, H. M.,** Chemical hydrolysis of some organic phosphorus and carbamate pesticides in aquatic environments, *Environ. Lett.,* 3, 171, 1972.

90. **Wolfe, N. L., Zepp, R. G. and Paris, D. F.,** Use of structure-reactivity relationships to estimate hydrolytic persistence of carbamate pesticides, *Water Res.,* 12, 561, 1978.

91. **Deuel, L. E., Jr., Price, J. D., Turner, F. T., and Brown, K. W.,** Persistence of carbofuran and its metabolites, 3-keto and 3-hydroxy carbofuran, under flooded rice culture, *J. Environ. Qual.,* 8, 23, 1979.

92. **Seiber, J. N., Catahan, M. P., and Barril, C. R.,** Loss of carbofuran from rice paddy water: chemical and physical factors, *J. Environ. Sci. Health,* B13, 131, 1978.

93. **Siddaramappa, R., Tirol, A. C., Seiber, J. N., Heinrichs, E. A., and Watanabe, I.,** The degradation of carbofuran in paddy water and flooded soil of untreated and retreated rice fields, *J. Environ. Sci. Health,* B13, 369, 1978.

94. **Venkateswarlu, K., Gowda, T. K. S., and Sethunathan, N.,** Persistence and biodegradation of carbofuran in flooded soils, *J. Agric. Food Chem.,* 25, 533, 1977.

95. **Soderquist, C. J., Bowers, J. B., and Crosby, D. G.,** Dissipation of molinate in a rice field, *J. Agric. Food Chem.,* 25, 940, 1977.

96. **Minakawa, O., Ishii, S., and Konno, H.,** Analytical method of residue of molinate, a herbicide in paddy field, and actions of molinate to living bodies, *Nippon Koshu Eisei Zasshi,* 25, 645, 1978, *Pestic. Abstr.* 79-0792.

97. **Grover, R.,** personal communication, 1979.

98. **DeMarco, A. C. and Hayes, E. R.,** Photodegradation of thiolcarbamate herbicides, *Chemosphere,* 8, 321, 1979.

99. **Frank, R. and Ripley, B. D.,** Land Use Activities in Eleven Agricultural Watersheds in Southern Ontario, Canada, 1975-76, International Joint Commission, Windsor, Ontario, 1977.

100. **Frank, R., Braun, H. E., Sirons, G., Holdrinet, M. V. H., Ripley, B. D., Onn, D., and Coote, R.,** Stream Flow Quality — Pesticides in Eleven Agricultural Watersheds in Southern Ontario, Canada, 1974-77, International Joint Commission, Windsor, Ontario, 1978.

101. **Frank, R., Braun, H. E., Holdrinet, M. V. H., Sirons, G. J., and Ripley, B. D.,** Agricultural and water quality in Canadian Great Lakes Basin. V. Use of pesticides in eleven agricultural watersheds and their persistence in stream water, 1975-77, *J. Environ. Qual.,* in press.

102. **Caro, J. H., Freeman, H. P., Glotfelty, D. C., Turner, B. C., and Edwards, W. M.,** Dissipation of soil-incorporated carbofuran in the field, *J. Agric. Food Chem.,* 21, 1010, 1973.

103. **Ripley, B. D.,** unpublished data, 1978.

104. **Frank, R., Sirons, G. J., and Ripley, B. D.,** Herbicide contamination and decontamination of well waters in Ontario, 1969-1978, *Pestic. Monit. J.,* 13, 120, 1979.

105. New and Revised Great Lakes Water Quality Objectives, Vol. 1, International Joint Commission, Windsor, Ontario, 1977.

106. Canadian Drinking Water Standards and Objectives 1968, Health and Welfare Canada Publication H48-1969, Department of Supply and Service, Ottawa, 1972.

106a. Guidelines for Canadian Drinking Water Quality 1978, Health and Welfare Canada Publication H48-10/1978, Department of Supply and Service, Ottawa, 1979.

106b. Pesticide Safety Handbook, Ontario Ministry of the Environment, Toronto, 1979.

106c. Guidelines and Criteria for Water Quality Management in Ontario, Ontario Ministry of the Environment, Toronto, July 1974.

106d. Water Management — Goals, policies, objectives and implementation procedures of the Ministry of the Environment, Ontario Ministry of Environment, Toronto, November 1978.

107. **Johnson, D. P. and Stansbury, H. A.,** Adaptation of Sevin insecticide (carbaryl) residue method to various crops, *J. Agric. Food Chem.,* 13, 235, 1965.

108. Carbofuran: Criteria for Interpreting the Effects of its Use on Environmental Quality, Publ. No. NRCC 16740, National Research Council of Canada, Ottawa, Ontario, 1979.

109. Williams, I. H., Brown, M. J., and Whitehead, P., Persistence of carbofuran residues in some British Columbia soils, *Bull. Environ. Contam. Toxicol.*, 15, 242, 1976.
110. Getzin, L. W., Persistence and degradation of carbofuran in soil, *Environ. Entomol.*, 2, 461, 1973.
111. Miles, J. R. W. and Harris, C. R., Insecticide residues in organic soils of six vegetable growing areas in southwestern Ontario, 1976, *J. Environ. Sci. Health*, B13, 199, 1978.
112. Proceedings of 11th Eastern Canada Pesticide Residue Workshop, Provincial Pesticide Residue Testing Laboratory, Ontario Ministry of Agriculture and Food, Guelph, Ontario, May 15 to 19, 1979.
113. Tucker, B. V. and Pack, D. E., Bux insecticide soil metabolism, *J. Agric. Food Chem.*, 20, 412, 1972.
114. Asai, R. I., Gunther, F. A., and Westlake, W. E., Influence of some soil characteristics on the dissipation rate of Landrin insecticide, *Bull. Environ. Contam. Toxicol.*, 11, 352, 1974.
115. Coppedge, J. R., Lindquist, D. A., Bull, D. L., and Dorough, H. W., Fate of 2-methyl-2-(methylthio)propionaldehyde *O*-(methylcarbamoyl)oxime(Temik) in cotton plants and soil, *J. Agric. Food Chem.*, 15, 902, 1967.
116. Andrawes, N. R., Bagley, W. P., and Herrett, R. A., Fate and carryover properties of Temik aldicarb pesticide [2-methyl-2-(methylthio)propionaldehyde *O*-(methylcarbamoyl)oxime] in soil, *J. Agric. Food Chem.*, 19, 727, 1971.
117. Leistra, M., Smelt, J. H., and Lexmond, T. M., Conversion and leaching of aldicarb in soil columns, *Pestic. Sci.*, 7, 471, 1976.
118. Smelt, J. H., Leistra, M., Houx, N. W. H., and Dekker, A., Conversion rates of aldicarb and its oxidation products in soil. I. Aldicarb sulfone, *Pestic. Sci.*, 9, 279, 1978.
119. Smelt, J. H., Leistra, M., Houx, N. W. H., and Dekker, A., Conversion rates of aldicarb and its oxidation products in soils. II. Aldicarb sulfoxide, *Pestic. Sci.*, 9, 286, 1978.
119a. Smelt, J. H., Leistra, M., Houx, N. W. H., and Dekker, A., Conversion rates of aldicarb and its oxidation products in soil. III. Aldicarb, *Pestic. Sci.*, 9, 293, 1978.
120. Severo, R., *The New York Times*, Tuesday, March 4, 1980, C1.
121. Sharom, M. S., Miles, J. R. W., Harris, C. R., and McEwen, F. L., Behaviour of 12 insecticides in soil and aqueous suspensions of soil and sediment, *Water Res.*, 14, 1095, 1980.
121a. Sharom, M. S., Miles, J. R. W., Harris, C. R., and McEwen, F. L., Persistence of 12 insecticides in water, *Water Res.*, 14, 1089, 1980.
122. Wauchope, R. D., The pesticide content of surface water draining from agricultural fields — a review, *J. Environ. Qual.*, 7, 459, 1978.
123. Felsot, A. and Dahm, P. A., Sorption of organophosphorus and carbamate insecticides by soil, *J. Agric. Food Chem.*, 27, 557, 1979.
124. Chiou, C. T., Freed, V. H., Schmedding, D. W., and Kohnert, R. L., Partition coefficient and bioaccumulation of selected organic chemicals, *Environ. Sci. Technol.*, 11, 475, 1977.
125. Briggs, G. G., A Simple Relationship Between Soil Adsorption of Organic Chemicals and their Octanol/Water Partition Coefficients, Proc. 7th Brit. Insecticide and Fungicide Conf., British Crop Protection Council, Brighton, England, 1973.
126. Kenaga, E. E., Guidelines for environmental study of pesticides: determination of bioconcentration potential, *Residue Rev.*, 44, 73, 1972.
127. Davies, J. E., Barquet, A., Freed, V. H., Haque, R., Morgade, C., Sonneborne, R. E., and Vaclavek, C., Human pesticide poisonings by a fat-soluble organophosphate insecticide, *Arch. Environ. Health*, 30, 608, 1975.
128. Lu, P. -Y. and Metcalf, R. L., Environmental fate and biodegradability of benzene derivatives as studied in a model ecosystem, *Environ. Health Perspect.*, 10, 269, 1975.
129. Metcalf, R. L., Kapoor, I. P., Lu, P. -Y., Schuth, C. K., and Sherman, P., Model ecosystem studies of the environmental fate of six organochlorine pesticides, *Environ. Health Perspect.*, 4, 35, 1973.
130. Metcalf, R. L., Sanborn, J. R., Lu, P. -Y., and Nye, D., Laboratory model ecosystem studies of the degradation and fate of radiolabeled tri-, tetra-, and pentachlorobiphenyl compared with DDE, *Arch. Environ. Contam. Toxicol.*, 3, 151, 1975.
131. Neely, W. B., Branson, D. R., and Blau, G. E., Partition coefficient to measure bioconcentration potential of organic chemicals in fish, *Environ. Sci. Technol.*, 8, 1113, 1974.
132. Tulp, M. Th. M. and Hutzinger, O., Some thoughts on aqueous solubilities and partition coefficients of PCB, and the mathematical correlation between bioaccumulation and physico-chemical properties, *Chemosphere*, 7, 849, 1978.
133. Kaufman, D. D., Degradation of carbamate herbicides in soil, *J. Agric. Food Chem.*, 15, 582, 1967.
134. Smith, A. E. and Hayden, B. J., Field persistence studies with eight herbicides commonly used in Saskatchewan, *Can. J. Plant Sci.*, 56, 769, 1976.
135. Crosby, D. G., Herbicide photodecomposition, in *Herbicides: Chemistry, Degradation, and Mode of Action*, Vol. 2, 2nd ed., Kearney, P. C. and Kaufman, D. D., Eds., Marcel Dekker, New York, 1976, chap. 18.

136. **Kaufman, D. D.**, Degradation of pesticides by soil microorganisms, in *Pesticides in Soil and Water,* Guenzi, W. D., Ed., Soil Science Society of America, Madison, Wis., 1974, chap. 8.

137. **Fang, S. C.**, Thiocarbamates, in *Herbicides: Chemistry, Degradation, and Mode of Action,* Vol 1, 2nd ed., Kearney, P. C. and Kaufman, D. D., Eds., Marcel Dekker, New York, 1975, chap. 5.

138. **Parochetti, J. V. and Warren, G. F.**, Biological activity and dissipation of IPC and CIPC, *Weed Sci.,* 16, 13, 1968.

139. **Still, G. G. and Herrett, R. A.**, Methylcarbamates, carbanilates, and acylanilides, in *Herbicides: Chemistry, Degradation, and Mode of Action,* Vol. 2, 2nd ed., Kearney, P. C. and Kaufman, D. D., Eds., Marcel Dekker, New York, 1976, chap. 12.

140. **Parochetti, J. V. and Warren, G. F.**, Vapor losses of IPC and CIPC, *Weeds,* 14, 281, 1966.

141. **Kearney, P. C.**, Influence of physiochemical properties on biodegradability of phenylcarbamate herbicides, *J. Agric. Food Chem.,* 15, 568, 1967.

142. **Kaufman, D. D., Kearney, P. C., Von Endt, D. W., and Miller, D. E.**, Methylcarbamate inhibition of phenylcarbamate metabolism in soil, *J. Agric. Food Chem.,* 18, 513, 1970.

143. **Liu, S. and Bollag, J.-M.**, Metabolism of carbaryl by a soil fungus, *J. Agric. Food Chem.,* 19, 487, 1971.

144. **Rodriguez, L. D. and Dorough, H. W.**, Degradation of carbaryl by soil microorganisms, *Arch. Environ. Contam. Toxicol.,* 6, 47, 1977.

145. **Williams, I. H., Pepin, H. S., and Brown, M. J.**, Degradation of carbofuran by soil microorganisms, *Bull. Environ. Contam. Toxicol.,* 15, 244, 1976.

146. **Fung, K. K. H. and Uren, N. C.**, Microbial transformation of S-methyl N-[(methylcarbamoyl)oxy]thioacetimidate (methomyl) in soils, *J. Agric. Food Chem.,* 25, 966, 1977.

147. **Kazano, H., Kearney, P. C., and Kaufman, D. D.**, Metabolism of methylcarbamate insecticides in soils, *J. Agric. Food Chem.,* 20, 975, 1972.

148. **Kadoum, A. M. and Mock, D. E.**, Herbicide and insecticide residues in tailwater pits: water and pit bottom soil from irrigated corn and sorghum fields, *J. Agric. Food Chem.,* 26, 45, 1978.

149. **Korn, S.**, The uptake and persistence of carbaryl in channel catfish, *Trans. Am. Fish Soc.,* 102, 137, 1973.

150. **Mitchell, L. E.**, Pesticides: properties and prognosis, in *Organic Pesticides in the Environment,* Advances in Chemistry Series 60, Gould, R. F., Ed., American Chemical Society, Washington, D. C., 1966, chap. 1.

151. **Eto, M.**, *Organophosphorus Pesticides: Organic and Biological Chemistry,* CRC Press, Boca Raton, Fla., 1974, 275.

152. **Edgington, L. V. and Peterson, C. A.**, Systemic fungicides: theory, uptake, and translocation, in *Antifungal Compounds,* Vol. 2, Siegel, M. R. and Sisler, H. D., Eds., Marcel Dekker, New York, 1977, chap. 2.

153. **Edgington, L. V., Martin, R. A., Bruin, G. C., and Parson, I. M.**, Systemic fungicides: a perspective after 10 years, *Plant Dis.,* 1, 19, 1980.

154. **Gunther, F. A., Blinn, R. C., and Carman, G. E.**, Residues of Sevin on and in lemons and oranges, *J. Agric. Food Chem.,* 10, 222, 1962.

155. **Polizu, A., Greger, H., and Alexandri, A. V.**, Studies on the persistency of carbaryl residues in fruit, *Qual. Plant Mater. Veg.,* 20, 215, 1971.

156. **Pieper, G. R.**, Residue analysis of carbaryl on forest foliage and in stream water using HPLC, *Bull. Environ. Contam. Toxicol.,* 22, 167, 1979.

157. **Pieper, G. R. and Miskus, R. P.**, Determination of Zectran residues in aerial forest spraying, *J. Agric. Food Chem.,* 15, 915, 1967.

158. **Fahey, J. E., Wilson, M. C., and Armbrust, E. J.**, Residues of Supracide and carbofuran in green and dehydrated alfalfa, *J. Econ. Entomol.,* 63, 589, 1970.

159. **Ashworth, R. J. and Sheets, T. J.**, Uptake and translocation of carbofuran in tobacco plants, *J. Econ. Entomol.,* 63, 1301, 1970.

160. **Cook, R. F., Stanovick, R. P., and Cassil, C. C.**, Determination of carbofuran and its carbamate metabolite residues in corn using a nitrogen-specific gas chromatographic detector, *J. Agric. Food Chem.,* 17, 277, 1969.

160a. **FMC Corporation**, unpublished data, Middleport, N. Y., 1976.

161. **Pree, D. J. and Saunders, J. L.**, Metabolism of carbofuran in mugho pine, *J. Agric. Food Chem.,* 22, 620, 1974.

162. **Pree, D. J. and Saunders, J. L.**, Bioactivity and translocation of carbofuran residues in mugho pine, *Environ. Entomol.,* 2, 262, 1973.

163. **Andrawes, N. R., Romine, R. R., and Bagley, W. P.**, Metabolism and residues of Temik aldicarb pesticide in cotton foliage and seed under field conditions, *J. Agric. Food Chem.,* 21, 379, 1973.

164. **Baron, R. L.**, IUPAC reports on pesticides. III. Terminal residues of carbamate insecticides, *Pure Appl. Chem.,* 50, 503, 1978.

165. Greenhalgh, R., IUPAC Commission on terminal pesticide residues, *J. Assoc. Off. Anal. Chem.*, 61, 841, 1978.

166. Hill, K. R., IUPAC Commission on terminal residues, *J. Assoc. Off. Anal. Chem.*, 58, 1256, 1975.

167. Hill, K. R., IUPAC Commission on terminal residues, *J. Assoc. Off. Anal. Chem.*, 54, 1316, 1971.

168. Hill, K. R., IUPAC Commission on terminal residues, *J. Assoc. Off. Anal. Chem.*, 53, 987, 1970.

169. Egan, H., IUPAC Commission on terminal residues, *J. Assoc. Off. Anal. Chem.*, 52, 299, 1969.

170. Egan, H., IUPAC Commission on terminal residues, *J. Assoc. Off. Anal. Chem.*, 51, 372, 1968.

171. Egan, H., IUPAC Commission on terminal residues, *J. Assoc. Off. Anal. Chem.*, 50, 1071, 1967.

172. Metcalf, R. L., Sangha, G. K., and Kapoor, I. P., Model ecosystem for the evaluation of pesticide biodegradability and ecological magnification, *Environ. Sci. Technol.*, 5, 709, 1971.

173. Yu, C.-C., Booth, G. M., Hansen, D. J., and Larsen, J. R., Fate of carbofuran in a model ecosystem, *J. Agric. Food Chem.*, 22, 431, 1974.

174. Yu, C.-C., Booth, G. M., Hansen, D. J., and Larsen, J . R., Fate of Bux insecticide in a model ecosystem, *Environ. Entomol.*, 3, 975, 1974.

175. Kanazawa, J., Isensee, A. R., and Kearney, P. C., Distribution of carbaryl and 3,5-xylyl methylcarbamate in an aquatic model ecosystem, *J. Agric. Food Chem.*, 23, 760, 1975.

176. Georghiou, G. P. and Metcalf, R. L., Carbamate insecticides: comparative insect toxicity of Sevin, Zectran, and other new materials, *J. Econ. Entomol.*, 55, 125, 1962.

177. Stenersen, J., Gilman, A., and Vardanis, A., Carbofuran: its toxicity to and metabolism by earthworms (*Lumbricus terrestris*), *J. Agric. Food Chem.*, 21, 166, 1973.

178. Sundaram, K. M. S., Volpe, Y., Smith, G. G., and Duffy, J. R., A Preliminary Study on the Persistence and Distribution of Matacil in a Forest Environment, Report CC-X-116, Chemical Control Research Institute, Ottawa, Ontario, 1976.

179. Sundaram, K. M. S. and Szeto, S. Y., A study on the lethal toxicity of aminocarb to freshwater crayfish and its *in vivo* metabolism, *J. Environ. Sci. Health*, B14, 589, 1979.

180. Roller, N. F., Survey of Pesticide Use in Ontario, 1973, Economics Branch, Ontario Ministry of Agriculture and Food, Toronto, Ontario, 1973.

180a. Carey, A. E., Gowen, J. A., and Wiersma, G. B., Pesticide application and cropping data from 37 states, 1971 — national soils monitoring program, *Pestic. Monit. J.*, 12, 137, 1978.

181. Casida, J. E., Radiotracer studies on metabolism, degradation, and mode of action of insecticidal chemicals, *Residue Rev.*, 25, 149, 1969.

182. Roberts, R. B., Look, M., Haddon, W. F., and Dickerson, T. C., A new degradation product of the insecticide mexacarbate found in freshwater, *J. Agric. Food Chem.*, 26, 55, 1978.

183. Casida, J. E., Mode of action of carbamates, *Annu. Rev. Entomol.*, 8, 39, 1963.

184. Abdel-Wahab, A. M., Kuhr, R. J., and Casida, J. E., Fate of C14-carbonyl-labeled aryl methylcarbamate insecticide chemicals in and on bean plants, *J. Agric. Food Chem.*, 14, 290, 1966.

185. Dorough, H. W., Fate of Furadan (NIA-10242) in bean plants, *Bull. Environ. Contam. Toxicol.*, 3, 164, 1968.

186. Kuhr, R. J., Metabolism of carbamate insecticide chemicals in plants and insects, *J. Agric. Food Chem.*, 18, 1023, 1970.

187. Dorough, H. W., Metabolism of insecticidal methylcarbamates in animals, *J. Agric. Food Chem.*, 18, 1015, 1970.

188. Bull, D. L., Lindquist, D. A., and Coppedge, J. R., Metabolism of 2-methyl-2-(methylthio)propionaldehyde O-(methylcarbamoyl)oxime (Temik, UC-21149) in insects, *J. Agric. Food Chem.*, 15, 610, 1967.

189. Bull, D. L., Stokes, R. A., Coppedge, J. R., and Ridgway, R. L., Further studies of the fate of aldicarb in soil, *J. Econ. Entomol.*, 63, 1283, 1970.

190. Knaak, J. B., Tallant, M. J., and Sullivan, L. J., The metabolism of 2-methyl-2-(methylthio)propionaldehyde O-(methylcarbamoyl)oxime in the rat, *J. Agric. Food Chem.*, 14, 573, 1966.

191. Metcalf, R. L., Fukuto, T. R., Collins, C., Borck, K., Burke, J., Reynolds, H. T., and Osman, M. F., Metabolism of 2-methyl-2-(methylthio)propionaldehyde O-(methylcarbamoyl)oxime (Temik) in plants and insects, *J. Agric. Food Chem.*, 14, 579, 1966.

192. Andrawes, N. R., Dorough, H. W., and Lindquist, D. A., Degradation and elimination of Temik in rats, *J. Econ. Entomol.*, 60, 979, 1967.

193. Bartley, W. J., Andrawes, N. R., Chancey, E. L., Bagley, W. P., and Spurr, H. W., The metabolism of Temik aldicarb pesticide [2-methyl-2-(methylthio)propionaldehyde O-(methylcarbamoyl)oxime] in the cotton plant, *J. Agric. Food Chem.*, 18, 446, 1970.

194. Bull, D. L., Metabolism of UC-21149 [2-methyl-2-(methylthio)propionaldehyde O-(methylcarbamoyl)oxime] in cotton plants and soil in the field, *J. Econ. Entomol.*, 61, 1598, 1968.

195. Andrawes, N. R., Bagley, W. P., and Herrett, R. A., Metabolism of 2-methyl-2-(methylthio)propionaldehyde O-(methylcarbamoyl)oxime (Temik aldicarb pesticide) in potato plants, *J. Agric. Food Chem.*, 19, 731, 1971.

196. **Whitten, C. J. and Bull, D. L.**, Fate of 3,3-dimethyl-1-(methylthio)-2-butanone *O*-(methylcarbamoyl)oxime (Diamond Shamrock DS-15647) in cotton plants and soils, *J. Agric. Food Chem.*, 22, 234, 1974.

197. **Holm, R. E., Chin, W. T., Wagner, D. H., and Stallard, D. E.**, Metabolism of thiofanox in cotton plants, *J. Agric. Food Chem.*, 23, 1056, 1975.

198. **Chin, W. T., Duane, W. C., Ballee, D. L., and Stallard, D. E.**, Mechanism of degradation of thiofanox in aqueous solutions, *J. Agric. Food Chem.*, 24, 1071, 1976.

199. **Chin, W. T., Duane, W. C., Meeks, J. R., and Stallard, D. E.**, Changes in thiofanox residue in potatoes resulting from storage and cooking, *J. Agric. Food Chem.*, 24, 1001, 1976.

200. **Harvey, J., Jr., Jelinek, A. G., and Sherman, H.**, Metabolism of methomyl in the rat, *J. Agric. Food Chem.*, 21, 769, 1973.

201. **Harvey, J., Jr. and Reiser, R. W.**, Metabolism of methomyl in tobacco, corn, and cabbage, *J. Agric. Food Chem.*, 21, 775, 1973.

202. **Harvey, J., Jr. and Pease, H. L.**, Decomposition of methomyl in soil, *J. Agric. Food Chem.*, 21, 784, 1973.

203. **Harvey, J., Jr., Han, J.C.-Y., and Reiser, R. W.**, Metabolism of oxamyl in plants, *J. Agric. Food Chem.*, 26, 529, 1978.

204. **Harvey, J., Jr. and Han, J. C. -Y.**, Decomposition of oxamyl in soil and water, *J. Agric. Food Chem.*, 26, 536, 1978.

205. **Harvey, J., Jr. and Han, J. C.-Y.**, Metabolism of oxamyl and selected metabolites in the rat, *J. Agric. Food Chem.*, 26, 902, 1978.

206. **Menzie, C. M.**, Metabolism of Pesticides, Special Scientific Report, Wildlife No. 127, Bureau of Sport Fisheries and Wildlife, Washington, D.C., 1969.

206a. **Menzie, C. M.**, Metabolism of Pesticides — An Update, Special Scientific Report, Wildlife No. 184, Bureau of Sport Fisheries and Wildlife, Washington, D.C., 1974.

206b. **Menzie, C. M.**, Metabolism of Pesticides — Update II, Special Scientific Report, Wildlife No. 212, Bureau of Sport Fisheries and Wildlife, Washington, D. C., 1978.

207. **Fukuto, T. R. and Sims, J. J.**, Metabolism of insecticides and fungicides, in *Pesticides in the Environment*, Vol. 1, (Part 1), White-Stevens, R., Ed., Marcel Dekker, New York, 1971, 145.

207a. **Matsumura, F.**, Degradation of pesticide residues in the environment, in *Environmental Pollution by Pesticides*, Edwards, C. A., Ed., Plenum Press, New York, 1973, chap. 13.

207b. **Paulson, G. D.**, Metabolic fates of herbicides in animals, *Residue Rev.*, 58, 1, 1975.

207c. **Dahm, P. A.**, Chemistry and metabolism of insecticides, in *Agricultural Practices and Water Quality*, Willrich, T. L. and Smith, G. E., Eds., Iowa State University Press, Ames, 1970, 167.

208. **Kuhr, R.**, The formation and importance of carbamate insecticide metabolites as terminal residues, in *International Symposium on Pesticide Terminal Residues, Tel-Aviv, 1971*, Pure Appl. Chem. Suppl., Tahori, A. S., Ed., Butterworth, New York, 1971, 199.

209. **Krishna, J. G. and Casida, J. E.**, Fate in rats of the radiocarbon from ten variously labeled methyl- and dimethylcarbamate-C^{14} insecticide chemicals and their hydrolysis products, *J. Agric. Food Chem.*, 14, 98, 1966.

210. **Braun, H. E., Ritcey, G. M., Frank, R., McEwen, F. L., and Ripley, B. D.**, Dissipation rates of insecticides in six minor vegetable crops grown on organic soils in Ontario, Canada, *Pestic. Sci.*, 11, 605, 1980.

211. **Hubbell, J. P. and Casida, J. E.**, Metabolic fate of the *N,N*-dialkylcarbamoyl moiety of thiocarbamate herbicides in rats and corn, *J. Agric. Food Chem.*, 25, 404, 1977.

212. **Engst, R.**, (IUPAC), Ethylenethiourea, *Pure Appl. Chem.*, 49, 675, 1977.

213. **Ripley, B. D. and Simpson, C. M.**, Residues of zineb and ethylene thiourea in orchard treated pears and commercial pear products, *Pestic. Sci.*, 8, 487, 1977.

214. **Ripley, B. D., Cox, D. F., Wiebe, J., and Frank, R.**, Residues of Dikar and ethylenethiourea in treated grapes and commercial grape products, *J. Agric. Food Chem.*, 26, 134, 1978.

215. **Ripley, B. D. and Cox, D. F.**, Residues of ethylenebis(dithiocarbamate) and ethylenethiourea in treated tomatoes and commercial tomato products, *J. Agric. Food Chem.*, 26, 1137, 1978.

216. **Kuhr, R. J. and Casida, J. E.**, Persistent glycosides of metabolites of methylcarbamate insecticide chemicals formed by hydroxylation in bean plants, *J. Agric. Food Chem.*, 15, 814, 1967.

217. **Knaak, J. B., Eldridge, J. M., and Sullivan, L. J.**, Systemic approach to preparation and identification of glucuronic acid conjugates, *J. Agric. Food Chem.*, 15, 605, 1967.

218. **Thompson, R. M. and Gerber, N.**, Separation of permethylated isomeric glucuronides by gas chromatography and analysis of the mass spectra, *J. Chromatogr.*, 124, 321, 1976.

219. **Stanley, C. W., Thornton, J. S., and Katague, D. B.**, Gas chromatographic method for residues of Baygon and metabolites in plant tissues, *J. Agric. Food Chem.*, 20, 1265, 1972.

220. **Stanley, C. W. and Thornton, J. S.**, Gas chromatographic method for residues of Baygon and its major metabolites in animal tissue and milk, *J. Agric. Food Chem.*, 20, 1269, 1972.

221. **Robbins, J. D., Bakke, J. E., and Feil, V. J.**, Metabolism of 4-benzothienyl *N*-methylcarbamate (Mobam) in rats. Balance study and urinary metabolite separation, *J. Agric. Food Chem.*, 17, 236, 1969.

222. **Paulson, G. D. and Zehr, M. V.**, Metabolism of *p*-chlorophenyl *N*-methylcarbamate in the chicken, *J. Agric. Food Chem.*, 19, 471, 1971.

223. **Cardona, R. A. and Dorough, H. W.**, Syntheses of the β-D-glucosides of 4- and 5-hydroxy-1-naphthyl *N*-methylcarbamate, *J. Agric. Food Chem.*, 21, 1065, 1973.

224. **Kilsheimer, J. R., Kaufman, H. A., Foster, H. M., Driscoll, P. R., Glick, L. A., and Napier, R. P.**, Benzo[b]thienyl carbamate insecticides, *J. Agric. Food Chem.*, 17, 91, 1969.

225. **Balba, M. H., Singer, M. S., Slade, M., and Casida, J. E.**, Synthesis of possible metabolites of methylcarbamate insecticide chemicals. Substituted-aryl *N*-hydroxymethylcarbamates, *J. Agric. Food Chem.*, 16, 821, 1968.

226. **Durden, J. A., Jr., Stollings, H. W., Casida, J. E., and Slade, M.**, The synthesis of hydroxymethylcarbamates, *J. Agric. Food Chem.*, 18, 459, 1970.

227. **Fahmy, M. A. H. and Fukuto, T. R.**, Convenient method for synthesis of hydroxymethylcarbamates, *J. Agric. Food Chem.*, 20, 168, 1972.

228. **Durden, J. A., Jr.**, The synthesis of nuclearly hydroxylated 1-naphthyl methylcarbamates. The 6- and 7-isomers, *J. Agric. Food Chem.*, 19, 432, 1971.

229. **Richey, F. A., Jr., Bartley, W. J., Fitzpatrick, J. T., and Kurtz, A. P.**, Chemical synthesis of the carbaryl metabolite trans-5,6-dihydro-5,6-dihydroxy-1-naphthyl methylcarbamate, *J. Agric. Food Chem.*, 20, 825, 1972.

230. **Balba, M. H. and Casida, J. E.**, Synthesis of possible metabolites of methylcarbamate insecticide chemicals. Hydroxyaryl and hydroxyalkylphenyl methylcarbamates, *J. Agric. Food Chem.*, 16, 561, 1968.

231. **Slade, M. and Casida, J. E.**, Metabolic fate of 3,4,5- and 2,3,5 -trimethylphenyl methylcarbamates, the major constituents in Landrin insecticide, *J. Agric. Food Chem.*, 18, 467, 1970.

232. **Durden, J. A., Jr., Bartley, W. J., and Stephen, J. F.**, The preparation of standards for aldicarb (Temik) metabolism study, *J. Agric. Food Chem.*, 18, 454, 1970.

233. **Casida, J. E., Kimmel, E. C., Lay, M., Ohkawa, H., Rodebush, J. E., Gray, R. A., Tseng, C. K., and Tilles, H.**, Thiocarbamate sulfoxide herbicides, in *Pesticides, Lectures held at the IUPAC 3rd Int. Congr. Pestic. Chem., Helsinki, Environ. Qual. Saf., Suppl. Vol. 3*, Coulston, F. and Korte, F., Eds., Georg Thieme, Stuttgart, 1975, 675.

234. **Ecke, G. G.**, Synthesis of glucosides of isopropyl-3-chlorocarbanilate metabolites, *J. Agric. Food Chem.*, 21, 792, 1973.

235. **Paulson, G. D., Jacobsen, A. M., Zaylskie, R. G., and Feil, V. J.**, Isolation and identification of propham (isopropyl carbanilate) metabolites from the rat and the goat, *J. Agric. Food Chem.*, 21, 804, 1973.

236. **Dorough, H. W., McManus, J. P., Kumar, S. S., and Cardona, R. A.**, Chemical and metabolic characteristics of 1-naphthyl β-D-glucoside, *J. Agric. Food Chem.*, 22, 642, 1974.

237. **McCully, K. A.**, Analysis of multiple pesticide residues — chemical identity and confirmation of results, *World Rev. Pest Control*, 8, 59, 1969.

238. **Williams, I. H.**, Carbamate insecticide residues in plant material: determination by gas chromatography, *Residue Rev.*, 38, 1, 1971.

239. **Kanazawa, J.**, Annual review of analysis of pesticides for agriculture, *Bunseki Kogaku*, 22, 155R, 1973.

240. **Sherma, J.**, Chromatographic analysis of pesticide residues, *Crit. Rev. Anal. Chem.*, 3, 299, 1973.

241. **Ruzicka, J. H. A.**, Organophosphorus and carbamate detection system, *Pestic. Sci.*, 4, 417, 1973.

242. **Dorough, H. W. and Thorstensen, J. H.**, Analysis for carbamate insecticides and metabolites, *J. Chromatogr. Sci.*, 13, 214, 1975.

243. **Magallona, E. D.**, Gas chromatographic determination of residues of insecticidal carbamates, *Residue Rev.*, 56, 1, 1975.

244. **Maier-Bode, H. and Riedmann, M.**, Gas chromatographic determination of nitrogen-containing pesticides using the nitrogen flame ionization detector (N-FID), *Residue Rev.*, 54, 113, 1975.

245. **Fishbein, L.**, *Chromatography of Environmental Hazards*, Vol. 3, Elsevier, Amsterdam, 1975, chap. 15.

246. **Grob, R. L., Ed.**, *Chromatographic Analysis of the Environment*, Marcel Dekker, New York, 1975.

247. **Van Middelem, C. H.**, Assay procedures for pesticide residues, in *Pesticides in the Environment*, Vol. 1 (Part 2), White-Stevens, R., Ed., Marcel Dekker, New York, 1971, 309.

248. **Chesters, G., Pionke, H. B., and Daniel, T. C.**, Extraction and analytical techniques for pesticides in soil, sediment, and water, in *Pesticides in Soil and Water*, Guenzi, W. D., Ed., Soil Science Society of America, Madison, Wis., 1974, chap. 16.

249. **Burchfield, H. P. and Storrs, E. E.**, Residue analysis, in *Antifungal Compounds,* Vol. 1, Siegel, M. R. and Sisler, H. D., Eds., Marcel Dekker, New York, 1977, chap. 14.

250. **Cochrane, W. P. and Purkayastha, R.**, Analysis of herbicide residues by gas chromatography, *Toxicol. Environ. Chem. Rev.,* 1, 137, 1973.

251. **Thornburg, W.**, Pesticide residues, *Anal. Chem.,* Biennial Applied Reviews, April 1971, 1973, 1975, 1977, 1979.

252. **Benson, W. R. and Jones, H. A.**, The literature of pesticide chemistry, *J. Assoc. Off. Anal. Chem.,* 50, 22, 1967.

253. **Benson, W. R. and Blalock, C. R.**, The literature of pesticide chemistry. II, *J. Assoc. Off. Anal. Chem.,* 54, 192, 1971.

254. **Duggan, R. E.**, Ed., Pesticide Analytical Manual, Vol. 1 and 2, U.S. Food and Drug Administration, Department of Health, Education and Welfare, Washington, D.C., 1967.

255. **Horwitz, W.**, Ed., *Official Methods of Analysis of the Association of Official Analytical Chemists,* 12th ed., Association of Official Analytical Chemists, Washington, D.C., 1975.

256. **McLeod, H. A. and Ritcey, W. R.**, Eds., Analytical Methods for Pesticide Residues in Foods, Department of Supply and Service, Ottawa, Ontario, 1969.

257. **Thompson, J. F.**, Ed., Analysis of Pesticide Residues in Human and Environmental Samples, Pesticides and Toxic Substances Effect Laboratory, National Environmental Research Center, Research Triangle Park, N.C., 1972.

258. **Storherr, R. W.**, Report on carbamate pesticides, fumigants, and miscellaneous, *J. Assoc. Off. Anal. Chem.,* 62, 376, 1979.

259. **Newsome, W. H.**, Report on fungicides, herbicides, and plant growth regulators, *J. Assoc. Off. Anal. Chem.,* 62, 379, 1979.

260. **Slade, P.**, IUPAC Commission on the development, improvement, and standardization of methods of pesticide residue analysis, *J. Assoc. Off. Anal. Chem.,* 59, 894, 1976.

261. Guidelines on Sampling and Statistical Methodologies for Ambient Pesticide Monitoring, Federal Working Group on Pest Management, Washington, D.C., October 1964.

262. Guidelines on Analytical Methodology for Pesticide Residue Monitoring, Federal Working Group on Pest Management, Washington, D.C., June 1975.

263. **Taras, M. J., Greenberg, A. E., Hoak, R. D., and Rand, M. C.**, Eds., *Standard Methods for the Examination of Water and Waste Water,* American Public Health Association, Washington, D.C., 1971.

264. **Zweig, G. and Sherma, J.**, Sample Preparation, in *Analytical Methods for Pesticides and Plant Growth Regulators,* Vol. 6, Zweig, G., Ed., Academic Press, New York, 1972, chap. 1.

265. **Kawar, N. S., DeBatista, G. C., and Gunther, F. A.**, Pesticide stability in cold-stored plant parts, soils, and dairy products, and in cold-stored extractives solutions, *Residue Rev.,* 48, 45, 1973.

266. **Naftel, C., Robinson, P., Salahub, E., and Wong, M.**, Stability of pesticides in extracting solvent, paper presented at 11th Eastern Canada Pestic. Residue Workshop, Guelph, Ontario, May 1979.

267. **Farrow, R. P., Lamb, F. C., Cook, R. W., Kimball, J. R., and Elkins, E. R.**, Removal of DDT, malathion, and carbaryl from tomatoes by commercial and home preparative methods, *J. Agric. Food Chem.,* 16, 65, 1968.

268. **Elkins, E. R., Lamb, F. C., Farrow, R. P., Cook, R. W., Kawai, M., and Kimball, J. R.**, Removal of DDT, malathion, and carbaryl from green beans by commercial and home preparative procedures, *J. Agric. Food Chem.,* 16, 962, 1968.

269. **Kiigemagi, U. and Deinzer, M. L.**, Dislodgeable and total residues of methomyl on mint foliage, *Bull. Environ. Contam. Toxicol.,* 22, 517, 1979.

270. **Howard, S. F. and Yip, G.**, Stability of metallic ethylene bisdithiocarbamates in chopped kale, *J. Assoc. Off. Anal. Chem.,* 54, 1371, 1971.

271. **Ripley, B. D.**, unpublished data, 1979.

272. **Koivistoinen, P. and Karinpää, A.**, Stability of isopropyl N-phenylcarbamate (IPC) and isopropyl N-(3-chlorophenyl) carbamate (CIPC) residues on fruit treated after harvest, *J. Agric. Food Chem.,* 13, 459, 1965.

273. **Holmstead, R. L., Allsup, T. L., and Fullmer, O. H.**, Characterization of a plant constituent interfering in the analysis of residues of 2,3-dihydro-2,2-dimethyl-3,7-benzofurandiol, a metabolite of carbofuran, *J. Assoc. Off. Anal. Chem.,* 62, 89, 1979.

274. **Suffet, I. H.**, The p-value approach to quantitative liquid-liquid extraction of pesticides and herbicides from water. II. Selection of water:solvent ratios and number of extractions, *J. Agric. Food Chem.,* 21, 288, 1973.

275. **Wheeler, W. B. and Frear, D. E. H.**, Extraction of chlorinated hydrocarbon pesticides from plant materials, *Residue Rev.,* 16, 86, 1966.

276. **Gunther, F. A.**, Instrumentation in pesticide residue determinations, *Adv. Pest Control Res.,* 5, 191, 1962.

277. **Coburn, J. A., Ripley, B. D., and Chau, A. S. Y.,** Analysis of pesticide residues by chemical derivatization. II. *N*-methylcarbamates in natural water and soils, *J. Assoc. Off. Anal. Chem.*, 59, 188, 1976.

278. **Thompson, J. F., Reid, S. J., and Kantor, E. J.,** A multiclass, multiresidue analytical method for pesticides in water, *Arch. Environ. Contam. Toxicol.*, 6, 143, 1977.

279. **Bowman, M. C. and Beroza, M.,** Extraction p-values of pesticides and related compounds in six binary solvent systems, *J. Assoc. Off. Agric. Chem.*, 48, 943, 1965.

280. **Suffet, I. H. and Faust, S. D.,** The p-value approach to quantitative liquid-liquid extraction of pesticides from water. I. Organophosphates: choice of pH and solvent, *J. Agric. Food Chem.*, 20, 52, 1972.

281. **Suffet, I. H.,** The p-value approach to quantitative liquid-liquid extraction of pesticides and herbicides from water. III. Liquid-liquid extraction of phenoxy acid herbicides from water, *J. Agric. Food Chem.*, 21, 591, 1973.

282. **Suffet, I. H. and Faust, S. D.,** Liquid-liquid extraction of organic pesticides from water: the p-value approach to quantitative extraction, in *Fate of Organic Pesticides in the Aquatic Environment*, Advances in Chemistry, Series 111, Gould, R. F., Ed., American Chemical Society, Washington, D.C., 1972, chap. 2.

283. **Suffet, I. H.,** personal communication, 1974.

284. **Gesser, H. D., Chow, A., Davis, F. C., Uthe, J. F., and Reinke, J.,** The extraction and recovery of polychlorinated biphenyls (PCB) using porous polyurethane foam, *Anal. Lett.*, 4, 883, 1971.

285. **Uthe, J. F., Reinke, J., and Gesser, H.,** Extraction of organochlorine pesticides from water by porous polyurethane, coated with selective adsorbent, *Environ. Lett.*, 3, 117, 1972.

286. **Bidleman, T. F. and Olney, C. E.,** High-volume collection of atmospheric polychlorinated biphenyls, *Bull. Environ. Contam. Toxicol.*, 11, 442, 1974.

287. **Bedford, J. W.,** The use of polyurethane foam plugs for extraction of polychlorinated biphenyl (PCB's) from natural waters, *Bull. Environ. Contam. Toxicol.*, 12, 622, 1974.

288. **Turner, B. C. and Glotfelty, D. E.,** Field air sampling of pesticide vapors with polyurethane foam, *Anal. Chem.*, 49, 7, 1977.

289. **Lewis, R. G., Brown, A. R., and Jackson, M. D.,** Evaluation of polyurethane foam for sampling of pesticides, polychlorinated biphenyls and polychlorinated naphthalenes in ambient air, *Anal. Chem.*, 49, 1668, 1977.

290. **Burnham, A. K., Calder, G. V., Fritz, J. S., Junk, G. A., Svec, H. J., and Willis, R.,** Identification and estimation of neutral organic contaminants in potable water, *Anal. Chem.*, 44, 139, 1972.

291. **Musty, P. R. and Nickless, G.,** Use of amberlite XAD-4 for extraction and recovery of chlorinated insecticides and polychlorinated biphenyls from water, *J. Chromatogr.*, 89, 185, 1974.

292. **Junk, G. A., Richard, J. J., Grieser, M. D., Witiak, D., Witiak, J. L., Arguello, M. D., Vick, R., Svec, H. J., Fritz, J. S., and Calder, G. V.,** Use of macroreticular resins in the analysis of water for trace organic contaminants, *J. Chromatogr.*, 99, 745, 1974.

293. **Richard, J. J. and Fritz, J. S.,** Adsorption of chlorinated pesticides from river water with XAD-2 resin, *Talanta*, 21, 91, 1974.

294. **Richard, J. J., Junk, G. A., Avery, M. J., Nehring, N. L., Fritz, J. S., and Svec, H. J.,** Analysis of various Iowa waters for selected pesticides: atrazine, DDE, and dieldrin — 1974, *Pestic. Monit. J.*, 9, 117, 1975.

295. **Daughton, C. G., Crosby, D. G., Garnas, R. L., and Hsieh, D. P. H.,** Analysis of phosphorus-containing hydrolytic products of organophosphorus insecticides in water, *J. Agric. Food Chem.*, 24, 236, 1976.

296. **Paschal, D. C., Bicknell, R., and Dresbach, D.,** Determination of ethyl and methyl parathion in runoff water with high performance liquid chromatography, *Anal. Chem.*, 49, 1551, 1977.

297. **Coburn, J. A., Valdmanis, I. A., and Chau, A. S. Y.,** Evaluation of XAD-2 for multiresidue extraction of organochlorine pesticides and polychlorinated biphenyls from natural waters, *J. Assoc. Off. Anal. Chem.*, 60, 224, 1977.

298. **Yamato, Y., Suzuki, M., and Watanabe, T.,** Extraction of benzene hexachloride isomers from water samples, using a macroreticular resin, *J. Assoc. Off. Anal. Chem.*, 61, 1135, 1978.

299. **Mallet, V. N., Brun, G. L., MacDonald, R. N., and Berkane, K.,** A comparative study of the use of XAD-2 resin and the conventional serial solvent extraction procedure for the analysis of fenitrothion and some derivatives in water preservation techniques, *J. Chromatogr.*, 160, 81, 1978.

300. **Woodrow, J. E. and Seiber, J. N.,** Portable device with XAD-4 resin trap for sampling airborne residues of some organophosphorus pesticides, *Anal. Chem.*, 50, 1229, 1978.

301. **Rees, G. A. V. and Au, L.,** Use of XAD-2 macroreticular resin for the recovery of ambient trace levels of pesticides and industrial organic pollutants from water, *Bull. Environ. Contam. Toxicol.*, 22, 561, 1979.

302. **Wolkoff, A. and Creed, C.**, Clean-up and concentration of environmental samples for HPLC, paper presented at 11th Eastern Canada Pestic. Residue Workshop, Provincial Pesticide Residue Testing Laboratory, Ontario Ministry of Agriculture and Food, Guelph, May 15 to 17, 1979.

303. **Werner, W.**, High-performance liquid chromatography in environmental analysis, *GIT Fachz. Lab.*, 22, 785, 1978; *Anal. Abstr.*, 36, 4H30, 1979.

304. **Waters Associates, Inc.**, Sep-Pak Cartridges for Rapid Sample Preparation, Application Note F82, (83265), Milford, Mass., May 1978.

305. **Waters Associates, Inc.**, Rapid Sample Preparation for Analysis of PCBs in Water, Application Note J37, (82237), Milford, Mass., September 1978.

305a. **Waters Associates, Inc.**, Pesticide Residues in Drinking Water, Application Note, H63, (83319), Milford, Mass., November 1976.

306. **Sundaram, K. M. S., Szeto, S. Y., and Hindle, R.**, Evaluation of Amberlite XAD-2 as the extractant for carbamate insecticides from natural water, *J. Chromatogr.*, 177, 29, 1979.

306a. **Sundaram, K. M. S., Szeto, S., and Hindle, R.**, Isolation and Analysis of Aminocarb and its Phenol from Environmental Waters, Report FPM-X-18, Forest Pest Management Institute, Sault Ste. Marie, Ontario, June 1978.

307. **Abbott, D. C., Blake, K. W., Tarrant, K. R., and Thomson, J.**, Thin-layer chromatographic separation, identification and estimation of residues of some carbamate and allied pesticides in soil and water, *J. Chromatogr.*, 30, 136, 1967.

308. **El-Dib, M. A.**, Thin-layer chromatographic detection of carbamate and phenylurea pesticide residues in natural waters, *J. Assoc. Off. Anal. Chem.*, 53, 756, 1970.

309. **Tucker, B.**, Bux insecticide, in *Analytical Methods for Pesticides and Plant Growth Regulators*, Vol. 7, Sherma, J. and Zweig, G., Eds., Academic Press, New York, 1973, chap. 6.

310. **Cook, R. F.**, Carbofuran, in *Analytical Methods for Pesticides and Plant Growth Regulators*, Vol. 7, Sherma, J. and Zweig, G., Eds., Academic Press, New York, 1973, chap. 7.

311. **Aly, O. M.**, Spectrophotometric determination of Sevin in natural waters, *J. Am. Water Work Assoc.*, 59, 906, 1967.

312. **Nagasawa, K., Uchiyama, H., Ogamo, A., and Shinozuka, T.**, Gas-chromatographic determination of microamounts of carbaryl and 1-naphthol in natural water as sources of water supplies, *J. Chromatogr.*, 144, 77, 1977.

313. **Cohen, I. C., Norcup, J., Ruzicka, J. H. A., and Wheals, B. B.**, An electron-capture gas chromatographic method for the determination of some carbamate insecticides as 2,4-dinitrophenyl derivatives of their phenol moieties, *J. Chromatogr.*, 49, 215, 1970.

314. **Seiber, J. N., Crosby, D. G., Fouda, H., and Soderquist, C. J.**, Ether derivatives for the determination of phenols and phenol-generating pesticides by electron capture gas chromatography, *J. Chromatogr.*, 73, 89, 1972.

315. **Ripley, B. D.**, unpublished data, 1978.

316. **Gutenmann, W. H. and Lisk, D. J.**, Gas chromatographic residue determination of Sevin as brominated 1-naphthyl acetate, *J. Agric. Food Chem.*, 13, 48, 1965.

317. **Kuhr, R. J., Davis, A. C., and Bourke, J. B.**, Dissipation of Guthion, Sevin, Polyram, Phygon and Systox from apple orchard soil, *Bull. Environ. Contam. Toxicol.*, 11, 224, 1974.

318. **Venkateswarlu, K. and Sethunathan, N.**, Degradation of carbofuran in rice soils as influenced by repeated applications and exposure to aerobic conditions following anaerobiosis, *J. Agric. Food Chem.*, 26, 1148, 1978.

319. **Ragab, M. T. H., Kimball, E. R., and Chaisson, C. A.**, Direct electron capture gas chromatographic determination of 3-oxocarbofuran in soil, *Chemosphere*, 6, 487, 1977.

320. **Butler, L. I. and McDonough, L. M.**, Determination of residues of carbofuran and its toxic metabolites by electron capture gas chromatography after derivative formation, *J. Assoc. Off. Anal. Chem.*, 54, 1357, 1971.

321. **Caro, J. H., Glotfelty, D. E., Freeman, H. P., and Taylor, A. W.**, Acid ammonium acetate extraction and electron capture gas chromatographic determination of carbofuran in soils, *J. Assoc. Off. Anal. Chem.*, 56, 1319, 1973.

322. **Holland, P. T.**, Routine methods for analysis of organophosphorus and carbamate insecticides in soil and ryegrass, *Pestic. Sci.*, 8, 354, 1977.

323. **Westlake, W. E., Ittig, M., and Gunther, F. A.**, Determination of m-sec-butylphenyl N-methyl-N-thiophenylcarbamate (RE-11775) in water, soil and vegetation, *Bull. Environ. Contam. Toxicol.*, 8, 109, 1972.

324. **Chiba, M.**, Factors affecting the extraction of organochlorine insecticides from soil, *Residue Rev.*, 30, 63, 1969.

325. **Kurtz, D. A. and Studholme, C. R.**, Recovery of trichlorfon (Dylox) and carbaryl (Sevin) in songbirds following spraying of forest for gypsy moth, *Bull. Environ. Contam. Toxicol.*, 11, 78, 1974.

326. **Claborn, H. V., Roberts, R. H., Mann, H. D., Bowman, M. C., Ivey, M. C., Weidenbach, C. P., and Radeleff, R. D.**, Residues in body tissues of livestock sprayed with Sevin or given Sevin in the diet, *J. Agric. Food Chem.*, 11, 74, 1963.

327. **Johnson, D. P., Critchfield, F. E., and Authur, B. W.**, Determination of Sevin insecticide and its metabolites in poultry tissue and eggs, *J. Agric. Food Chem.*, 11, 77, 1963.

328. **Wong, L. and Fisher, F. M.**, Determination of carbofuran and its toxic metabolites in animal tissue by gas chromatography of their *N*-trifluoroacetyl derivatives, *J. Agric. Food Chem.*, 23, 315, 1975.

329. **Cook, R. F., Jackson, J. E., Shuttleworth, J. M., Fullmer, O. H., and Fujie, G. H.**, Determination of the phenolic metabolites of carbofuran in plant and animal matrices by gas chromatography of their 2,4-dinitrophenyl ether derivatives, *J. Agric. Food Chem.*, 25, 1013, 1977.

330. **Anderson, C. A.**, Baygon, in *Analytical Methods for Pesticides and Plant Growth Regulators*, Vol. 7, Sherma, J. and Zweig, G., Eds., Academic Press, New York, 1973, chap. 5.

331. **Sundaram, K. M. S.**, Analytical Evaluation of Matacil in Fish Collected from the Insecticide Treated Area in Quebec, unpublished report, Chemical Control Research Institute, Department of the Environment, Ottawa, 1976.

332. **Argauer, R. J., Shimanuki, H., and Alvarez, C. C.**, Fluorometric determination of carbaryl and 1-naphthol in honeybees (*Apis mellifera* L.) with confirmation by gas chromatography, *J. Agric. Food Chem.*, 18, 688, 1970.

333. **Winterlin, W., Walker, G., and Luce, A.**, Carbaryl residues in bees, honey, and bee bread following exposure to carbaryl via the food supply, *Arch. Environ. Contam. Toxicol.*, 1, 362, 1973.

334. **Morse, R. A., St. John, L. E., and Lisk, D. J.**, Residue analysis of Sevin in bees and pollen, *J. Econ. Entomol.*, 56, 415, 1963.

335. **Johnson, D. P. and Stansbury, H. A., Jr.**, Determination of carbaryl residues in honey bees, *J. Assoc. Off. Agric. Chem.*, 48, 771, 1965.

336. **Butler, L. I. and McDonough, L. M.**, Specific GLC method for determining residues of carbaryl by electron capture detection after derivative formation, *J. Assoc. Off. Anal. Chem.*, 53, 495, 1970.

337. **Shafik, M. T., Sullivan, H. C., and Enos, H. F.**, A method for the determination of 1-naphthol in urine, *Bull. Environ. Contam. Toxicol.*, 6, 34, 1971.

338. **Argauer, R. J.**, Determination of residues of Banol and other carbamate pesticides after hydrolysis and chloroacetylation, *J. Agric. Food Chem.*, 17, 888, 1969.

339. **Bowman, M. C. and Beroza, M.**, Determination of Niagara NIA-10242 and its phenol degradation product in corn silage and milk and determination of other carbamates by GLC of their thiophosphoryl derivatives, *J. Assoc. Off. Anal. Chem.*, 50, 926, 1967.

340. **Bowman, M. C. and Beroza, M.**, Determination of residues of Mobil MC-A-600 (benzo[b]thien-4-yl methylcarbamate) and its hydrolysis product (benzo[b]thiophene-4-ol) in coastal Bermuda grass and milk, *J. Agric. Food Chem.*, 15, 894, 1967.

341. **Chasar, A. G. and Lucchesi, C. A.**, Methods for the determination of Mobam (4-benzothienyl *N*-methylcarbamate), *J. Agric. Food Chem.*, 15, 1030, 1967.

342. **Miskus, R., Gordon, H. T., and George, D. A.**, Colorimetric determination of 1-naphthyl *N*-methylcarbamate in agricultural crops, *J. Agric. Food Chem.*, 7, 613, 1959.

343. **Elessawi, M. A. and El-Refai, A. R.**, Persistence of Sevin residues on some vegetable crops after various harvest times, *J. Assoc. Off. Anal. Chem.*, 50, 1109, 1967.

344. **Johnson, D. P.**, Determination of Sevin insecticide residues in fruits and vegetables, *J. Assoc. Off. Agric. Chem.*, 46, 234, 1963.

345. **Johnson, D. P.**, Determination of Sevin insecticide residues in fruits and vegetables, *J. Assoc. Off. Agric. Chem.*, 47, 283, 1964.

346. **Benson, W. R. and Finocchiaro, J. M.**, Rapid procedure for carbaryl residues: modification of the official colorimetric method, *J. Assoc. Off. Agric. Chem.*, 48, 676, 1965.

347. **Rangaswamy, J. R. and Majumder, S. K.**, Colorimetric method for estimation of carbaryl and its residues on grains, *J. Assoc. Off. Anal. Chem.*, 57, 592, 1974.

348. **Finocchiaro, J. M. and Benson, W. R.**, Thin-layer chromatographic determination of carbaryl (Sevin) in some foods, *J. Assoc. Off. Agric. Chem.*, 48, 736, 1965.

349. **Palmer, N. J. and Benson, W. R.**, Collaborative study of the thin layer chromatographic method for carbaryl residues in apples and spinach, *J. Assoc. Off. Anal. Chem.*, 51, 679, 1968.

350. **Wales, P. J., McLeod, H. A., and McKinley, W. P.**, TLC-enzyme inhibition procedure to detect some carbamate standards and carbaryl in food extracts, *J. Assoc. Off. Anal. Chem.*, 51, 1239, 1968.

351. **Gajan, R. J., Benson, W. R., and Finocchiaro, J. M.**, Determination of carbaryl in crops by oscillographic polarography, *J. Assoc. Off. Agric. Chem.*, 48, 958, 1965.

352. **Porter, M. L., Gajan, R. J., and Burke, J. A.**, Acetonitrile extraction and determination of carbaryl in fruits and vegetables, *J. Assoc. Off. Anal. Chem.*, 52, 177, 1969.

353. **Faucheux, L. J., Jr.**, Rapid cleanup for carbaryl, using channel layer chromatography, *J. Assoc. Off. Anal. Chem.*, 51, 676, 1968.

354. **Argauer, R. J. and Webb, R. E.**, Rapid fluorometric evaluation of the deposition and persistence of carbaryl in the presence of an adjuvant on bean and tomato leaves, *J. Agric. Food Chem.*, 20, 732, 1972.

355. **Ware, G. W., Estesen, B., and Cahill, W. P.**, Dislodgable insecticide residues on cotton, *Bull. Environ. Contam. Toxicol.*, 14, 606, 1975.

356. **Ralls, J. W. and Cortes, A.**, Determination of Sevin in green beans by bromination and electron capture gas chromatography, *J. Gas Chromatogr.*, 2, 132, 1964.

357. **Van Middelem, C. H., Norword, T. L., and Waites, R. E.**, Electron affinity GLC residue determination of Sevin and other carbamates following hydrolysis and bromination, *J. Gas Chromatogr.*, 3, 310, 1965.

358. **Lawrence, J. F. and Leduc, R.**, High pressure liquid chromatography with ultraviolet absorbance or fluorescence detection of carbaryl in potato and corn, *J. Assoc. Off. Anal. Chem.*, 61, 872, 1978.

359. **Ware, G. W., Estesen, B., and Cahill, W. P.**, Dislodgable insecticide residues on cotton (1975), *Bull. Environ. Contam. Toxicol.*, 20, 17, 1978.

360. **Kojima, M., Shiga, N., Matano, O., and Goto, S.**, Microdetermination of carbaryl residue, *J. Pestic. Sci.*, 2, 311, 1977; *Pestic. Abstr.* 77-2575.

361. **Lau, S. C. and Marxmiller, R. L.**, Residue determination of Landrin insecticide by trifluoroacetylation and electron-capture gas chromatography, *J. Agric. Food Chem.*, 18, 413, 1970.

362. **Boyack, G. A.**, Banol, in *Analytical Methods for Pesticides, Plant Growth Regulators, and Food Additives*, Volume 5, Zweig, G., Ed., Academic Press, New York, 1967, chap. 8.

363. **Bache, C. A. and Lisk, D. J.**, Microwave emission residue analysis of carbamate and triazine pesticides, *J. Gas Chromatogr.*, 6, 301, 1968.

364. **Bowman, M. C. and Beroza, M.**, Determination of Mesurol and five of its metabolites in apples, pears, and corn by gas chromatography, *J. Assoc. Off. Anal. Chem.*, 52, 1054, 1969.

365. **Greenhalgh, R., Wood, G. W., and Pearce, P. A.**, A rapid GC method of monitoring Mesurol [4-(methylthio)-3,5-xylyl-N-methyl carbamate] and its sulfoxide and sulfone metabolites and their persistence in lowbush blueberries, *J. Environ. Sci. Health*, B12, 229, 1977.

366. **Moye, H. A.**, Reaction gas chromatographic analysis of pesticides. I. On-column transesterification of N-methylcarbamates by methanol, *J. Agric. Food Chem.*, 19, 452, 1971.

367. **Cassil, C. C., Stanovick, R. P., and Cook, R. F.**, A specific gas chromatographic method for residues of organic nitrogen pesticides, *Residue Rev.*, 26, 63, 1969.

368. **Hawk, R. E., Sheets, T. J., Cassil, C. C., Fullmer, O. H., Fujie, G. H., and McCarthy, J. F.**, Effect of application method on the transfer of carbofuran and 3-hydroxycarbofuran residues into cigarette mainstream smoke, *Tob. Sci.*, XX, 3, 1976; *Tobacco*, 178, 32, 1976.

369. **Van Middelem, C. H., Moye, H. A., and Janes, M. J.**, Carbofuran and 3-hydroxycarbofuran determination in lettuce by alkali-flame gas chromatography, *J. Agric. Food Chem.*, 19, 459, 1971.

370. **Williams, I. H. and Brown, M. J.**, Determination of carbofuran and 3-hydroxycarbofuran residues in small fruits, *J. Agric. Food Chem.*, 21, 399, 1973.

371. **Finlayson, D. G., Williams, I. H., Brown, M. J., and Campbell, C. J.**, Distribution of insecticide residues in carrots at harvest, *J. Agric. Food Chem.*, 24, 606, 1976.

372. **Archer, T. E.**, Effects of light on the fate of carbofuran during the drying of alfalfa, *J. Agric. Food Chem.*, 24, 1057, 1976.

373. **Archer, T. E., Stokes, J. D., and Bringhurst, R. S.**, Fate of carbofuran and its metabolites on strawberries in the environment, *J. Agric. Food Chem.*, 25, 536, 1977.

374. **Lawrence, J. F. and Leduc, R.**, Direct analysis of carbofuran and two nonconjugated metabolites in crops by high-pressure liquid chromatography with UV absorption detection, *J. Agric. Food Chem.*, 25, 1362, 1977.

375. **Lawrence, J. F., Lewis, D. A., and McLeod, H. A.**, Detection of carbofuran and metabolites directly or as their heptafluorobutyryl derivatives using gas-liquid or high pressure liquid chromatography with different detectors, *J. Chromatogr.*, 138, 143, 1977.

376. **Ernst, G. F., Röder, S. J., Tjan, G. H., and Jansen, J. T. A.**, Thin layer chromatographic detection and indirect gas chromatographic determination of three carbamate pesticides, *J. Assoc. Off. Anal. Chem.*, 58, 1015, 1975.

377. **Johansson, C. E.**, A multiresidue analytical method for determining organochlorine, organophosphorus, dinitrophenyl and carbamate pesticides in apples, *Pestic. Sci.*, 9, 313, 1978.

378. **Butler, L. I. and McDonough, L. M.**, Method for the determination of residues of carbamate insecticides by electron-capture gas chromatography, *J. Agric. Food Chem.*, 16, 403, 1968.

379. **Holden, E. R.**, Gas chromatographic determination of residues of methylcarbamate insecticides in crops as their 2,4-dinitrophenyl ether derivatives, *J. Assoc. Off. Anal. Chem.*, 56, 713, 1973.

380. **Holden, E. R.**, Collaborative study of the 2,4-dinitrophenyl ether multiresidue method for use in determining four carbamate pesticides in crops, *J. Assoc. Off. Anal. Chem.*, 58, 562, 1975.

381. Klein, A. K., Report on extraction procedures for chloro-organic pesticides, *J. Assoc. Off. Agric. Chem.*, 41, 551, 1958.

382. Klein, A. K., Report on extraction procedures for chloro-organic insecticides, *J. Assoc. Off. Agric. Chem.*, 42, 539, 1959.

383. Storherr, R. W., Reports on carbamate pesticides and on fumigants and miscellaneous pesticides, *J. Assoc. Off. Anal. Chem.*, 54, 324, 1971.

384. Watts, R. R., Extraction efficiency study — examination of three procedures for extracting ^{14}C-labeled and unlabeled residues of organophosphorus pesticides and carbaryl from bean leaves and kale, *J. Assoc. Off. Anal. Chem.*, 54, 953, 1971.

385. Wheeler, W. B., Thompson, N. P., Andrade, P., and Krause, R. T., Extraction efficiencies for pesticides in crops. I. [^{14}C]Carbaryl extraction from mustard greens and radishes, *J. Agric. Food Chem.*, 26, 1333, 1978.

386. Van Middelem, C. H. and Peplow, A. J., Evaluation of extraction procedures for the removal of ^{14}C-carbofuran and its toxic metabolites from cabbage leaves, *J. Agric. Food Chem.*, 21, 100, 1973.

387. Marquardt, R. P. and Luce, E. N., Determination of residual 4-dimethylamino-3,5-xylyl methylcarbamate and 4-dimethylamino-3,5-xylenol by use of luteoarsenotungstic acid, *J. Agric. Food Chem.*, 11, 418, 1963.

388. Marquardt, R. P., Zectran, in *Analytical Methods for Pesticides, Plant Growth Regulators, and Food Additives*, Vol. 2, Zweig, G., Ed., Academic Press, New York, 1964, chap. 47.

389. Holden, E. R., Jones, W. M., and Beroza, M., Determination of residues of methyl- and dimethyl-carbamate insecticides by gas chromatography of their 2,4-dinitroaniline derivatives, *J. Agric. Food Chem.*, 17, 56, 1969.

390. Johnson, D. P. and Stansbury, H. A., Jr., Determination of Temik residues in raw fruits and vegetables, *J. Assoc. Off. Anal. Chem.*, 49, 399, 1966.

391. Meagher, W. R., Hendrickson, R., and Shively, B. G., Spectrophotometric determination of Temik residues in citrus, *J. Assoc. Off. Anal. Chem.*, 50, 1242, 1967.

392. Hendrickson, R. and Meagher, W. R., Residues of 2-methyl-2-(methylthio)propionaldehyde *O*-(methylcarbamoyl)-oxime (Temik) in citrus following soil applications, *J. Agric. Food Chem.*, 16, 903, 1968.

393. Maitlen, J. C., McDonough, L. M., and Beroza, M., Determination of residues of 2-methyl-2-(methylthio)propionaldehyde *O*-(methylcarbamoyl)oxime (UC-21149, Temik), its sulfoxide, and its sulfone by gas chromatography, *J. Agric. Food Chem.*, 16, 549, 1968.

394. Beckman, H., Giang, B. Y., and Qualia, J., Preparation and detection of Temik and its metabolites as residues, *J. Agric. Food Chem.*, 17, 70, 1969.

395. Carey, W. F. and Helrich, K., Improved quantitative method for the determination of aldicarb and its oxidation products in plant materials, *J. Assoc. Off. Anal. Chem.*, 53, 1296, 1970.

396. Maitlen, J. C., McDonough, L. M., and Beroza, M., Rapid method for the extraction, cleanup, and GLC determination of toxic residues of Temik, *J. Assoc. Off. Anal. Chem.*, 52, 786, 1969.

397. Lindquist, R. K., Krueger, H. R., Spadafora, R. R., and Mason, J. F., Application of aldicarb to greenhouse tomatoes: plant growth, fruit yields, greenhouse whitefly control, and residues in fruits, *J. Econ. Entomol.*, 65, 862, 1972.

398. Krueger, H. R. and Mason, J. F., Effects of plant growth regulators on levels of phorate and aldicarb in soybeans, *J. Agric. Food Chem.*, 22, 338, 1974.

399. Woodham, D. W., Edwards, R. R., Reeves, R. G., and Schutzmann, R. L., Total toxic aldicarb residues in soil, cottonseed, and cotton lint following a soil treatment with the insecticide on the Texas high plains, *J. Agric. Food Chem.*, 21, 303, 1973.

400. Woodham, D. W., Reeves, R. G., and Edwards, R. R., Total toxic aldicarb residues in weeds, grasses, and wildlife from the Texas high plains following a soil treatment with the insecticide, *J. Agric. Food Chem.*, 21, 604, 1973.

401. Pease, H. L. and Kirkland, J. J., Determination of methomyl residues using microcoulometric gas chromatography, *J. Agric. Food Chem.*, 16, 554, 1968.

402. Tappan, W. B., Wheeler, W. B., and Lundy, H. W., Methomyl residues on cigar-wrapper and flue-cured tobaccos in Florida, *J. Econ. Entomol.*, 66, 197, 1973.

403. Krueger, H. R., Lindquist, R. K., Mason, J. F., and Spadafora, R. R., Application of methomyl to greenhouse tomatoes: greenhouse whitefly control and residues in foliage and fruits, *J. Econ. Entomol.*, 66, 1223, 1973.

404. Fung, K. K. H., Determination and identification of *S*-methyl *N*-[(methyl-carbamoyl)oxy]thioacetimidate (methomyl) residues in tobacco, *J. Agric. Food Chem.*, 23, 695, 1975.

405. Leidy, R. B., Domanski, J. J., Haire, P. L., and Sheets, T. J., Effects of weathering and flue-curing on methomyl residues on tobacco, *Arch. Environ. Contam. Toxicol.*, 5, 199, 1977.

406. **Reeves, R. G. and Woodham, D. W.**, Gas chromatographic analysis of methomyl residues in soil, sediment, water, and tobacco utilizing the flame photometric detector, *J. Agric. Food Chem.*, 22, 76, 1974.

407. **Mendoza, C. E., McLeod, H. A., Shields, J. B., and Phillips, W. E. J.**, Determination of methomyl in rape seeds, oils and meals, *Pestic. Sci.*, 5, 231, 1974.

408. **McLeod, H. A. and Wales, P. J.**, A low temperature cleanup procedure for pesticides and their metabolites in biological samples, *J. Agric. Food Chem.*, 20, 624, 1972.

409. **McLeod, H. A.**, Low temperature apparatus for cleanup of biological sample extracts, *Anal. Chem.*, 44, 1328, 1972.

410. **McLeod, H. A., Mendoza, C. E., and McCully, K. A.**, Analysis of foods for methomyl using thin-layer chromatography after low temperature clean-up, *Pestic. Sci.*, 6, 11, 1975.

411. **Chin, W. -T., Duane, W. C., Szalkowski, M. B., and Stallard, D. E.**, Gas chromatographic determination of thiofanox residues in soil, plants, and water, *J. Agric. Food Chem.*, 23, 963, 1975.

412. **Holt, R. F. and Pease, H. L.**, Determination of oxamyl residues using flame photometric gas chromatography, *J. Agric. Food Chem.*, 24, 263, 1976.

413. **Singhal, J. P., Khan, S., and Bansal, O. P.**, Spectrophotometric determination of oxamyl as copper dithiocarbamate, *J. Agric. Food Chem.*, 25, 377, 1977.

414. **Davis, P. L., O'Bannon, J. H., and Munroe, K. A.**, Extraction and analysis of oxamyl from citrus leaves, *J. Agric. Food Chem.*, 26, 777, 1978.

415. **Thean, J. E., Fong, W. G., Lorenz, D. R., and Stephens, T. L.**, High pressure liquid chromatographic determination of methomyl and oxamyl on vegetable crops, *J. Assoc. Off. Anal. Chem.*, 61, 15, 1978.

416. **Bissinger, W. E. and Fredenburg, R. H.**, The determination of micro quantities of isopropyl *N*-phenylcarbamate (IPC) in head lettuce, *J. Assoc. Off. Agric. Chem.*, 34, 812, 1951.

417. **Gard, L. N. and Rudd, N. G.**, Isopropyl *N*-(3-chlorophenyl)carbamate (CIPC) in soil and crop, *J. Agric. Food Chem.*, 1, 630, 1953.

418. **Gard, L. N.**, Determination of isopropyl *N*-(3-chlorophenyl)carbamate residues in potatoes treated for sprout inhibition, *J. Agric. Food Chem.*, 7, 339, 1959.

419. **Montgomery, M. and Freed, V. H.**, Determining micro amounts of isopropyl *N*-phenylcarbamate, *J. Agric. Food Chem.*, 7, 617, 1959.

420. **Ferguson, C. E. and Gard, L.**, IPC and CIPC residue analysis, *J. Agric. Food Chem.*, 17, 1062, 1969.

421. **Ferguson, C. E., Jr., Gard, L. N., Shupe, R. H., and Dress, K. S.**, Infrared analysis for residues of isopropyl *N*-(3-chlorophenyl)carbamate (CIPC) in white potatoes, *J. Agric. Food Chem.*, 11, 428, 1963.

422. **Riden, J. R. and Hopkins, T. R.**, Decline and residue studies on 4-chloro-2-butynyl *N*-(3-chlorophenyl)carbamate, *J. Agric. Food Chem.*, 9, 47, 1961.

423. **Gutenmann, W. H. and Lisk, D. J.**, Electron affinity residue determination of CIPC, monuron, diuron, and linuron by direct hydrolysis and bromination, *J. Agric. Food Chem.*, 12, 46, 1964.

424. **Cohen, I. H. and Wheals, B. B.**, An electron-capture gas chromatographic method for the determination of substituted urea and carbamate herbicides as 2,4-dinitrophenyl derivatives of their amine moieties, *J. Chromatogr.*, 43, 233, 1969.

425. **Onley, J. H. and Yip, G.**, Herbicidal carbamates: extraction, cleanup, and gas chromatographic determination by thermionic, electron capture, and flame photometric detectors, *J. Assoc. Off. Anal. Chem.*, 54, 1366, 1971.

426. **Lawrence, J. F.**, Evaluation and confirmation of an alkylation-gas-liquid chromatographic method for the determination of carbamate and urea herbicides in foods, *J. Assoc. Off. Anal. Chem.*, 59, 1061, 1976.

427. **Lawrence, J. F. and Laver, G. W.**, Analysis of carbamate and urea herbicides in foods by gas-liquid chromatography after alkylation, *J. Agric. Food Chem.*, 23, 1106, 1975.

428. **Batchelder, G. H. and Patchett, G. G.**, A colorimetric method for the determination of EPTC residues in crops and soils, *J. Agric. Food Chem.*, 8, 214, 1960.

429. **Hughes, R. E., Jr. and Freed, V. H.**, The determination of ethyl *N,N*-di-n-propylthiolcarbamate (EPTC) in soil by gas chromatography, *J. Agric. Food Chem.*, 9, 381, 1961.

430. **Crosby, D. G. and Bowers, J. B.**, Amine derivatives for pesticide residue analysis, *J. Agric. Food Chem.*, 16, 839, 1968.

431. **Koren, E., Foy, C. L., and Ashton, F. M.**, Adsorption, volatility, and migration of thiocarbamate herbicides in soil, *Weed Sci.*, 17, 148, 1969.

432. **Lawrence, J. F.**, Confirmation of some organonitrogen herbicides and fungicides by chemical derivatization and gas chromatography, *J. Agric. Food Chem.*, 24, 1236, 1976.

433. **Smith, A. E.**, Factors affecting the loss of tri-allate from soils, *Weed Res.*, 9, 306, 1969.

434. **Smith, A. E.**, Estimation of *S*-2,3-dichloroallyl *N,N*-diisopropylthiolcarbamate (diallate) residues in soils by electron-capture gas chromatography, *J. Agric. Food Chem.*, 17, 1052, 1969.

435. McKone, C. E. and Hance, R. J., Estimation of S-2,3,3-trichloroallyl N,N-diisopropylthiolcarbamate (triallate) residues in soil, barley straw, and grain by electron-capture gas chromatography, *J. Agric. Food Chem.*, 15, 935, 1967.

436. Hermanson, H. P., Siewierski, M., and Helrich, K., Extraction and gas-liquid chromatographic determination of vernolate in soil, *J. Assoc. Off. Anal. Chem.*, 52, 175, 1969.

437. Hermanson, H. P., Siewierski, M., and Winnett, A. G., Thin-layer chromatographic separation of vernolate, suspected metabolites and derivatives, *Bull. Environ. Contam. Toxicol.*, 5, 38, 1970.

438. Hermanson, H. P., Helrich, K., and Carey, W. F., Gas-liquid chromatography of propanol, propylamine, and dipropylamine, *Anal. Lett.*, 1, 941, 1968.

439. Long, J. W. and Thompson, L., Jr., Herbicide residues in air-cured burley tobacco, *J. Agric. Food Chem.*, 22, 82, 1974.

440. Ja, W. Y., Patchett, G. G., and Smith, W. J., Eptam, in *Analytical Methods for Pesticides and Plant Growth Regulators*, Vol. 6, Zweig, G. and Sherma, J., Eds., Academic Press, New York, 1972, chap. 95.

441. Schulten, H. -R. and Stöber, I., Determination and identification of biocides of the phenylurea, carbamate and thiocarbamate type in surface waters by high pressure liquid chromatography and field desorption mass spectrometry, *Fresenius Z. Anal. Chem.*, 293, 370, 1978.

442. Gray, R. A., A vapor trapping apparatus for determining the loss of EPTC and other herbicides from soils, *Weeds*, 13, 138, 1965.

443. Gray, R. A. and Weierich, A. J., Factors affecting the vapor loss of EPTC from soils, *Weeds*, 13, 141, 1965.

444. Lowen, W. K. and Pease, H. L., Dithiocarbamates, in *Analytical Methods for Pesticides, Plant Growth Regulators, and Food Additives*, Vol. 3, Zweig, G., Ed., Academic Press, New York, 1964, chap. 7.

445. Burke, J. A., Mills, P. A., and Bostwick, D. C., Experiments with evaporation of solutions of chlorinated pesticides, *J. Assoc. Off. Anal. Chem.*, 49, 999, 1966.

446. Chiba, M. and Morley, H. V., Studies of losses of pesticides during sample preparation, *J. Assoc. Off. Anal. Chem.*, 51, 55, 1968.

447. McKinley, W. P., Coffin, D. E., and McCully, K. A., Cleanup processes for pesticide residue analysis, *J. Assoc. Off. Agric. Chem.*, 47, 863, 1964.

448. Jones, L. R. and Riddick, J. A., Separation of organic insecticides from plant and animal tissue, *Anal. Chem.*, 24, 569, 1952.

449. Morley, H. V., Adsorbents and their application to column cleanup of pesticide residues, *Residue Rev.*, 16, 1, 1966.

450. Mills, P. A., Variation of Florisil activity: simple method for measuring adsorbent capacity and its use in standardizing Florisil columns, *J. Assoc. Off. Anal. Chem.*, 51, 29, 1968.

451. Florisil, The Preferred Adsorbent for Chromatography: Properties, Applications, Bibliography, Floridin Company, Pittsburgh, Pa.

452. Mills, P. A., Bong, B. A., Kamps, L. R., and Burke, J. A., Elution solvent system for Florisil column cleanup in organochlorine pesticide residue analyses, *J. Assoc. Off. Anal. Chem.*, 55, 39, 1972.

453. Mills, P. A., Onley, J. H., and Gaither, R. A., Rapid method for chlorinated pesticide residues in nonfatty foods, *J. Assoc. Off. Agric. Chem.*, 46, 186, 1963.

454. Law, L. M. and Goerlitz, D. F., Microcolumn chromatographic cleanup for the analysis of pesticides in water, *J. Assoc. Off. Anal. Chem.*, 53, 1276, 1970.

455. Sherma, J. and Shafik, T. M., A multiclass, multiresidue analytical method for determining pesticide residues in air, *Arch. Environ. Contam. Toxicol.*, 3, 55, 1975.

456. Mendoza, C. E. and Shields, J. B., A microcolumn apparatus for rapid cleanup of 2,4-dinitrophenylmethylamine extracts for carbamate pesticide analysis, *J. Agric. Food Chem.*, 22, 528, 1974.

457. Johnson, L. G., Formation of pentafluorobenzyl derivatives for the identification and quantitation of acid and phenol pesticide residues, *J. Assoc. Off. Anal. Chem.*, 56, 1503, 1973.

458. Brown, M. J., personal communication, 1979.

459. Lovelock, J. E. and Lipsky, S. R., Electron affinity spectroscopy — a new method for the identification of functional groups in chemical compounds separated by gas chromatography, *J. Am. Chem. Soc.*, 82, 431, 1960.

460. Harley, J., Nel, W., and Pretorius, V., Flame ionization detector for gas chromatography, *Nature (London)*, 181, 177, 1958.

461. McWilliam, I. G. and Dewar, R. A., Flame ionization detector for gas chromatography, *Nature (London)*, 181, 760, 1958.

462. Karmen, A. and Giuffrida, L., Enhancement of the response of the hydrogen flame ionization detector to compounds containing halogen and nitrogen, *Nature (London)*, 201, 1204, 1964.

463. Giuffrida, L., A flame ionization detector highly selective and sensitive to phosphorus — a sodium thermionic detector, *J. Assoc. Off. Agric. Chem.*, 47, 293, 1964.

464. Aue, W. A., Gehrke, C. W., Tindle, R. C., Stalling, D. L., and Rugle, C. D., Application of the alkali-flame detector to nitrogen containing compounds, *J. Gas Chromatogr.,* 5, 381, 1967.
465. Brody, S. S. and Chaney, J. E., Flame photometric detector. The application of a specific detector for phosphorus and for sulfur compounds—sensitive to subnanogram quantities, *J. Gas Chromatogr.,* 4, 42, 1966.
466. Coulson, D. M. and Cavanagh, L. A., Automatic chloride analyzer, *Anal. Chem.,* 32, 1245, 1960.
467. Coulson, D. M., Cavanagh, L. A., De Vries, J. E., and Walther, B., Microcoulometric gas chromatography of pesticides, *J. Agric. Food Chem.,* 8, 399, 1960.
468. Coulson, D. M., Electrolytic conductivity detector for gas chromatography, *J. Gas Chromatogr.,* 3, 134, 1965.
469. Coulson, D. M., Selective detection of nitrogen compounds in electrolytic conductivity gas chromatography, *J. Gas Chromatogr.,* 4, 285, 1966.
470. Hall, R. C., A highly sensitive and selective microelectrolytic conductivity detector for gas chromatography, *J. Chromatogr. Sci.,* 12, 152, 1974.
471. Burgett, C. A., Smith, D. H., and Bente, H. B., The nitrogen-phosphorus detector and its applications in gas chromatography, *J. Chromatogr.,* 134, 57, 1977.
472. Kolb, B., Auer, M., and Pospisil, P., Ionization detector for gas chromatography with switchable selectivity for carbon, nitrogen and phosphorus, *J. Chromatogr.,* 134, 65, 1977.
473. Kolb, B., Auer, M., and Pospisil, P., Reaction mechanism in an ionization detector with tunable selectivity for carbon, nitrogen and phosphorus, *J. Chromatogr. Sci.,* 15, 53, 1977.
474. Hall, R. C., The nitrogen detector in gas chromatography, *Crit. Rev. Anal. Chem.,* 7, 323, 1978.
475. Westlake, W. E. and Gunther, F. A., Advances in gas chromatographic detectors illustrated from applications to pesticide residue evaluations, *Residue Rev.,* 18, 175, 1967.
476. Lawrence, J. F., Ryan, J. J., and Leduc, R., Comparison of electron capture and electrolytic conductivity for the gas chromatographic detection of some perchloro derivatives of diethylstibesterol, *J. Chromatogr.,* 147, 398, 1978.
477. Ryan, J. J. and Lawrence, J. F., Comparison of electron-capture and electrolytic conductivity detection for the gas-liquid chromatographic analysis of some perfluoro derivatives of four agricultural chemicals, *J. Chromatogr.,* 135, 117, 1977.
478. Lawrence, J. F. and Ryan, J. J., Comparison of electron-capture and electrolytic-conductivity detection for the gas-liquid chromatographic analysis of heptafluorobutyryl derivatives of some agricultural chemicals, *J. Chromatogr.,* 130, 97, 1977.
479. Papers of the Symposium on Substance-Selective Detectors in Chromatography, ACS, San Francisco, September 1976, in *J. Chromatogr.,* 134, 1, 1977.
480. Lawrence, J. F., Effect of some operating paameters on the response of the Coulson conductivity detector, *J. Chromatogr.,* 87, 333, 1973.
481. Cochrane, W. P., Wilson, B. P., and Greenhalgh, R., Determination of sulfur- and chlorine-containing pesticides with an electrolytic conductivity detector, *J. Chromatogr.,* 75, 207, 1973.
482. Cochrane, W. P. and Wilson, B. P., Electrolytic conductivity detection of some nitrogen-containing herbicides, *J. Chromatogr.,* 63, 364, 1971.
483. Wilson, B. P. and Cochrane, W. P., Comparison of the Coulson and Hall electrolytic conductivity detectors for the determination of nitrogen-containing pesticides, *J. Chromatogr.,* 106, 174, 1975.
484. Greenhalgh, R. and Cochrane, W. P., Comparative gas chromatographic response of organophosphorus compounds containing nitrogen and nitrogen compounds with alkali flame and electrolytic conductivity detectors, *J. Chromatogr.,* 70, 37, 1972.
485. Greenhalgh, R. and Wilson, M. A., Optimization and response of the Pye flame photometric detector to some insecticides in the phosphorus and sulfur modes, *J. Chromatogr.,* 128, 157, 1976.
486. Cochrane, W. P., Maybury, R. B., and Greenhalgh, R. G., Comparative study of the linearity and sensitivity of electron capture and flame photometric detectors using a pesticide standard, *J. Environ. Sci. Health,* B14, 197, 1979.
487. Burgett, C. A. and Green, L. E., Improved flame photometric detection without solvent flameout, *J. Chromatogr. Sci.,* 12, 356, 1974.
488. Kapila, S. and Vogt, C. R., FPD: Burner configurations and the response to hetero-organics, *J. Chromatogr. Sci.,* 17, 327, 1979.
489. Cochrane, W. P. and Greenhalgh, R., Evaluation and comparison of selective gas chromatographic detectors for the analysis of pesticide residues, *Chromatographia,* 9, 255, 1976.
490. Burchfield, H. P., Johnson, D. E., Rhoades, J. W., and Wheeler, R. J., Selective detection of phosphorus, sulfur, and halogen compounds in the gas chromatography of drugs and pesticides, *J. Gas Chromatogr.,* 3, 28, 1965.
491. Patchett, G. G., Evaluation of the electrolytic conductivity detector for residue analysis of nitrogen-containing pesticides, *J. Chromatogr. Sci.,* 8, 155, 1970.
492. Laski, R. R. and Watts, R. R., Gas chromatography of organonitrogen pesticides, using a nitrogen specific detection system, *J. Assoc. Off. Anal. Chem.,* 56, 328, 1973.

493. Ekchardt, J. G., Denton, M. B., and Moyers, J. L., Sulfur FPD flow optimization and response normalization with a variable exponential function device, *J. Chromatogr. Sci.*, 13, 133, 1975.

494. McLeod, H. A., Butterfield, A. G., Lewis, D., Phillips, W. E. J., and Coffin, D. E., Gas-liquid chromatography system with flame ionization, phosphorus, sulfur, nitrogen and electron capture detectors operating simultaneously for pesticide residue analysis, *Anal. Chem.*, 47, 674, 1975.

495. McLeod, H. A., Systems for automated multiple pesticide residue analysis, *J. Chromatogr. Sci.*, 13, 302, 1975.

496. Aue, W. A., Detectors for use in GC analysis of pesticides, *J. Chromatogr. Sci.*, 13, 329, 1975.

497. Wheeler, L. and Strother, A., Chromatography of N-methylcarbamates in the gaseous phase, *J. Chromatogr.*, 45, 362, 1969.

498. Zielinski, W. L., Jr. and Fishbein, L., Gas chromatography of carbamate derivatives. I. Simple carbamates, *J. Gas Chromatogr.*, 3, 142, 1965.

499. Zielinski, W. L., Jr. and Fishbein, L., Structural relation to chromatographic behaviour of carbamates, *J. Gas Chromatogr.*, 3, 260, 1965.

500. Zielinski, W. L., Jr. and Fishbein, L., Gas chromatography of carbamate derivatives. II. N-substituted carbamates, *J. Gas Chromatogr.*, 3, 333, 1965.

501. McNair, H. M. and Bonelli, E. J., *Basic Gas Chromatography*, Varian Aerograph, Walnut Creek, Calif., 1967.

502. Strother, A., Gas chromatography of various phenyl N-methylcarbamates, *J. Gas Chromatogr.*, 6, 110, 1968.

503. Brun, G. L., personal communication, 1979.

504. Bache, C. A. and Lisk, D. J., Note on the versatility of OV-17 substrate for gas chromatography of pesticides, *J. Assoc. Off. Anal. Chem.*, 51, 1270, 1968.

505. Eberle, D. O. and Gunther, F. A., Chromatographic, spectrometric, and irradiation behavior of five carbamate insecticides, *J. Assoc. Off. Agric. Chem.*, 48, 927, 1965.

506. Wisniewski, J. V., Sample decomposition during analysis. Gas chromatography of aromatic carbamates, *Facts Methods Sci. Res.*, 7, 5, 1966.

507. Gudzinowicz, B. J., *Gas Chromatographic Analysis of Drugs and Pesticides*, Marcel Dekker, New York, 1967, chap. 9.

508. Peck, J. M. and Harkiss, K. J., Gas chromatographic analysis of some carbamate derivatives, *J. Chromatogr. Sci.*, 9, 370, 1971.

509. Ebing, W., Gas chromatographische analyse insektizider carbamate, *Chimia*, 19, 501, 1965.

510. Riva, M. and Carisano, A. Direct gas chromatographic determination of carbaryl, *J. Chromatogr.*, 42, 464, 1969.

511. Lewis, D. L. and Paris, D. F., Direct determination of carbaryl by gas-liquid chromatography using electron capture detection, *J. Agric. Food Chem.*, 22, 148, 1974.

512. Weyer, L. G., Gas-liquid chromatographic method for the analysis of carbaryl insecticide formulations, *J. Assoc. Off. Anal. Chem.*, 57, 778, 1974.

513. Argauer, R. J. and Warthen, J. D., Jr., Separation of 1- and 2-naphthols and determination of trace amounts of 2-naphthyl methylcarbamate in carbaryl formulations by high pressure liquid chromatography with confirmation by spectrofluorometry, *Anal. Chem.*, 47, 2472, 1975.

514. Greenhalgh, R., Marshall, W. D., and King, R. R., Trifluoroacetylation of Mesurol [4-methylthio-3,5-xylyl-N-methylcarbamate], its sulfoxide, sulfone, and phenol analogs for analysis by gas chromatography, *J. Agric. Food Chem.*, 24, 266, 1976.

515. Pierce Handbook and General Catalog, Pierce Chemical Company, Rockford, Ill., 1979—80.

516. Lorah, E. J. and Hemphill, D. D., Direct chromatography of some N-methylcarbamate pesticides, *J. Assoc. Off. Anal. Chem.*, 57, 570, 1974.

517. Ragab, M. T. H., Direct electron capture gas chromatography of 2,3-dihydro-2,2-dimethyl-3-oxo-7-benzofuranyl N-methylcarbamate; a metabolite of carbofuran insecticide, *Anal. Lett.*, 10, 551, 1977.

518. Luke, M. A., Froberg, J. E., and Masumoto, H. T., Extraction and cleanup of organochlorine, organophosphate, organonitrogen, and hydrocarbon pesticides in produce for determination by gasliquid chromatography, *J. Assoc. Off. Anal. Chem.*, 58, 1020, 1975.

519. Aue, W. A., Hastings, C. R., and Kapila, S., On the unexpected behavior of a common gas chromatographic phase, *J. Chromatogr.*, 77, 299, 1973.

520. Oda, M., Shida, N., and Kashiwa, T., Direct gas chromatography of carbamate insecticides, *Bull. Agric. Chem. Inspect. Stn.*, 16, 60, 1976; *Anal. Abst.*, 34, 2G 32, 1978.

521. Hall, R. C. and Harris, D. E., Direct gas chromatographic determination of carbamate pesticides using Carbowax 20M-modified supports and the electrolytic conductivity detector, *J. Chromatogr.*, 169, 245, 1979.

522. Gutenmann, W. H. and Lisk, D. J., Gas chromatographic determination of phenolic pesticides and residues, *J. Assoc. Off. Anal. Chem.*, 48, 1173, 1965.

523. Yip, G. and Howard, S. F., Extraction and cleanup procedure for the gas chromatographic determination of four dinitrophenolic pesticides, *J. Assoc. Off. Anal. Chem.*, 51, 24, 1968.

524. **Baker, R. A.,** Phenolic analyses by direct aqueous injection gas chromatography, *J. Am. Water Works Assoc.,* 58, 751, 1966.

525. **Bowman, M. C. and Beroza, M.,** Apparatus combining gas chromatography with spectrofluorometry by means of a flowing liquid interface, *Anal. Chem.,* 40, 535, 1968.

526. **Bache, C. A. and Lisk, D. J.,** Selective emission spectrometric determination of nanogram quantities of organic bromine, chlorine, iodine, phosphorus, and sulfur compounds in a helium plasma, *Anal. Chem.,* 39, 786, 1967.

527. **Beroza, M. and Coad, R. A.,** Reaction gas chromatography, *J. Gas Chromatogr.,* 4, 199, 1966.

528. **Kvalvag, J., Iwata, Y., and Gunther, F. A.,** High-pressure liquid chromatographic separation of o,p'- and p,p'-dicofol and their dichlorobenzophenone degradation products, *Bull. Environ. Contam. Toxicol.,* 21, 25, 1979.

529. **Zweig, G. and Sherma, J.,** Kelthane (Dicofol), in *Analytical Methods for Pesticides and Plant Growth Regulators,* Vol. 6, Zweig, G., Ed., Academic Press, New York, 1972, chap. 34.

530. **Bowman, M. C. and Beroza, M.,** Rapid GLC method for determining residues of fenthion, disulfoton, and phorate in corn, milk, grass, and feces, *J. Assoc. Off. Anal. Chem.,* 52, 1231, 1969.

531. **Fishbein, L. and Zielinski, W. L., Jr.,** Structural transformations during the gas chromatography of carbamates, *Chromatographia,* 2, 38, 1969.

532. **Bullock, D. J. W.,** Pirimicarb, in *Analytical Methods for Pesticides and Plant Growth Regulators,* Vol. 7, Academic Press, New York, 1973, 399.

533. **Williams, I. H.,** Direct gas chromatographic determination of methomyl (Lannate) with microcoulometric nitrogen detection, *Pestic. Sci.,* 3, 179, 1972.

534. **Chapman, R. A. and Harris, C. R.,** Determination of residues of methomyl and oxamyl and their oximes in crops by gas-liquid chromatography of oxime trimethylsilyl ethers, *J. Chromatogr.,* 171, 249, 1979.

535. **Masuda, T., Kanazawa, J., and Inoue, H.,** Determination of methomyl residues in potatoes, *Pestic. Technol.,* 28, 22, 1972.

535a. **Kawahara, T. and Maeda, H.,** Residue analysis of Mesomil in crops, *Noyaku Kensasho Hokoku,* 11, 89, 1971.

536. **Kiigemagi, U., Wellman, D., Cooley, E. J., and Terriere, L. C.,** Residues of the insecticides phorate and methomyl in mint hay and oil, *Pestic. Sci.,* 4, 89, 1973.

537. **Lee, Y. W., Ford, R. J., McDonald, H., McKinlay, K. S., Putnam, L. G., and Saha, J. G.,** Residues of methomyl in rape plant and seed following its application for the control of bertha army worm, *Mamestra configurata* (Lepidoptera:Noctuidae), *Can. Entomol.,* 104, 1745, 1972.

538. **Fung, K. K. H.,** Determination and confirmation of methomyl residues in soil and water, *Pestic. Sci.,* 7, 571, 1976.

539. **Ogata, J. N., Yanagihara, K. H., Hylin, J. W., and Bevenue, A.,** Use of the alkali-flame ionization detector for the determination of methomyl residues in plant materials, *J. Chromatogr.,* 157, 401, 1978.

540. **Halfhill, J. E. and Maitlen, J. C.,** Residues of aldicarb in fresh and dried alfalfa, *J. Econ. Entomol.,* 66, 557, 1973.

541. **Benson, W. R.,** Report on carbamate pesticides, *J. Assoc. Off. Anal. Chem.,* 52, 266, 1969.

542. **Barney, J. E., II.** Collaborative study of the gas chromatographic determination of six thiocarbamate herbicides in formulations, *J. Assoc. Off. Anal. Chem.,* 57, 53, 1974.

543. **Barney, J. E., II.** Second collaborative study of a gas-liquid chromatographic method for the assay of thiocarbamate formulations, *J. Assoc. Off. Anal. Chem.,* 59, 213, 1976.

544. **Lawrence, J. F.,** Gas chromatographic separation of herbicides of major interest in Canada, with electrolytic conductivity detection in the nitrogen and chlorine modes, *J. Chromatogr.,* 121, 85, 1976.

545. **Zielinski, W. L., Jr. and Fishbein, L.,** Gas chromatography of metallic derivatives of ethylenebis(dithiocarbamic acids), *J. Chromatogr.,* 23, 302, 1965.

546. **McLeod, H. A. and McCully, K. A.,** Head space gas procedure for screening food samples for dithiocarbamate pesticide residues, *J. Assoc. Off. Anal. Chem.,* 52, 1226, 1969.

547. **Bighi, C.,** Microdetermination of dithiocarbamates by gas chromatography, *J. Chromatogr.,* 14, 348, 1964.

548. **Bighi, C. and Saglietto, G.,** Gas-chromatographic study of the decomposition of thiocarbamic compounds as a function of temperature, *J. Chromatogr.,* 17, 13, 1965.

549. **Onuska, F. I. and Boos, W. R.,** Gas chromatographc and mass spectrometric studies of S-alkyl derivatives of N,N-dialkyl dithiocarbamates, *Anal. Chem.,* 45, 967, 1973.

550. **Newsome, W. H.,** A method for determining ethylenebis(dithiocarbamate) residues on food crops as bis(trifluoroacetamido)ethane, *J. Agric. Food Chem.,* 22, 886, 1974.

551. **Keppel, G. F.,** Modification of the carbon disulfide evolution method for dithiocarbamate residues, *J. Assoc. Off. Anal. Chem.,* 52, 162, 1969.

552. **King, G. S. and Blau, K.,** in *Handbook of Derivatives for Chromatography,* Blau, K. and King, G. S., Eds., Heydon and Sons, London, 1977, chap. 1.

553. **Sweeley, C. C.**, in *Handbook of Derivatives for Chromatography*, Blau, K. and King, G. S., Eds., Heydon and Sons, London, 1977, xi.

554. **Blau, K. and King, G. S., Eds.**, *Handbook of Derivatives for Chromatography*, Heydon and Sons, London, 1977.

555. **Lawrence, J. F. and Frei, R. W.**, *Chemical Derivatization in Liquid Chromatography*, Elsevier, Amsterdam, 1976.

556. **Lawrence, J. F., Ed.**, Derivatization in chromatography. I, *J. Chromatogr. Sci.*, 17, 113, 1979.

557. **Cochrane, W. P.**, Application of chemical derivatisation techniques for pesticide analysis, *J. Chromatogr. Sci.*, 17, 124, 1979.

558. **Epstein, A. J., Gaskill, D. R., and Lucchesi, C. A.**, Gas chromatographic determination of 4-benzothienyl *N*-methylcarbamate, *Anal. Chem.*, 39, 721, 1967.

559. **Greenhalgh, R. and Kovacicova, J.**, A chemical confirmatory test for organophosphorus and carbamate insecticides and triazine and urea herbicides with reactive NH moieties, *J. Agric. Food. Chem.*, 23, 325, 1975.

560. **Cochrane, W. P.**, Confirmation of pesticide residue identity by chemical derivatization, *Int. Conf. Environ. Sensing Assess., 1975*, Institute of Electrical and Electronic Engineers, New York, 1, 41, 1976.

561. **Pierce, A. E.**, Silylation of Organic Compounds, Pierce Chemical Co., Rockford, Ill., 1968.

562. **Fishbein, L. and Zielinski, W L., Jr.**, Gas chromatography of trimethylsilyl derivatives. I. Pesticidal carbamates and ureas, *J. Chromatogr.*, 20, 9, 1965.

563. **Bache, C. A., St. John, L. E., Jr., and Lisk, D. J.**, Gas chromatographic analysis of insensitive pesticides as their halomethyldimethylsilyl derivatives, *Anal. Chem.*, 40, 1241, 1968.

564. **Khalifa, S. and Mumma, R. O.**, Gas chromatographic separation of the aglycone metabolites of carbaryl, *J. Agric. Food Chem.*, 20, 632, 1972.

565. **Mathur, S. B., Iwata, Y., and Gunther, F. A.**, Gas-liquid chromatographic analysis of carbaryl as its *N*-thiomethyl derivative, *J. Agric. Food Chem.*, 26, 768, 1978.

566. **Black, A. L., Chiu, Y. C., Fahmy, M. A. H., and Fukuto, T. R.**, Selective toxicity of *N*-sulfenylated derivatives of insecticidal methylcarbamate esters, *J. Agric. Food Chem.*, 21, 747, 1973.

567. **Boulton, J. J. K., Boyce, C. B. C., Jewess, P. J., and Jones, R. F.**, Comparative properties of *N*-acetyl derivatives of oxime *N*-methylcarbamates and aryl *N*-methylcarbamates as insecticides and acetylcholinesterase inhibitors, *Pestic. Sci.*, 2, 10, 1971.

568. **Fraser, J., Harrison, I. R., and Wakerley, S. B.**, Synthesis and insecticidal activity of *N*-acyl-*N*-methylcarbamates, *J. Sci. Food Agric., Suppl.*, p.8, 1968.

569. **Fraser, J., Clinch, P. G., and Reay, R. C.**, *N*-acylation of *N*-methylcarbamate insecticides and its effect on biological activity, *J. Sci. Food Agric.*, 16, 615, 1965.

570. **Robertson, W. A. H., Fraser, J., and Clinch, P. G.**, Boots Pure Drug Co. Ltd., British Patent 982,235, February 3, 1965.

571. **Look, M.**, Synthesis of *N*-acetyl-1-^{14}C Zectran (4-dimethylamino-3,5-xylyl-*N*-acetyl-1'-^{14}C methylcarbamate) by microacetylation, *J. Agric. Food Chem.*, 16, 893, 1968.

572. **Sullivan, L. J., Eldridge, J. M., and Knaak, J. B.**, Determination of carbaryl and some other carbamates by gas chromatography, *J. Agric. Food Chem.*, 15, 927, 1967.

573. **Blau, K. and King, G. S.**, in *Handbook of Derivatives for Chromatography*, Blau, K. and King, G. S., Eds., Heydon and Sons, London, 1977, chap. 3.

574. **Magallona, E. D., Gunther, F. A., and Iwata, Y.**, Synthesis and spectroscopic characterization of acetyl-Cl, (x = 0,1,2 or 3) and nitroso derivatives of carbaryl and three possible metabolites, *Arch. Environ. Contam. Toxicol.*, 5, 177, 1977.

575. **Magallona, E. D. and Gunther, F. A.**, Gas chromatographic evaluation of some *N*-derivatives of intact carbaryl, *Arch. Environ. Contam. Toxicol.*, 5, 185, 1977.

576. **Greenhalgh, R., King, R. R., and Marshall, W. D.**, Trifluoroacetylation of pesticides and metabolites containing a sulfoxide moiety for quantitation by gas chromatography and chemical confirmatory purposes, *J. Agric. Food Chem.*, 26, 475, 1978.

577. **Seiber, J. N.**, *N*-Perfluoroacyl derivatives for methylcarbamate analysis by gas chromatography, *J. Agric. Food Chem.*, 20, 443, 1972.

578. **Lawrence, J. F.**, Gas chromatographic analysis of heptafluorobutyryl derivatives of some carbamate insecticides, *J. Chromatogr.*, 123, 287, 1976.

579. **Shafik, T. M., Bradway, D., and Mongan, P. F.**, Electron-capture gas chromatography of picogram levels of aromatic *N*-methylcarbamate insecticides, presented at the 163rd Natl. ACS Meet., Pesticide Section, Boston, April 1972.

580. **Sherma, J. and Shafik, T. M.**, A multiclass, multiresidue analytical method for determining pesticide residue in air, *Arch. Environ. Contam. Toxicol.*, 3, 55, 1975.

581. **Chapman, R. A. and Robinson, J. R.**, Simplified method for the determination of residues of carbofuran and its metabolites in crops using gas-liquid chromatography-mass fragmentography, *J. Chromatogr.*, 140, 209, 1977.

582. Holmstead, R. L. and Casida, J. E., Chemical ionization mass spectrometry of N-methylcarbamate insecticides, some of their metabolites, and related compounds, *J. Assoc. Off. Anal. Chem.*, 58, 541, 1975.

583. Fieser, L. F. and Fieser, M., *Reagents for Organic Synthesis,* John Wiley & Sons, New York, 1967.

584. Bose, R. J., Formation of the N-trifluoroacetate of carbofuran, *J. Agric. Food Chem.*, 25, 1209, 1977.

585. Mumma, R. O. and Khalifa, S., Mass spectra of trifluoroacetyl derivatives of carbofuran and its aglycone metabolites, *J. Agric. Food Chem.*, 20, 1090, 1972.

586. Skinner, S. I. M. and Greenhalgh, R., Mass Spectra of Insecticides, Herbicides and Fungicides and Metabolites, Agriculture Canada Publication, Department of Supply and Service, Ottawa, Ontario, 1977.

587. Moye, H. A., Esters of sulfonic acids as derivatives for the gas chromatographic analysis of carbamate pesticides, *J Agric. Food Chem.*, 23, 415, 1975.

588. Landowne, R. A. and Lipsky, S. R., The electron capture spectrometry of haloacetates: a means of detecting ultramicro quantities of sterols by gas chromatography, *Anal. Chem.*, 35, 532, 1963.

589. Argauer, R. J., Rapid procedure for the chloracetylation of microgram quantities of phenols and detection by electron-capture gas chromatography, *Anal. Chem.*, 40, 122, 1968.

590. Miller, C. W., Shafik, M. T., and Biros, F. J., A method for sampling and analysis of propoxur in air, *Bull. Environ. Contam. Toxicol.*, 8, 339, 1972.

591. Paulson, G. D. and Portnoy, C. E., Sulfate ester conjugates: a one-step method for replacing the sulfate with an acetyl group, *J. Agric. Food Chem.*, 18, 180, 1970.

592. Paulson, G. D., Zaylskie, R. G., Zehr, M. V., Portnoy, C. E., and Feil, V. J., Metabolism of carbaryl(1-naphthyl N-methylcarbamate) in chicken urine, *J. Agric. Food Chem.*, 18, 110, 1970.

593. Kato, S., Multi-residue analytical method for detecting N-methylcarbamate insecticides by gas-chromatography of their monochloroacetyl derivatives, *Bunseki*, 3, 145, 1979; *Pestic. Abst.*, 12, 79-1472, 1979.

594. Lamparski, L. L. and Nestrick, T. J., Determination of trace phenols in water by gas chromatographic analysis of heptafluorobutyryl derivatives, *J. Chromatogr.*, 156, 143, 1978.

595. Fenimore, D. C., Freeman, R. R., and Loy, P. R., Determination of Δ⁹-tetrahydrocannabinol in blood by electron capture gas chromatography, *Anal. Chem.*, 45, 2331, 1973.

596. Reinheimer, J. D., Douglass, J. P., Leister, H., and Voelkel, M. B., Aromatic nucleophilic substitution reaction in qualitative organic chemistry: the reaction of 2,4-dinitrofluorobenzene with phenols, *J. Org. Chem.*, 22, 1743, 1957.

597. Cohen, I. C., Norcup, J., Ruzicka, J. H. A., and Wheals, B. B., Trace determination of phenols by gas chromatography as their 2,4-dinitrophenyl ethers, *J. Chromatogr.*, 44, 251, 1969.

598. Caro, J. H., Taylor, A.W., and Freeman, H. P., Comparative behavior of dieldrin and carbofuran in the field, *Arch. Environ. Contam. Toxicol.*, 3, 437, 1976.

599. Clegg, D. E. and Martin, P. R., Residues of the carbamate acaricide, 3-methyl-5-isopropylphenyl-N-(n-butanoyl)-N-methylcarbamate (Promacyl) and two metabolites in the tissue and milk of cattle, *Pestic. Sci.*, 4, 447, 1973.

600. Kawahara, F. K., Microdetermination of derivatives of phenols and mercaptans by means of electron capture gas chromatography, *Anal. Chem.*, 40, 1009, 1968.

601. Kawahara, F. K., Microdetermination of pentafluorobenzyl ester derivatives of organic acids by means of electron capture gas chromatography, *Anal. Chem.*, 40, 2073, 1968.

602. Kawahara, F. K., Gas chromatographic analysis of mercaptans, phenols, and organic acids in surface waters with use of pentafluorobenzyl derivatives, *Environ. Sci. Technol.*, 5, 235, 1971.

603. Johnson, L. G., Analysis of pesticides in water using silica gel column clean-up, *Bull. Environ. Contam. Toxicol.*, 5, 542, 1971.

604. Coburn, J. A. and Chau, A. S. Y., Confirmation of pesticide residue identity. VIII. Organophosphorus pesticides, *J. Assoc. Off. Anal. Chem.*, 57, 1272, 1974.

605. Chau, A. S. Y. and Terry, K., Analysis of pesticides by chemical derivatization. III. Gas chromatographic characteristics and conditions for the formation of pentafluorobenzyl derivatives of ten herbicidal acids, *J. Assoc. Off. Anal. Chem.*, 59, 633, 1976.

606. Agemian, H. and Chau, A. S. Y., Determination of pesticides by derivative formation. IV. A sensitive gas-chromatographic method for the determination of MCPA and MCPB herbicides after esterification with 1-bromomethyl-2,3,4,5,6-pentafluorobenzene, *Analyst (London)*, 101, 732, 1976.

607. Agemian, H. and Chau, A. S. Y., Analysis of pesticide residues by chemical derivatization. V. Multiresidue analysis of eight phenoxyalkanoic acid herbicides in natural waters, *J. Assoc. Off. Anal. Chem.*, 60, 1070, 1977.

608. De Beer, J., Van Peteghem, C., and Heyndrickx, A., Comparative study of gas-liquid chromatographic behavior of the pentafluorobenzyl esters and the methyl esters of ten chlorophenoxyalkyl acids, *J. Chromatogr.*, 157, 97, 1978.

609. De Beer, J.`O., Van Peteghem, C. H., and Heyndrickx, A. M., Pentafluorobenzyl ester derivatives of nine chlorophenoxy acid herbicides: spectroscopic properties and utility for identification purposes, *J. Assoc. Off. Anal. Chem.*, 61, 1140, 1978.

610. Coburn, J. A., personal communication, 1979.

611. Cummins, L. M., in *Recent Advances in Gas Chromatography*, Domsky, I. I. and Perry, J. A., Eds., Marcel Dekker, New York, 1971, 329.

612. Burchfield, H. P. and Storrs, E. E., *Biochemical Applications of Gas Chromatography*, Academic Press, New York, 1972, 279.

612a. Syzmanski, H. A., *Biochemical Applications of Gas Chromatography*, Plenum Press, New York, 1964, 39.

613. Ripley, B. D., French, B. J., and Edgington, L. V., unpublished data, 1980.

614. Sumida, S., Takaki, M., and Miyamoto, J., Rapid gas chromatographic determination of microquantities of N-methyl carbamates as their 2,4-dinitrophenyl derivatives, *Agric. Biol. Chem.*, 34, 1576, 1970.

615. Day, E. W., Jr., Golab, T., and Koons, J. R., Determination of micro quantities of C_1-C_4 primary and secondary amines by electron affinity detection, *Anal. Chem.*, 38, 1053, 1966.

616. Turner, B. C. and Caro, J. H., Uptake and distribution of carbofuran and its metabolites in field-grown corn plants, *J. Environ. Qual.*, 2, 245, 1973.

617. Sumida, S., Takaki, M., and Miyamoto, T., Method for the determination of residue of Meobal (3,4-dimethylphenyl N-methylcarbamate) in rice grain, *Botyu-Kagaku*, 35, 72, 1970.

618. Mendoza, C. E. and Shields, J. B., Determination of methomyl by using 1-fluoro-2,4-dinitrobenzene reaction and gas-liquid chromatography, *J. Agric. Food Chem.*, 22, 255, 1974.

619. Frei, R. W. and Lawrence, J. F., Determination of Matacil and Zectran by fluorigenic labeling, thin layer chromatography, and *in situ* fluorimetry, *J. Assoc. Off. Anal. Chem.*, 55, 1259, 1972.

620. Clarke, D. D., Wilk, S., and Gitlow, S. E., Electron capture properties of halogenated amine derivatives, *J. Gas. Chromatogr.*, 4, 310, 1966.

621. McCurdy, W. H., Jr. and Reiser, R. W., Trace analysis of fatty amines by gas chromatography, *Anal. Chem.*, 38, 795, 1966.

622. Tilden, R. L. and Van Middelem, C. H., Determination of carbaryl as an amide derivative by electron-capture gas chromatography, *J. Agric. Food Chem.*, 18, 154, 1970.

623. Knaak, J. B. and Sullivan, L. J., Metabolism of 3,4-dichlorobenzyl N-methylcarbamate in the rat, *J. Agric. Food Chem.*, 16, 454, 1968.

624. Drawert, F., Felgenhauer, R., and Kupfer, G., Reaction gas chromatography, *Agnew. Chem.*, 72, 555, 1960.

625. Tanaka, F. S. and Wien, R. G., Gas chromatography of substituted phenylureas by flash-heater methylation with trimethylanilinium hydroxide, *J. Chromatogr.*, 87, 85, 1973.

626. Wien, R. G. and Tanaka, F. S., Gas chromatography of N-methyl and N-acyl carbamates by flash-heater reaction with trimethylanilinium hydroxide, *J. Chromatogr.*, 130, 55, 1977.

627. Bromilow, R. H. and Lord, K. A., Analysis of sulfur-containing carbamates by formation of derivatives in the gas-liquid chromatograph using trimethylphenylammonium hydroxide, *J. Chromatogr.*, 125, 495, 1976.

628. Gaylord, N. G. and Sroog, C. E., The reaction of carbamates with alcohols, *J. Org. Chem.*, 18, 1632, 1953.

629. Hubbell, D. H., Rothwell, D. F., Wheeler, W. B., Tappan, W. B., and Rhoads, F. M., Microbiological effects and persistence of some pesticide combinations in soil, *J. Environ. Qual.*, 2, 96, 1973.

630. Moye, H. A., Reaction gas chromatographic analysis of pesticides. II. On-column transesterification of organophosphates by methanol, *J. Agric. Food Chem.*, 21, 621, 1973.

631. Khan, S. U., Chemical derivatization of herbicide residues for gas liquid chromatographic analysis, *Residue Rev.*, 59, 21, 1975.

632. Thier, H. P., Analysengang zur ermittlung von pestizidrueckstaenden in pflanzenmaterial. I. Teil., *Deut. Lebensms. Rundsch.*, 68, 345, 1972.

633. Thier, H. P., Analysengang zur ermittlung von pestizidrueckstaenden in pflanzenmaterial (Schluss), *Deut. Lebensm. Rundsch.*, 68, 397, 1972.

634. Thier, H. P., Analytik der herbizide, *Agnew. Chem.*, 86, 244, 1974.

635. Spengler, D. and Hamroll, B., Separation and determination of carbamate and urea herbicides by reaction gas chromatography, *J. Chromatogr.*, 49, 205, 1970.

636. Henkel, H. G., Separation of alkylated and chlorinated anilines by gas chromatography, *J. Gas Chromatogr.*, 3, 320, 1965.

637. Moffat, A. C., Horning, E. C., Martin, S. B., and Rowland, M., Perfluorobenzene derivatives as derivatizing agents for the gas chromatography of primary and secondary amines using electron capture detection, *J. Chromatogr.*, 66, 255, 1972.

638. **Hartvig, P., Handl, W., Vessman, J., and Svahr, C. M.,** Electron capture gas chromatography of tertiary amines as pentafluorobenzyl carbamates, *Anal. Chem.,* 48, 390, 1976.
639. **Bradway, D. E. and Shafik, T.,** Electron capture gas chromatographic analysis of the amine metabolites of pesticides: derivatization of anilines, *J. Chromatogr. Sci.,* 15, 322, 1977.
640. **Mitten, M. E.,** unpublished results, PPG Industries, Inc., Barberton, Ohio, 1969.
641. *Fed. Regist.,* 42, 40618, 1977.
642. **Newsome, W. H.,** A method for the determination of ethylenethiuram monosulfide on food crops, *J. Agric. Food Chem.,* 23, 348, 1975.
643. **Newsome, W. H.,** A method for the determination of ethylenebis(isothiocyanate) on food crops, *J. Agric. Food Chem.,* 24, 420, 1976.
644. **Newsome, W. H. and Panopio, L. G.,** A method for the determination of 2-imidazoline residues in food crops, *J. Agric. Food Chem.,* 26, 638, 1978.
645. **Newsome, W. H.,** A method for the determination of ethyleneurea in foods as the pentafluorobenzamide, *J. Agric. Food Chem.,* 26, 1325, 1978.
646. **Bontoyan, W. R., Looker, J. B., Kaiser, T. E., Giang, P., and Olive, B. M.,** Survey of ethylenethiourea in commercial ethylenebisdithiocarbamate formulations, *J. Assoc. Off. Anal. Chem.,* 55, 923, 1972.
647. **Bontoyan, W. R. and Looker, J. B.,** Degradation of commercial ethylene bisdithiocarbamate formulations to ethylenethiourea under elevated temperature and humidity, *J. Agric. Food Chem.,* 21, 338, 1973.
648. **Ludwig, R. A., Thorn, G. D., and Miller, D. M.,** Studies on the mechanism of fungicidal action of disodium ethylene bisthiocarbamate (Nabam), *Can. J. Bot.,* 32, 48, 1954.
649. **Czeglédi-Jankó, G. and Holló, A.,** Determination of the degradation products of ethylenebis (dithiocarbamates) by thin-layer chromatography and some investigations of their decomposition *in vitro, J. Chromatogr.,* 31, 89, 1967.
650. **Newsome, W. H., Shields, J. B., and Villeneuve, D. C.,** Residues of maneb, ethylenethiuram monosulfide, ethylenethiourea, and ethylenediamine on beans and tomatoes field treated with maneb, *J. Agric. Food Chem.,* 23, 756, 1975.
651. **Newsome, W. H.,** Residues of four ethylenebis(dithiocarbamates) and their decomposition products on field-sprayed tomatoes, *J. Agric. Food Chem.,* 24, 999, 1976.
652. **Pease, H. L. and Holt, R. F.,** Manganese ethylenebis(dithiocarbamate) (maneb)/ethylenethiourea(ETU) residue studies on five crops treated with ethylenebis(dithiocarbamate) (EBDC) fungicides, *J. Agric. Food Chem.,* 25, 561, 1977.
653. **Newsome, W. H. and Laver, G. W.,** Effect of boiling on the formation of ethylenethiourea in zineb-treated foods, *Bull. Environ. Contam. Toxicol.,* 10, 151, 1973.
654. **Watts, R. R., Storherr, R. W., and Onley, J. H.,** Effects of cooking on ethylenebisdithiocarbamate degradation to ethylene thiourea, *Bull. Environ. Contam. Toxicol.,* 12, 224, 1974.
655. **Graham, S. L. and Hansen, W. H.,** Effects of short-term administration of ethylenethiourea upon thyroid function of the rat, *Bull. Environ. Contam. Toxicol.,* 7, 19, 1972.
656. **Graham, S. L., Hansen, W. H., Davis, K. J., and Perry, C. H.,** Effects of one-year administration of ethylenethiourea upon the thyroid of the rat, *J. Agric. Food Chem.,* 21, 324, 1973.
657. **Ulland, B. M., Weisburger, J. H., Weisburger, E. K., Rice, J. M., and Cypher, R.,** Thyroid cancer in rats from ethylene thiourea intake, *J. Natl. Cancer Inst.,* 49, 583, 1972.
658. **Khera, K. S.,** Ethylenethiourea: teratogenicity study in rats and rabbits, *Teratology,* 7, 243, 1973.
659. **Nash, R. G.,** Uptake of ethylenebis(dithiocarbamate) fungicides and ethylenethiourea by soybeans, *J. Agric. Food Chem.,* 24, 596, 1976.
660. **Hoagland, R. E. and Frear, D. S.,** Behavior and fate of ethylenethiourea in plants, *J. Agric. Food Chem.,* 24, 129, 1976.
661. **Rhoades, R. C.,** Studies with manganese [^{14}C] ethylenebis(dithiocarbamate) ([^{14}C]maneb) fungicide and [^{14}C] ethylenethiourea ([^{14}C]ETU) in plants, soil, and water, *J. Agric. Food Chem.,* 25, 528, 1977.
662. **Blazquez, C. H.,** Residue determination of ethylenethiourea (2-imidazolidinethione) from tomato foliage, soil, and water, *J. Agric. Food Chem.,* 21, 330, 1973.
663. **Onley, J. H. and Yip, G.,** Determination of ethylene thiourea residues in foods, using thin-layer and gas chromatography, *J. Assoc. Off. Anal. Chem.,* 54, 165, 1971.
664. **Haines, L. D. and Adler, I. L.,** Gas chromatographic determination of ethylene thiourea residues, *J. Assoc. Off. Anal. Chem.,* 56, 333, 1973.
665. **Otto, S., Keller, W., and Drescher, N.,** A new gas chromatographic determination of ethylenethiourea residues without derivatization, *J. Environ. Sci. Health,* B12, 179, 1977.
666. **Onley, J. H., Giuffrida, L., Ives, N. F., Watts, R. R., and Storherr, R. W.,** Gas-liquid chromatography and liquid chromatography of ethylenethiourea in fresh vegetable crops, fruits, milk, and cooked foods, *J. Assoc. Off. Anal. Chem.,* 60, 1105, 1977.

667. **Onley, J. H.,** Gas-liquid chromatographic method for determining ethylenethiourea in potatoes, spinach, applesauce, and milk: collaborative study, *J. Assoc. Off. Anal. Chem.,* 60, 1111, 1977.
668. **Newsome, W. H.,** Determination of ethylenethiourea residues in apples, *J. Agric. Food Chem.,* 20, 967, 1972.
669. **Nash, R. G.,** Improved gas-liquid chromatographic method for determining ethylenethiourea in plants, *J. Assoc. Off. Anal. Chem.,* 57, 1015, 1974.
670. **Nash, R. G.,** Gas-liquid chromatographic method for determining ethylenethiourea in plants, *J. Assoc. Off. Anal. Chem.,* 58, 566, 1975.
671. **King, R. R.,** Derivatization of ethylenethiourea with *m*-trifluoromethylbenzyl chloride for analysis by electron-capture gas chromatography, *J. Agric. Food Chem.,* 25, 73, 1977.
672. **Engst, R. and Schnaak, W.,** Chromatographic-polarographic determination of ethylene thiourea in foods, *Nahrung,* 18, 597, 1974.
673. **Engst, R. and Schnaak, W.,** Analysis of residues of degradation products of ethylene-bis-dithiocarbamates, in *Pesticides, Lectures held at the IUPAC 3rd Int. Congr. Pest. Chem., Helsinki (Environ. Qual. Saf., Suppl. Vol. 3),* Coulston, F. and Korte, F., Eds., Georg Thieme, Stuttgart, 1975, 62.
674. **Singh, J., Cochrane, W. P., and Scott, J.,** Extractive acylation of ethylenethiourea from water, *Bull. Environ. Contam. Toxicol.,* 23, 470, 1979.
675. **Engst, R. and Schnaak, W.,** Residues of dithiocarbamate fungicides and their metabolites on plant foods, *Residue Rev.,* 52, 45, 1974.
676. **Hadden, N., Baumann, F., MacDonald, F., Munk, M., Stevenson, R., Gere, D., Zamaroni, F., and Major R.,** *Basic Liquid Chromatography,* Varian Aerograph, Walnut Creek, Calif., 1971.
677. **Snyder, L. R. and Kirkland, J. J.,** *Introduction to Modern Liquid Chromatography,* John Wiley & Sons, New York, 1974.
678. **Brown, P. R.,** *High Pressure Liquid Chromatography: Biochemical and Biomedical Applications,* Academic Press, New York, 1973.
679. **Kirkland, J. J., Ed.,** *Modern Practice of Liquid Chromatography,* John Wiley & Sons, New York, 1971.
680. **Simpson, C. F., Ed.,** *Practical High Performance Liquid Chromatography,* Heydon and Sons, London, 1976.
681. **Moye, H. A.,** High speed liquid chromatography of pesticides, *J. Chromatogr. Sci.,* 13, 268, 1975.
682. **Horgan, D. F., Jr.,** High speed liquid chromatography, in *Analytical Methods for Pesticides and Plant Growth Regulators,* Vol. 7, Zweig, G., Ed., Academic Press, New York, 1973, chap. 2.
683. Pesticides by High Performance Liquid Chromatography, Bulletin No. 122, Whatman, Inc., Clinton, N. J., 1978.
684. **Walton, H. F.,** Ion exchange and liquid column chromatography, Biennial Fundamental Reviews, *Anal. Chem.,* 1978.
685. **Sparacino, C. M. and Hines, J. W.,** High-performance liquid chromatography of carbamate pesticides, *J. Chromatogr. Sci.,* 14, 549, 1976.
686. **Bakalyar, S. R.,** Mobile phases for high performance liquid chromatography, *Chromatogr. Rev.,* 4(1), 1, 1978.
687. **Lawrence, J. F. and Turton, D.,** High-performance liquid chromatographic data for 166 pesticides, *J. Chromatogr.,* 159, 207, 1978.
688. **Colvin, B. M., Engdahl, B. S., and Hanks, A. R.,** Determination of carbaryl in pesticide formulations and fertilizers by high-pressure liquid chromatography, *J. Assoc. Off. Anal. Chem.,* 57, 648, 1974.
689. **Leitch, R. E.,** Precise quantitative analysis with a stable high-speed liquid-liquid chromatography column, *J. Chromatogr. Sci.,* 9, 531, 1971.
690. **Sidwell, J. A.,** Recent developments in the use of high performance liquid chromatography for multiresidue pesticide analysis, *Med. Fac. Landbouww. Rijksuniv. Gent,* 42, 1803, 1977.
691. **Thurston, A. D., Jr.,** Liquid Chromatography of Carbamate Pesticides, EPA-R2-72-079, National Environmental Research Center, U. S. Environmental Protection Agency, Corvallis, Ore., 1972.
692. **Ishii, Y. and Otake, T.,** Studies on the analysis of pesticides by high-performance liquid chromatography. I. Carbamate insecticides, *Bull. Agric. Chem. Insp. Stn. (Tokyo),* 13, 32, 1973.
693. **Seiber, J. N.,** Reverse-phase liquid chromatography of some pesticides and related compounds, *J. Chromatogr.,* 94, 151, 1974.
694. **Lawrence, J. F. and Leduc, R.,** High-pressure liquid chromatographic analysis of carbofuran and two non-conjugated metabolites in crops as fluorescent dansyl derivatives, *J. Chromatogr.,* 152, 507, 1978.
695. **Frei, R. W., Lawrence, J. F., Hope, J., and Cassidy, R. M.,** Analysis of carbamate insecticides by fluorigenic labelling and high-speed liquid chromatography, *J. Chromatogr. Sci.,* 12, 40, 1974.
696. **Lawrence, J. F.,** Direct analysis of some carbamate pesticides in foods by high-pressure liquid chromatography, *J. Agric. Food Chem.,* 25, 211, 1977.

697. **Farrington, D. S. and Hopkins, R. G.**, Determination of ethylenethiourea in ethylenebisdithiocarbamate fungicides: comparison of high-performance liquid chromatography and gas-liquid chromatography, *Analyst (London)*, 104, 111, 1979.

698. **Kirkland, J. J.**, Preferred experimental conditions for trace analysis by modern liquid chromatography, *Analyst (London)*, 99, 859, 1974.

699. **McNair, H. M. and Chandler, C. D.**, High pressure liquid chromatography equipment. II, *J. Chromatogr. Sci.*, 12, 425, 1974.

700. **Veening, H.**, High resolution liquid chromatography, *Crit. Rev. Anal. Chem.*, 5, 165, 1975.

701. **Argauer, R. J.**, Fluorometric analysis of carbaryl insecticide in mixed formulations, *J. Assoc. Off. Anal. Chem.*, 53, 1166, 1970.

702. **Lawrence, J. F.**, Fluorimetric derivatization in high performance liquid chromatography, *J. Chromatogr. Sci.*, 17, 147, 1979.

703. **Frei, R. W. and Lawrence, J. F.**, Fluorigenic labelling in high-speed liquid chromatography, *J. Chromatogr.*, 83, 321, 1973.

704. **Kissinger, P. T., Bratin, K., Davis, G. C., and Pachla, L. A.**, The potential utility of pre- and post-column chemical reactions with electrochemical detection in liquid chromatography, *J. Chromatogr. Sci.*, 17, 137, 1979.

705. **Frei, R. W. and Scholten, A. H. M. T.**, Reaction detectors in HPLC, *J. Chromatogr. Sci.*, 17, 152, 1979.

706. **Jupille, T.**, UV-visible absorption derivatization in liquid chromatography, *J. Chromatogr. Sci.*, 17, 160, 1979.

707. **Moye, H. A.**, Dynamic fluorogenic labelling (DFL) of N-methylcarbamate pesticides for high speed liquid chromatographic analysis of residues, 89th Annu. Meet. AOAC, Abstract 28, Washington, D.C., 1975.

708. **Moye, H. A., Scherer, S. J., and St. John, P. A.**, Dynamic fluorogenic labeling of pesticides for high performance liquid chromatography: detection of N-methylcarbamates with o-phthaldehyde, *Anal. Lett.*, 10, 1049, 1977.

709. **Krause, R. T.**, Further characterization and refinement of an HPLC post-column fluorometric labeling technique for the determination of carbamate insecticides, *J. Chromatogr. Sci.*, 16, 281, 1978.

710. **DeStefano, J. J. and Beachell, H. C.**, Performance of large diameter columns for high speed liquid-solid chromatography, *J. Chromatogr. Sci.*, 10, 654, 1972.

710a. **Reece, P. A. and Cozamanis I.**, Injector loop sizes in HPLC, *Chromatogr. Rev.*, 5(2), 14, 1979.

711. **Kikta, E. J., Jr. and Stange, A. E.**, Phenones: a family of compounds broadly applicable to use as internal standards in high-performance liquid chromatography, *J. Chromatogr.*, 138, 41, 1977.

712. **Aten, C. F. and Bourke, J. B.**, Reverse-phase liquid chromatographic behavior of some carbamate and urea pesticides, *J. Agric. Food Chem.*, 25, 1428, 1977.

713. **Larose, R. H.**, High-speed liquid chromatographic cleanup of environmental samples prior to gas chromatographic determination of lindane, *J. Assoc. Off. Anal. Chem.*, 57, 1046, 1974.

714. **Coburn, J. A. and Sampson, R. C. J.**, personal communication, 1979.

715. **Frei, R. W.**, Trace enrichment and chemical derivatization in liquid chromatography; problems and potential in environmental analysis, *Int. J. Environ. Anal. Chem.*, 5, 143, 1978.

715a. **Frei, R. W.**, Trace enrichment in HPLC, *Chromatogr. Rev.*, 5(2), 7, 1979.

716. **McLafferty, F. W., Knutti, R., Venkataraghavan, R., Arpino, P. J., and Dawkins, B. G.**, Continuous mass spectrometric monitoring of a liquid chromatograph with subnanogram sensitivity using an on-line computer, *Anal. Chem.*, 47, 1503, 1975.

717. **McFadden, W. H., Schwartz, H. L., and Evans, S.**, Direct analysis of liquid chromatography effluents, *J. Chromatogr.*, 122, 389, 1976.

717a. **Arpino, P. J. and Guiochon, G.**, LC/MS coupling, *Anal. Chem.*, 51, 682A, 1979.

718. **Ramasamy, M.**, The identification and determination of organophosphorus and carbamate insecticides by thin-layer chromatography, *Analyst (London)*, 94, 1075, 1969.

719. **Stahl, E., Ed.**, *Thin-Layer Chromatography, A Laboratory Handbook,* Academic Press, New York, 1965.

720. **Aly, O. M. and Faust, S. D.**, Thin-layer chromatographic analysis in water pollution, in *Chromatographic Analysis of the Environment*, Grob, R. L., Ed., Marcel Dekker, New York, 1975, chap. 14.

721. **Chiba, M. and Morley, H. V.**, Thin-layer chromatography as a rapid screening method for the determination of carbaryl and 1-naphthol residues without cleanup, *J. Assoc. Off. Agric. Chem.*, 47, 667, 1964.

722. **Finocchiaro, J. M. and Benson, W. R.**, Thin layer chromatography of some carbamate and phenylurea pesticides, *J. Assoc. Off. Anal. Chem.*, 50, 888, 1967.

723. **Ogamo, A., Kasahara, H., and Nagasawa, K.**, Thin-layer chromatography of carbamate insecticides, *Eisei Kagaku*, 19, 340, 1973; *Pestic. Abstr.*, 74-0972.

724. **Nagasawa, K., Yoshidome, H., and Kamata, F.**, Separation and detection of carbamates and related compounds on polyamide layers, *J. Chromatogr.*, 52, 453, 1970.

725. **Hutzinger, O., Jamieson, W. D., MacNeil, J. D., and Frei, R. W.**, Electron-donor-acceptor complexing reagents for the analysis of pesticides. I. Survey of reagents and instrumental techniques, *J. Assoc. Off. Anal. Chem.*, 54, 1100, 1971.

726. **MacNeil, J. D., Frei, R. W., and Hutzinger, O.**, Electron-donor-acceptor complexing reagents for the detection of pesticides on TLC. II. Carbamates, anilides, and ureas. Spray reagents, chromatography, and instrumental techniques, *Int. J. Environ. Anal. Chem.*, 1, 205, 1972.

727. **Tewari, S. N. and Singh, R.**, Identification and detection of carbamate pesticides in autopsy tissues using thin-layer chromatography, *Fresenius Z. Anal. Chem.*, 294, 287, 1979.

728. **Davies, R. D.**, Detection limits for some carbamate and phenylurea pesticides by high-performance thin-layer chromatography, *J. Chromatogr.*, 170, 453, 1979.

729. **MacNeil, J. D. and Frei, R. W.**, Quantitative thin-layer chromatography of pesticides, *J. Chromatogr. Sci.*, 13, 279, 1975.

730. **Suzuki, K., Nagayoshi, H., and Kashiwa, T.**, Systematic separation and identification of 13 carbamate pesticides in their mixture, *Agric. Biol. Chem.*, 37, 218, 1973.

731. **Baron, R. L., Sphon, J. A., Chen, J. T., Lustig, E., Doherty, J. D., Hansen, E. A., and Kolbye, S. M.**, Confirmatory isolation and identification of a metabolite of carbaryl in urine and milk, *J. Agric. Food Chem.*, 17, 883, 1969.

732. **Mendoza, C. E.**, Analysis of pesticides by the thin-layer chromatographic-enzyme inhibition technique, *Residue Rev.*, 43, 105, 1972.

733. **Mendoza, C. E.**, Analysis of pesticides by the thin-layer chromatographic enzyme-inhibition technique. II, *Residue Rev.*, 50, 43, 1974.

734. **Mendoza, C. E.**, Thin-layer chromatography and enzyme inhibition techniques, *J. Chromatogr.*, 78, 29, 1973.

735. **Villeneuve, D. C.**, A review of enzymatic techniques used for pesticide residue analysis, *Adv. Chem. Ser.*, 104, 27, 1971.

736. **Zweig, G. and Archer, T. E.**, Residue determination of Sevin (1-naphthyl N-methylcarbamate) in wine by cholinesterase inhibition and paper chromatography, *J. Agric. Food Chem.*, 6, 910, 1958.

737. **Mendoza, C. E., Wales, P. J., McLeod, H. A., and McKinley, W. P.**, Enzymatic detection of ten organophosphorus pesticides and carbaryl on thin-layer chromatograms: an evaluation of indoxyl, substituted indoxyl and 1-naphthyl acetate as substrates of esterases, *Analyst (London)*, 93, 34, 1968.

738. **Mendoza, C. E. and Shields, J. B.**, Esterase specificity and sensitivity to organophosphorus and carbamate pesticides: factors affecting determination by thin layer chromatography, *J. Assoc. Off. Anal. Chem.*, 54, 507, 1971.

739. **Mendoza, C. E. and Shields, J. B.**, Determination of some carbamates by enzyme inhibition techniques using thin-layer chromatography and colorimetry, *J. Agric. Food Chem.*, 21, 178, 1973.

740. **Voss, G.**, Cholinesterase inhibition autoanalysis of insecticidal organophosphates and carbamates, *J. Assoc. Off. Anal. Chem.*, 52, 1027, 1969.

741. **Willard, H. H., Merritt, L. L., Jr., and Dean, J. A.**, *Instrumental Methods of Analysis*, 3rd ed., Van Nostrum, Princeton, N.J., 1958, chap. 3.

742. **Frei, R. W., Lawrence, J. F., and Belliveau, P. E.**, An *in situ* fluorimetric method for the determination of Sevin and α-naphthol on thin-layer chromatograms, *Fresinius Z. Anal. Chem.*, 254, 271, 1971.

743. **Frei, R. W. and Lawrence, J. F.**, Fluorigenic labelling of carbamates with dansyl chloride. IV. *In situ* quantitation of N-methyl carbamate insecticides on thin-layer chromatograms, *J. Chromatogr.*, 67, 87, 1972.

744. **Bowman, M. C. and Beroza, M.**, Spectrophotofluorescent and spectrophotophosphorescent data on insecticidal carbamates and the analysis of five carbamates in milk by spectrophotofluorometry, *Residue Rev.*, 17, 23, 1967.

745. **Ishii, Y.**, Determination of carbaryl (1-naphthyl N-methylcarbamate) in fish tissue by spectrophotofluorescence, *Bull. Agric. Chem. Insp. Stn. (Tokyo)*, 10, 51, 1970.

746. **Ott, D. E., Ittig, M., and Friestad, H. O.**, Automated steam distillation and fluorometry for screening for carbaryl as 1-naphthol in fruits and vegetables, *J. Assoc. Off. Anal. Chem.*, 54, 160, 1971.

747. **Belliveau, P. E., Mallet, V., and Frei, R. W.**, A spray method for the fluorescent detection of sulfur-containing organic compounds, *J. Chromatogr.* 48, 478, 1970.

748. **Belliveau, P. E. and Frei, R. W.**, pH sensitive fluorogenic spray reagent. I. General discussion and investigation of 1,2-dichloro-4,5-dicyanobenzoquinone, *Chromatographia*, 4, 189, 1971.

749. **Mallet, V. N., Belliveau, P. E., and Frei, R. W.**, *In situ* fluorescence spectroscopy of pesticides and other organic pollutants, *Residue Rev.*, 59, 51, 1975.

750. **Mallet, V. and Frei, R. W.**, An investigation of flavones as fluorogenic spray reagents for organic compounds on a cellulose matrix. I. General discussion of the method, *J. Chromatogr.*, 54, 251, 1971.

751. **Mallet, V. and Frei, R. W.**, An investigation of flavones as fluorogenic spray reagents for organic compounds on a cellulose matrix. II. Detection of pesticides, *J. Chromatogr.*, 56, 69, 1971.

752. **Frei, R. W. and Lawrence, J. F.**, Florigenic labelling of carbamates with dansyl chloride. I. Study of reaction conditions, *J. Chromatogr.*, 61, 174, 1971.

753. **Lawrence, J. F. and Frei, R. W.**, Fluorigenic labelling of carbamates using dansyl chloride. II. Fluorescence phenomena of the derivatives, *J. Chromatogr.*, 66, 93, 1972.

754. **Lawrence, J. F., Legay, D. S., and Frei, R. W.**, Fluorigenic labelling of carbamates using dansyl chloride. III. Thin-layer chromatographic properties of the derivatives, *J. Chromatogr.*, 66, 295, 1972.

755. **Lawrence, J. F. and Frei, R. W.**, Fluorimetric derivatization for pesticide residue analysis, *J. Chromatogr.*, 98, 253, 1974.

756. **Seiler, N. and Demisch, L.**, Fluorescent derivatives, in *Handbook of Derivatives for Chromatography*, Blau, K. and King, G., Eds., Heyden and Sons, London, 1977, chap. 9.

757. **Seiler, N. and Wiechmann, J.**, in *Progress in Thin-Layer Chromatography and Related Methods*, Vol. 1, Niederwieser, A. and Pataki, G., Eds., Ann Arbor Science Publishers, Ann Arbor, Mich., 1970, 94.

758. **Lawerence, J. F. and Frei, R. W.**, Fluorigenic labelling of N-methyl and N,N-dimethylcarbamates with 4-chloro-7-nitrobenzo-2,1,3-oxadiazole, *Anal. Chem.*, 44, 2046, 1972.

759. **Cyr, T., Cyr, N., and Haque, R.**, Spectrophotometric methods, in *Analytical Methods for Pesticides and Plant Growth Regulators*, Vol. 9, Zweig, G. and Sherma, J., Eds., Academic Press, New York, 1977, chap. 3.

760. **Post, A. P. and Stanley, C. W.**, Rapid screening test for carbaryl, *J. Agric. Food Chem.*, 15, 1124, 1967.

761. **McDermott, W. H. and DuVall, A. H.**, Carbaryl insecticide: analysis of formulations by colorimetry, *J. Assoc. Off. Anal. Chem.*, 53, 896, 1970.

762. **Ramasamy, M.**, A colorimetric method for the determination of eight carbamate insecticide residues, *Pestic. Sci.*, 5, 383, 1974.

763. **Van Gils, W. F.**, Spectrophotometric determination of propoxur residues on vegetable matter, *Analyst (London)*, 95, 88, 1970.

764. **Vonesch, E. E. and de Riveros, M. H. C. K.**, Colorimetric determination of carbaryl in wettable formulations, *J. Assoc. Off. Anal. Chem.*, 54, 128, 1971.

765. **de Riveros, M. H. C. K. and Vonesch, E. E.**, Colorimetric determination of carbaryl in apple, lettuce, chard, and tomato, *J. Assoc. Off. Anal. Chem.*, 54, 1083, 1971.

766. **Rangaswamy, J. R., Vijayashanker, Y. N., and Prakash, S. R.**, A simple spectrophotometric method for the determination of carbofuran residues, *J. Assoc. Off. Anal. Chem.*, 59, 1276, 1976.

767. **Blom, J.**, Sensitive and specific reaction for nitrate and for hydroxylamine, *Ber. Dtsch. Chem. Ges.*, 59, 121, 1926.

768. **Kearby, W. H., Ercegovich, C. D., and Bliss, M., Jr.**, Residue studies on aldicarb in soil and scotch pine, *J. Econ. Entomol.*, 63, 1317, 1970.

769. **Lee, D. F. and Roughan, J. A.**, Improvements to the nitrite-diazo dye (Blom's) method of determining hydroxylamine as used in the determination of residues of aldicarb, *Analyst (London)*, 96, 798, 1971.

770. **Clarke, D. G., Baum, H., Stanley, E. L., and Hester, W. F.**, Determination of dithiocarbamates, *Anal. Chem.*, 23, 1842, 1951.

771. **Lowen, W. K.**, Determination of dithiocarbamate residues on food crops, *Anal. Chem.*, 23, 1846, 1951.

771a. **Lowen, W. K.**, Analysis of manganese ethylenebisdithiocarbamate compositions and residues, *J. Assoc. Off. Agric. Chem.*, 36, 484, 1953.

772. **Viles, F. J.**, Field determination of carbon disulfide in air, *J. Ind. Hyg. Toxicol.*, 22, 188, 1940.

773. **Dickinson, D.**, Determination of Fermate (ferric dimethyldithiocarbamate, *Analyst (London)*, 71, 327, 1946.

774. **Cullen, T. E.**, Spectrophotometric determination of dithiocarbamate residues in food crops, *Anal. Chem.*, 36, 221, 1964.

775. **Stevenson, A.**, A simple color spot test for distinguishing between maneb, zineb, mancozeb, and selected mixtures, *J. Assoc. Off. Anal. Chem.*, 55, 939, 1972.

776. **Safe, S. and Hutzinger, O.**, *Mass Spectrometry of Pesticides and Pollutants*, CRC Press, Boca Raton, Fla., 1973.

777. **Garrison, A. W., Keith, L. H., and Alford, A. L.**, Confirmation of pesticide residues by mass spectrometry and NMR techniques, in *Fate of Organic Pesticides in the Aquatic Environment*, ACS Series 111, Gould, R. F., Ed., American Chemical Society, Washington, D.C., 1972, chap. 3.

778. **Silverstein, R. M. and Bassler, G. C.**, *Spectrometric Identification of Organic Compounds*, 2nd ed., John Wiley & Sons, New York, 1967, chap. 2.

779. McLafferty, F. W., *Interpretation of Mass Spectra,* W. A. Benjamin, New York, 1966.

780. Budzikiewicz, H., Djerassi, C., and Williams, D. H., *Mass Spectrometry of Organic Compounds,* Holden-Day, San Francisco, 1967.

781. Benyon, J. H., Saunders, R. A., and Williams, A. E., *The Mass Spectra of Organic Molecules,* Elsevier, Amsterdam, 1968.

782. Smith, S. R., Recent developments in ionization processes related to analytical methods in mass spectrometry, *Crit. Rev. Anal. Chem.,* 5, 243, 1975.

783. Vander Velde, G. and Ryan, J. F., Gas chromatography-mass spectrometry as applied to pesticide analysis, *J. Chromatogr. Sci.,* 13, 322, 1975.

784. Biros, F. J., Recent applications of mass spectrometry and combined gas chromatography-mass spectrometry to pesticide residue analysis, *Residue Rev.,* 40, 1, 1971.

785. Ryan, J. F., Residue analysis applications of mass spectrometry, in *Analytical Methods for Pesticides and Plant Growth Regulators,* Vol. 9, Zweig, G. and Sherma, J., Eds., Academic Press, New York, 1977, chap. 1.

785a. Gochman, N., Bowie, L. J., and Bailey, D. N., Specialized gas chromatography-mass spectrometry systems for clinical chemistry, *Anal. Chem.,* 51, 525A, 1979.

786. Damico, J. N. and Benson, W. R., The mass spectra of some carbamate pesticides, *J. Assoc. Off. Agric. Chem.,* 48, 344, 1965.

787. Benson, W. R. and Damico, J. N., Mass spectra of some carbamates and related ureas. II, *J. Assoc. Off. Anal. Chem.,* 51, 347, 1968.

788. Rankin, P. C., Negative ion mass spectra of some pesticidal compounds, *J. Assoc. Off. Anal. Chem.,* 54, 1340, 1971.

789. Lewis, C. P., Mass spectra of ethyl N-phenylcarbamate and ethyl N-ethylcarbamate, *Anal. Chem.,* 36, 176, 1964.

790. Lewis, C. P., Mass spectra of N-substituted ethyl carbamates, *Anal. Chem.,* 36, 1582, 1964.

791. Thompson, J. B., Brown, P., and Djerassi, C., Mass spectrometry in structural and stereochemical problems. CXIV. Electron impact induced rearrangements of thiocarbonates, carbamates, and thiocarbamates, *J. Am. Chem. Soc.,* 88, 4049, 1966.

792. Durden, J. A., Jr. and Bartley, W. J., The mass spectra of 4- , 5- , 6- , and 7-hydroxy-1-naphthyl methylcarbamates, *J. Agric. Food Chem.,* 19, 441, 1971.

793. Mumma, R. O., Khalifa, S., and Hamilton, R. H., Spectroscopic identification of metabolites of carbaryl in plants, *J. Agric. Food Chem.,* 19, 445, 1971.

794. Zweig, G. and Sherma, J., Eds., *Analytical Methods for Pesticides and Plant Growth Regulators,* Vol. 9, Academic Press, New York, 1977.

795. Gould, R. F., Ed., *Pesticide Identification at the Residue Level,* ACS Series 104, American Chemical Society, Washington, D.C., 1971.

796. Keith, L. H. and Alford, A. L., Review of the application of nuclear magnetic resonance spectroscopy in pesticide analysis, *J. Assoc. Off. Anal. Chem.,* 53, 1018, 1970.

797. Keith, L. H. and Alford, A. L., The high resolution NMR spectra of pesticides. III. The carbamates, *J. Assoc. Off. Anal. Chem.,* 53, 157, 1970.

798. Alford, A. L. and Keith, L. H., Catalog of Pesticide NMR Spectra, EPA Water Pollution Control Research Series, 16020 EWC 04/71, Environmental Protection Agency, Washington, D.C., 1971.

799. Giang, P. A., Library of infrared spectra of important pesticides, in *Analytical Methods for Pesticides and Plant Growth Regulators,* Vol. 9, Zweig, G. and Sherma, J., Eds., Academic Press, New York, 1977, chap. 6.

800. Giang, P. A., A bibliography on the use of infrared spectrometry in pesticides and pesticide residue studies, *Agric. Res. Ser.,* NE-91, 1978; *Pestic. Abstr.,* 79-0220.

801. Broderick, E. J., Bourke, J. B., Mattick, L. R., Tashenburg, E. F., and Avens, A. W., Determination of methylcarbamate pesticides in the presence of methyl anthranilate in concord grapes, *J. Assoc. Off. Anal. Chem.,* 49, 982, 1966.

802. Kovacs, M. F., Jr., Improved infrared method for carbaryl (Sevin) in commercial formulations, *J. Assoc. Off. Anal. Chem.,* 50, 566, 1967.

803. McDermott, W. H., Infrared analysis of carbaryl insecticide: modification of the extraction procedure to accommodate liquid suspension formulations, *J. Assoc. Off. Anal. Chem.,* 56, 576, 1973.

804. McDermott, W. H., Collaborative study of infrared spectrophotometric determination of aldicarb in formulations, *J. Assoc. Off. Anal. Chem.,* 57, 642, 1974.

Chapter 2

THE SUBSTITUTED UREA HERBICIDES

Allan E. Smith and Raj Grover

TABLE OF CONTENTS

I. Introduction 184
    A. History 184
    B. Use Patterns 184
    C. Formulations and Application Rates 184
    D. Synthesis 190
    E. Physical Properties 191
    F. Chemical Properties 191
    G. Toxicological Properties 192

II. Environmental Aspects 192

III. Review of Analytical Procedures 193
    A. Colorimetry 194
    B. Gas Chromatography (GC) 196
    C. High-Performance Liquid Chromatography (HPLC)......... 197
    D. Discussion of Analytical Procedures 198

IV. Analytical Methodology 199
    A. Extraction of Phenylurea Herbicides from Waters, Soils, and Plant
       Tissues.. 199
    B. GC Conditions for the Detection of Diuron 199
    C. HPLC Conditions for Urea Herbicides.................... 201
    D. Standards 205
    E. Sample Collection 205
    F. Sample Preparation..................................... 206
    G. Confirmatory Techniques................................ 206

V. Detailed Analytical Procedures 206
    A. Colorimetric Screening for Qualitative Phenylurea Determination in
       Water.. 206
    B. Analysis of Diuron in Water by GC...................... 207
    C. Analysis of Diuron in Sediment by GC 208
    D. Analysis of Diuron in Aquatic Vegetation and Biota by GC........ 208
    E. HPLC Procedure for Phenylureas in Water............... 208
    F. HPLC Procedure for Phenylureas in Sediments, Aquatic Plants, and
       Biota 208

References... 209

I. INTRODUCTION

A. History

The substituted urea herbicides are derivatives of urea (H_2NCONH_2), and the first compounds to be synthesized contained an amino moiety fully substituted with alkyl or alkoxy groups, while the second amino function contained a single mono- or di-halogenated phenyl group. Currently there are about 20 such aryl-dialkyl or aryl-alkyl-alkoxy urea derivatives (Table 1) and 10 structures based on heterocyclic substituents (Table 2) available as herbicides.

Monuron, the first phenylurea herbicide to be synthesized, was described by Bucha and Todd of E. I. du Pont de Nemours & Company in the U.S. in 1951. Du Pont quickly developed several other substituted ureas, such as fenuron, diuron, and neburon and thus set the commercial use of urea derivatives as herbicides on a practical basis. Since then, thousands of substituted ureas containing aliphatic, cycloaliphatic, aromatic, and heterocyclic substituents have been prepared and tested for biological activities against pests of all kinds. Systematic studies have established that substituted urea derivatives exhibit almost exclusively herbicidal activity, though diflubenzuron is used as an insecticide, and some of the thiourea compounds show fungicidal properties.

In general, the urea herbicides are taken up into the germinating weeds via the roots, from where they are translocated to the photosynthetic organs. Control of susceptible plant species is obtained by the inhibition of the Hill reaction in the photosynthesis process, resulting in foliar chlorosis and eventual death.

B. Use Patterns

Initially, the phenylureas were characterized as soil sterilants. However, their selective control of weeds in crops, especially at low dosages, was recognized soon thereafter (Table 3), culminating in the more recent development of chlortoluron and linuron as selective herbicides in cereals.[1,2] Fenuron and monuron, together with karbutilate, are still used for general weed control in noncrop land, while diuron, fluometuron, and norea are used as selective herbicides in crops, such as cotton and sugarcane. Diuron is also used as a pre-emergence herbicide in orchard and citrus crops. Chloroxuron, chlorbromuron, and norea are applied as selective herbicides in soybeans and the latter two also in potatoes. These three herbicides, together with linuron, are used as selective pre-emergence herbicides in several vegetable and small fruit crops. Siduron will control grassy weeds, such as crabgrass, foxtail, and barnyard grass in newly seeded and established grasses.

Diuron is the only urea herbicide recommended for direct application to water and is used for the control of certain aquatic weeds and algae in farm ponds and dugouts. In addition, diuron can be applied to banks of irrigation ditches and canals for general weed control thereby facilitating water flow.

C. Formulations and Application Rates

Since most of the urea herbicides have relatively low water solubilities (Table 4) they are generally formulated as wettable powders or granular dusts (Table 3). However, fenuron is also formulated as a liquid concentrate while cycluron and metobromuron are formulated as emulsifiable concentrates.

For selective weed control, the urea herbicides are applied as pre-emergence treatments to bare soil, though with certain crops diuron and linuron can be applied as post-emergence treatments to young emerged weeds.[1,2] For general vegetation control in noncropland, the herbicides may be applied as pre-emergence granular treatments.

Application rates of urea herbicides used for selective weed control in crops can vary

Table 1

COMMON AND CHEMICAL NAMES, FORMULAE, MOLECULAR WEIGHTS, AND STRUCTURES OF PHENYLUREA HERBICIDES

Common Name	Chemical Name	Formulae	Molecular weight	R_1	R_2	R_3	R_4
buturon	3-(p-chlorophenyl)-1-(1-methyl-2-propynyl)-1-methylurea	$C_{12}H_{13}ClN_2O$	236	-	-Cl	-CH$_3$	-CH-CH$_3$ / CH≡CH
chlorbromuron*	3-(4-bromo-3-chlorophenyl)-1-methoxy-1-methylurea	$C_9H_{10}BrClN_2O_2$	294	-Cl	-Br	-OCH$_3$	-CH$_3$
chloreturon	3-(3-chloro-4-ethoxyphenyl)-1,1-dimethylurea	$C_{11}H_{15}ClN_2O_2$	243	-Cl	-OC$_2$H$_5$	-CH$_3$	-CH$_3$
chloroxuron*	3-p-(p-chlorophenoxy)phenyl-1,1-dimethylurea	$C_{15}H_{15}ClN_2O_2$	291	-	(p-chlorophenoxy)	-CH$_3$	-CH$_3$
chlortoluron	3-(3-chloro-4-methylphenyl)-1,1-dimethylurea	$C_{10}H_{13}ClN_2O$	213	-Cl	-CH$_3$	-CH$_3$	-CH$_3$
difenoxuron	3-[4-methoxyphenoxy)-phenyl]-1,1-dimethylurea	$C_{16}H_{18}N_2O_3$	286	-	(methoxyphenoxy)	-CH$_3$	-CH$_3$
diflubenzuron	1-(4-chlorophenyl)-3-(2,6-difluorobenzoyl)urea	$C_{14}H_9ClF_2N_2O_2$	311	-	-Cl	-H	(2,6-difluorobenzoyl)
dimefuron	3-chloro-4-(3-tert-butyl-6-oxo-4,1,2-oxadiazinyl) phenyl-1,1-dimethyl)urea	$C_{16}H_{21}ClN_4O_3$	355	-Cl	(oxadiazinone with C(CH$_3$)$_3$)	-CH$_3$	-CH$_3$
diuron*	3-(3,4-dichlorophenyl)-1,1-dimethylurea	$C_9H_{10}Cl_2N_2O$	233	-Cl	-Cl	-CH$_3$	-CH$_3$

Table 1 (continued)

COMMON AND CHEMICAL NAMES, FORMULAE, MOLECULAR WEIGHTS, AND STRUCTURES OF PHENYLUREA HERBICIDES

Common name	Chemical name	Formula	MW				
fenuron*	1,1-dimethyl-3-phenylurea	C$_9$H$_{12}$N$_2$O	164	-	-	-CH$_3$	-CH$_3$
fluometuron*	1,1-dimethylurea-3-(a,a,a-trifluoro-m-tolyl)urea	C$_{10}$H$_{11}$F$_3$N$_2$O	232	-CH$_3$	-	-CH$_3$	-CH$_3$
isoproturon	3-(p-isopropylphenyl)-1,1-dimethylurea	C$_{12}$H$_{18}$N$_2$O	206	-	-CH(CH$_3$)$_2$	-CH$_3$	-CH$_3$
karbutilate*	3-(3,3-dimethylureido)phenyl-tert-butylcarbamate	C$_{14}$H$_{21}$N$_3$O$_3$	279	-	-OCONHC-(CH$_3$)$_3$	-CH$_3$	-CH$_3$
linuron*	3-(3,4-dichlorophenyl)-1-methoxy-1-methylurea	C$_9$H$_{10}$Cl$_2$N$_2$O$_2$	249	-Cl	-Cl	-OCH$_3$	-CH$_3$
metobromuron*	3-(p-bromophenyl)-1-methoxy-1-methylurea	C$_9$H$_{11}$BrN$_2$O$_2$	259	-	-Br	-OCH$_3$	-CH$_3$
metoxuron	3-(3-chloro-4-methoxyphenyl)-1,1-dimethylurea	C$_{10}$H$_{13}$ClN$_2$O$_2$	229	-Cl	-OCH$_3$	-CH$_3$	-CH$_3$
monuron*	3-(p-chlorophenyl)-1,1-dimethylurea	C$_9$H$_{11}$ClN$_2$O	199	-	-Cl	-CH$_3$	-CH$_3$
neburon*	3-(3,4-dichlorophenyl)-1-n-butyl-1-methylurea	C$_{12}$H$_{16}$Cl$_2$N$_2$O	275	-Cl	-Cl	-CH$_3$	-(CH$_2$)$_3$CH$_3$
siduron*	1-(2-methylcyclohexyl)-3-phenylurea	C$_{14}$H$_{20}$N$_2$O	232	-	-	-H	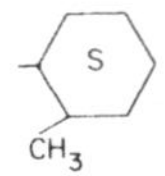

*WSSA common names

Table 2

COMMON AND CHEMICAL NAMES, FORMULAE, MOLECULAR WEIGHTS, AND STRUCTURES OF HETEROCYCLIC UREA HERBICIDES

Common Name	Chemical Name	Formula	Molecular weight	Structure
benzthiazuron	1-methyl-3-(2-benzthiazolyl)urea	$C_9H_9N_3SO$	207	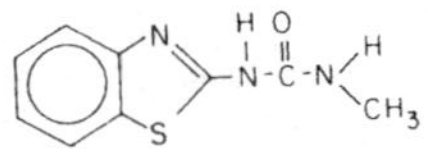
cycluron*	3-cyclooctyl-1,1-dimethylurea	$C_{11}H_{22}N_2O$	198	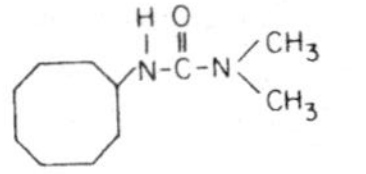
ethidimuron	1-(5-ethylsulfonyl-1,3,4-thiadiazol-2-yl)-1,3-dimethylurea	$C_7H_{12}N_4S_2O_3$	264	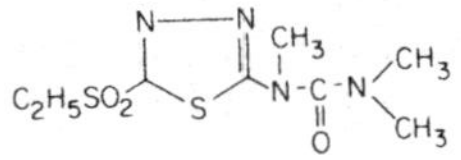
isonoruron	3-(hexahydro-4,7-methanoindan-2-yl)-1,1-dimethyl=urea	$C_{13}H_{22}N_2O$	222	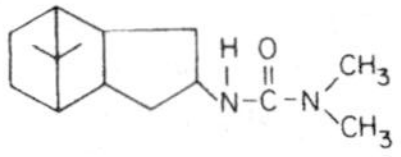
methabenzthia-zuron	3-(2-benzthiazolyl)-1,3-dimethylurea	$C_{10}H_{11}N_3SO$	221	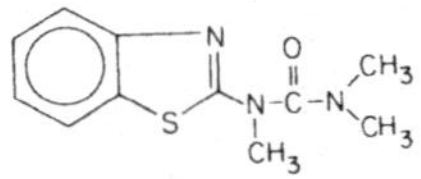
norea*	3-(hexahydro-4,7-methanoniondan-5-yl)-1,1-dimethyl=urea	$C_{13}H_{22}N_2O$	222	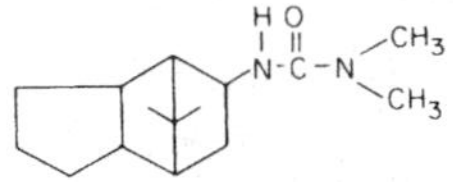
tebuthiuron*	1-(5-tert-butyl-1,3,4-thiadiazol-2-yl)-1,3-dimethylurea	$C_9H_{16}N_4SO$	228	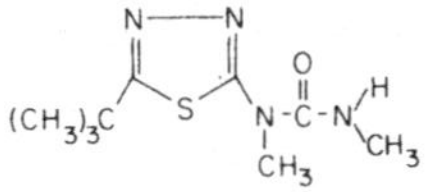
thiazfluron	1-(5-trifluoromethyl-1,3,4-thiadiazol-2-yl)-1,3-dimethylurea	$C_6H_7F_3N_4SO$	240	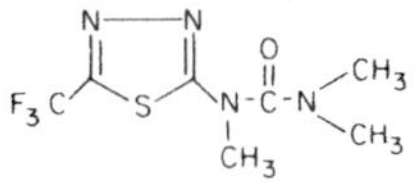

* WSSA common names

Table 3
USE PATTERNS OF UREA HERBICIDES[1,2]

Herbicide	Application method	Formulations[1]	Carrier	Rate (lb/ac)	Types of weeds controlled	Crops or other uses
Chlorbromuron	Pre- or post-emergence	WP	Water	0.75—4	Annual grasses and broadleaf weeds	Soybeans, potatoes, carrots, parsnips, winter wheat
Chloroxuron	Pre- or post-emergence	WP	Water	1—4	Annual grasses and broadleaf weeds	Soybeans, onions, strawberries, celery, chrysanthemums
Chortoluron	Pre- or post-emergence	WP	Water	2—4	General weed control, esp. *Alopecurus*	Winter barley, winter wheat
Diuron	Pre- or post emergence	WP/G	Water/oil	0.5—6	Annual grasses and broadleaf weeds	Cotton, sugarcane, pineapple, grapes, apples, pears, citrus, alfalfa
		WP/G	Water/oil	4—48	General weed control, soil sterilant	Noncrop land, industrial and rights-of-way areas, irrigation and drainage ditches
		WP/G	Water	0.5—2 (ppm)	Control of algae and aquatic weeds	Ponds, farm dug-outs
Fenuron (+ TCA)	Pre- or post-emergence	L/G	Water/oil	15—30	General weed control, brush control	Noncrop land
Fluometuron	Pre-plant, pre- or post-emergence	WP	Water	0.75—4	Annual grasses and broadleaf weeds	Cotton, sugarcane
Isoproturon	Pre- or post-emergence	WP	Water		Annual weeds, esp. *Alopecurus*	Winter wheat, winter barley
Karbutilate	Pre- or post-emergence	WP/G	Water	2—20	Annual and perennial broadleaf weeds and grasses, woody species	Railroad, highway, and utility rights-of-ways and other noncrop areas

Linuron	Pre- or post-emergence	WP	Water/surfactant	0.5—2	Germinating and newly established broadleaf weeds and grasses	Soybeans, cotton, corn, sorghum, wheat, potatoes, celery, carrots, parsnips, short-term control in non-crop land
Methabenz-thiazuron	Pre- or post-emergence	WP	Water		Annual weeds, esp. *Alopecurus* and *Poa*	Winter cereals, perennial ryegrass
Metobromuron	Pre-emergence	EC	Water		Annual grasses and broadleaf weeds	Potatoes
Metoxuron	Post-emergence	WP	Water		Annual weeds, esp. *Alopecurus* and *Matricaria*	Winter barley, winter wheat, carrots
Monolinuron	Pre-emergence, Post-planting	WP	Water	1—2	Annual grass and broadleaf weeds	Potatoes, leeks
Monuron (+TCA)	Pre- or post-emergence	WP OC/G	Water/oil	20—45	General weed killer, soil sterilant	Noncrop land
Norea	Pre- or post-emergence	WP	Water	1—5	Annual grasses and broadleaf weeds	Cotton, sorghum, sugarcane, lima beans, sweet potatoes, spinach, soybeans, potatoes, peas, woody and herbaceous ornamentals
Siduron	Pre-emergence	WP/G	Water	2—12	Germinating annual weed grasses	Newly-seeded and established bluegrass, fescue, redtop, smooth brome, perennial ryegrass, orchardgrass, zoysia, and some strains of bentgrass
Tebuthiuron	Pre- or post emergence	WP/G	Water	1—16	Annual and perennial grasses and broadleaf weeds, woody species	Noncrop land

ª WP, wettable powder; G, granular dust; L, liquid concentrate; EC, emulsifiable concentrate; OC, oil concentrate.

Table 4
SOME PHYSICAL AND TOXICOLOGICAL PROPERTIES OF COMMERCIALLY USED UREA HERBICIDES[1-4]

Herbicide	Melting point (°C)	Water solubility (°C)(ppm)	Vapor pressure (°C)(mmHg)	Extinction coefficient (E)(cm²/mg)	λmax (nm)	Acute oral LD$_{50}$ (rats)(mg/kg)
Chlorbromuron	94—96	50 (20)	4.0×10^{-7}(20)	42	249	2150
Chloroxuron	151—152	4 (20)	1.8×10^{-9}(20)	36	247	73000
Chlortoluron	147—148	70 (20)	3.0×10^{-8}(20)	78	242	—
Diuron	158—159	42 (25)	0.3×10^{-5}(50)	32	247	3400
Fenuron	133—134	3850 (25)	1.6×10^{-4}(60)	79	238	4000—5700
Fluometuron	163—164	90 (25)	5.0×10^{-7}(20)	66	242	78000
Isoproturon	—	—	—	77	240	—
Karbutilate	176—177	325 (20)	—	49	240	3000
Linuron	93—94	75 (25)	1.5×10^{-5}( )	42	248	1500
Methabenzthiazuron	119	—	$\times 10^{-5}$(20)	101	228	—
Metobromuron	95—96	330 (20)	3.0×10^{-6}(20)	39	247	2000—3000
Metoxuron	126—127	678 (24)	3.2×10^{-8}(20)	61	242	3200
Monolinuron	76—78	580 (20)	1.5×10^{-4}(22)	80	246	2250
Monuron	174—175	230 (25)	5.0×10^{-7}(25)	75	245	3600
Norea	171—172	150 (25)	—	33	196	1476
Siduron	133—138	18 (25)	$<8.0 \times 10^{-4}$(100)	—	—	75000
Tebuthiuron	159—161	2500 (25)	2.0×10^{-6}(25)	—	254	600

from 0.5 to 5.0 kg/ha and are dependent upon the crop and soil type (Table 3). For general vegetation control up to 50 kg/ha may be applied depending on the type of weed species present and the length of control desired.[1,2] Water is the usual carrier for selective control applications. For general vegetation control, the wettable powders can also be applied as suspensions in oil. Diuron can be sprayed directly to water surfaces of farm ponds and dugouts at rates sufficient to yield a herbicide concentration in the aquatic system of 0.5 to 2.0 ppm (Table 3). The lower rates are usually advocated for algal control, and the higher rates for the control of aquatic weeds and for extended algal control. For the control of weeds in ditchbanks or irrigation canal beds up to 35 kg/ha diuron may be used.

D. Synthesis

The most common method for the synthesis of phenyl-dialkyl ureas is based on the treatment of substituted aryl *iso*-cyanates with the appropriately substituted alkylamines.

aryl—N=C=O + HN—dialkyl ⟶ aryl—NH—CO—N—dialkyl

The reaction is carried out by combining the two reactants in an inert solvent, such as acetonitrile, and the ureas produced are obtained with a high degree of purity and in good yields.

A second method used for the manufacture of urea herbicides involves the reaction between dialkylcarbamic acid chlorides and substituted anilines.

aryl—NH$_2$ + dialkyl—N—CO—Cl ⟶ aryl—NH—CO—N—dialkyl

The arylalkoxy phenylureas are synthesized by the condensation of the aryl *iso*-cyanates with alkylalkoxyamines or, alternatively, may be prepared by alkylation of the appropriate hydroxamic acid derivatives.

aryl—N=C=O + NH—alkylalkoxy ⟶ aryl—NH—CO—N—Alkylalkoxy

$$\text{aryl—NH—CO—N}\overset{OH}{\underset{CH_3}{<}} \xrightarrow{(CH_3)_2SO_4} \text{aryl—NH—CO—N}\overset{OCH_3}{\underset{CH_3}{<}}$$

These and additional methods for the synthesis of urea herbicides are discussed in greater detail elsewhere.[3]

E. Physical Properties

Some of the physical properties of commonly encountered urea derivatives, such as melting points, water solubilities, vapor pressures, UV absorption maxima, and extinction coefficients are listed in Table 4.[1-4] In general, the urea herbicides are white crystalline solids of low vapor pressures and high melting points. Their solubility in water is relatively low; however, they are readily soluble in organic solvents such as ethanol, acetone, and benzene.

The simple phenylureas show basic properties, with pk_a values in the range of -1 to -2. With increasingly complex substituents the basicity decreases. In aqueous solutions, all urea herbicides show absorption maxima in the range 196 to 280 nm in the UV region of the spectrum.[4]

F. Chemical Properties

The substutited urea derivatives can undergo several chemical reactions, such as hydrolysis, halogenation, acylation, alkylation, salt formation, etc. Use of these reactions has been made for the development of analytical methodology for urea herbicides, while other reactions have been instrumental in the preparation of derivatives with modified or enhanced biological effectiveness. None of these chemical reactions play a significant role in the transformation of the urea herbicides in the environment.

Substituted ureas undergo hydrolysis in both acidic and alkaline media to the corresponding aryl and alkyl amines.

aryl—NH—CO—N—dialkyl ⟶ aryl—NH$_2$ + dialkyl—NH

Analysis of the substituted aniline hydrolysis products by colorimetric procedures can be used to quantitatively determine residues of urea herbicides in various substrates (see later). Halogenation of the aromatic ring of phenylureas in the *para* position has lead to the successful commercial development of bromo- derivatives, such as chlorbromuron and metobromuron.

The basicity of fenuron and monuron is sufficient to allow salt formation with the acidic herbicide TCA (trichloroacetic acid), and such salts have been commercially developed for total vegetation control.

$$\text{Cl}-\!\!\!\bigcirc\!\!\!-\overset{H}{\underset{H}{N}}-\overset{O}{C}-\overset{+}{N}\overset{CH_3}{\underset{CH_3}{<}} \cdot \ CCl_3CO_2^-$$

The amide nitrogen of the phenylurea structure can undergo *N*-acylation and *N*-alkylation reactions. Under the action of strong proton extracting agents, such as sodium hydride, the amide proton is removed. Treatment of this intermediate with alkyl halides results in formation of derivatives such as:

$$\text{aryl} - \text{N} - \text{CO} - \text{N} \bigg\langle \begin{array}{c} \\ \\ \end{array}$$
$$ | $$
$$ \text{R}$$

These products have better GC characteristics than the parent ureas (see later); thus analytical procedures based on the GC estimation of these compounds have been developed for residue analysis of several urea herbicides.[5,6] Other reactions of the urea herbicides are well illustrated elsewhere.[3] Aqueous solutions of monuron have been shown to decompose in sunlight[7] and several other reports have confirmed that many of the urea herbicides may be photochemically unstable.[8]

G. Toxicological Properties

The LD_{50} values for most of the urea herbicides, using rats as the test animal, range from 600 to 78,000 mg/kg, indicating low to moderate mammalian toxicity (Table 4).[1,9] The LC_{50} values for fish and other aquatic invertebrates range from 0.4 to 170 ppm for exposure periods varying from 48 to 96 hr (Table 5).[1,9-12]

There are only a few reports on the long-term toxic effects of urea herbicides or their mammalian mutagenicity potential.[13] Monuron is the only urea herbicide that has been implicated for possible carcinogenicity.[14]

II. ENVIRONMENTAL ASPECTS

With the exception of diuron, the urea herbicides are not recommended for the control of weeds in or near bodies of water. However, some urea herbicides have been detected in aquatic systems under a variety of situations which no doubt reflects their off-target transport from agricultural and other treated areas. The major routes of pesticidal transport from target areas to a nontarget body of water include the air, originating from drift and volatilization during and following application; by surface water run-off, both in solution and in the adsorbed state on the soil colloids; and by leaching and accidental spills near wells and water bodies.

There have been no reported cases of water contamination with urea herbicides resulting from off-target drift in the form of droplets, or as vapor. The latter is only to be expected since urea herbicides possess relatively low vapor pressures.

Off-target transport of diuron, linuron, fluometuron, and tebuthiuron in surface run-off waters has been reported.[15-19] The concentration of these herbicides in the run-off water varied from time to time depending upon the rate of application, rainfall, slope of the terrain, and solubility and adsorptive characteristics of the chemicals. In another study it was noted that the maximum concentration of diuron in the initial flush from irrigation ditches treated at a rate of 35.8 kg/ha was in the range of 200 ppb.[20]

Under conditions of high water table, ample rainfall or irrigation water, and favorable soil textural conditions, urea herbicides with moderate to high water solubilities may enter ground waters.[21] Once in the ground water, the degradation of these chemicals can be slow due to aerobic conditions and lack of microbial activity.

Urea herbicides are persistent chemicals and soil-based residues can remain for several months following application. Degradation is accomplished by biological processes and nonbiological breakdown is considered to be unimportant.[3] The biological breakdown or transformation of urea herbicides by plants, mammals, and by soil microorganisms has been well documented,[3] with N-dealkylation, N-dealkoxylation, and hydrolytic reactions being demonstrated in all three substrates. In addition, ring hydroxylation with oxidation of the aromatic moiety can also occur in plants and animals.[3]

Table 5
ACUTE TOXICITY TO FISH AND OTHER AQUATIC INVERTEBRATES FOR SELECTED UREA HERBICIDES[1,9-12]

Herbicide	Organism	LC$_{50}$ (ppm)	Duration of exposure (hr)
Chlorbromuron	Trout	0.56	96
	Catfish	11.0	96
	Bluegill	7.5	96
Chloroxuron	Killfish	15.0	
	Fathead minnows	0.4	
Diuron	Trout, rainbow	4.3	48
	Daphnia	1.4	48
	Coho salmon	16	24
	Bluegill	4—7.6	96
	Juvenile white mullet	6.3	48
Fluometuron	Rainbow trout	47	96
	Bluegill	96	96
	Catfish	55	96
	Carp	170	96
	Guppy	46	96
	Freshwater shrimp	60	100
	Killfish	50	100
Karbutilate	Bluegill	75	96
	Trout	135	96
Linuron	Rainbow trout	16	96
	Bluegill	16	96
Metobromuron	Rainbow trout	71	72
Monuron	Juvenile white mullet	16.3	48
Tebuthiuron	Trout	144	—
	Bluegill	112	—

The importance of substituted anilines, formed by biological hydrolytic mechanisms, with their possible transformation to toxic azo-benzenes, was considered a potential health hazard at one time. However, the presence of azo-benzene derivatives has not been noted in soils and plants receiving normal applications of urea herbicides.[3]

Little is known about the fate of pesticides, including the ureas, in the aquatic environment. It has been assumed that the processes that determine the behavior of pesticides in the terrestial system also operate in the aquatic situation. Once in the water, the dispersal of urea herbicides in the aquatic ecosystem will be a function of both the hydrological and limnological parameters. The availability and the degradation of these chemicals should therefore be determined by the familiar processes of adsorption and desorption, and photochemical and microbial degradation.

III. REVIEW OF ANALYTICAL PROCEDURES

There are no recommended procedures for the extraction and analysis of residues of urea herbicides in aquatic environments that can compare with those developed for the organochlorine pesticides. Only one or two procedures have been described which even allow for the estimation of more than one urea herbicide residue in the presence of each other, and seemingly analysts have been content to undertake analyses on samples containing only known urea treatments.

There are three major procedures available to the analyst for the determination of urea herbicide residues in water, soil, plant, and animal tissue. These are based on colorimetry, gas chromatography, and high-performance liquid chromatography. In this Section the principles of each procedure will be discussed, together with the advantages and disadvantages engendered by each method.

A. Colorimetry

It has been known for many years that aromatic amines can undergo diazotization with nitrous acid, and the resulting diazonium salts coupled with suitable reagents to form red-colored azo-dervatives. The analysis of microgram amounts of sulfanilamide using such a procedure was reported by Bratton and Marshall[22] in 1939, using *N*-(1-naphthyl)ethylenediamine as the coupling reagent. The overall reaction may be summarized as:

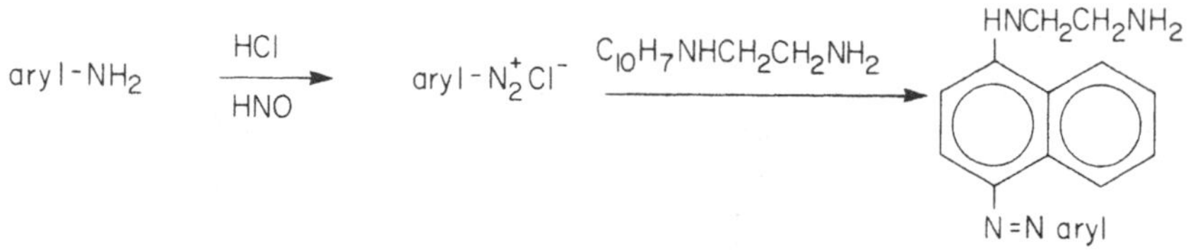

The absorptive maximum of the red azo-derivative is in the region of 550 nm, so that spectrophotometric determination of solution absorbance at this wavelength may be used to calculate the amount of aromatic amine originally present.

Substituted phenylureas undergo hydrolysis in both acidic and basic media to aromatic amines (anilines).[23] Thus a general analytical procedure has been developed for the quantitative determination of phenylurea herbicides which is based on the colorimetric quantitation of the aniline moiety formed by hydrolysis of the parent herbicide.[23-27]

Practically, the urea herbicides are solvent extracted from the samples to be analyzed,[23,26] then after evaporation of the solvent the extracts are boiled with either acid or alkali to effect hydrolysis of any phenylurea residues to their respective anilines. By using boiling alkali this hydrolysis can be carried out directly on the unextracted soil, plant, or animal tissue sample.[9,23-25,27] In this instance, the samples are heated with the alkali and the aromatic amine hydrolysis product is distilled off as it is formed into an organic solvent, such as *n*-hexane or *iso*-octane, from which it can be removed by treatment with dilute acid. Following hydrolysis, an acidic solution of the aromatic amine is treated with aqueous sodium or potassium nitrite to form the diazonium derivative. Excess nitrous acid is then decomposed by the addition of an aqueous solution of sulfamic acid or its ammonium salt. The solution containing the diazotized aniline is treated with *N*-(1-naphthyl)ethylenediamine, in water, to form the red or purple-colored azo-derivative which is quantitated spectrophotometrically. In general, about 15 or 20 min are required to complete formation of this azo-product, though longer times may be necessary.

Table 6 summarizes hydrolytic conditions that have been reported for some commonly encountered phenylurea herbicides, together with the times required for the formation of the azo-products from the diazotized amines. The wavelength of maximum absorbance of the azo-derivatives is also noted.

The sensitivity of this colorimetric procedure as applied to the analysis of some commonly used phenylurea herbicides in a variety of substrates is indicated in Table 7. Thus the method is capable of detecting phenylurea herbicide residues at a level of 0.1 μg/g or less.

Table 6

CONDITIONS FOR THE HYDROLYSIS OF PHENYLUREA HERBICIDES TO
AROMATIC AMINES, TOGETHER WITH THE TIMES FOR GENERATION OF
THE AZO-DERIVATIVES, AND THEIR ABSORPTION MAXIMA

Phenylurea herbicide	Aromatic amine product	Hydrolytic conditions	Time for generation of azo-product (min)	λ max of azo-product (nm)	Ref.
Fenuron	Aniline	6 *N*HCl, 1 hr	180	555	26
Buturon, monolinuron, monuron	4-Chloroaniline	5 *N*NaOH, 4 hr	15	555	23
Diuron, linuron, neburon	3,4-Dichloroaniline	5 *N*NaOH, 4 hr or 6 *N*HCl, 1 hr	15	555	25,26
Chlorbromuron	3-Chloro-4-bromoaniline	1.5 *N*NaOH, 3 hr	20	550	9
Fluometuron	3-Trifluoromethylaniline	1.5 *N*NaOH, 3 hr	20	525	9
Metobromuron	4-Bromoaniline	1.5 *N*NaOH, 3 hr	20	558	9

Table 7

SENSITIVITY OF DETECTION OF PHENYLUREA HERBICIDE
RESIDUES IN VARIOUS SUBSTRATES USING THE
COLORIMETRIC PROCEDURE

Phenylurea herbicide	Substrate	Extraction method	Sensitivity (µg/g)	Ref.
Diuron			0.02	
Fenuron		Solvent extraction followed	0.04	
Linuron	Water	by acidic hydrolysis	0.02	26
Monuron			0.03	
Neburon			0.02	
Monuron	Soil	Solvent extraction followed by acidic hydrolysis	0.05	23
Monuron	Soil	Direct hydrolysis of sample with alkali and distillation of aniline	0.1	23
Diuron	Soil, plant, and	Direct hydrolysis of sample	0.05	25
Monuron	animal tissue	with alkali and distillation	0.05	
Neburon		of aniline	0.05	
Chlorbromuron	Soil and plant	Direct hydrolysis of sample	0.05	
Fluometuron	tissue	with alkali and distillation	0.05	9
Metobromuron		of aniline	0.05	

Some substrates, notably plant tissues and soils,[25,27,28] contain small amounts of naturally occurring materials which also react with the color-forming reagent to give interfering red products. These interferences can, however, by successfully removed from the aniline complexes by chromatography using a cellulose column.[25,27,28]

The advantages to using the colorimetric analytical procedure are that it is simple, requires no sophisticated equipment, and is excellent for the analysis of known single residues of phenylurea herbicides in natural waters. The method is not applicable as a multi-residue analytical method, since all the azo-derivatives of the substituted anilines absorb in the same region (Table 6). Attempts have been made to introduce specificity into the colorimetric procedure by using chromatographic techniques to separate and identify the different azo-products.[28] However, it will be noted (Table 6) that the same aniline can be formed from more than one phenylurea. Thus 4-chloroaniline could arise via the hydrolysis of monuron, monolinuron, and buturon, while 3.4-dichloroaniline is formed from diuron, linuron, and neburon. The carbamate herbicides barban (4-chloro-2-butynyl-*m*-chlorocarbanilate), chlorpropham (isopropyl-*m*-chlorocarbanilate), propham (isopropyl carbanilate), and propanil (3′,4′-dichloropropionanilide) can also give rise to aromatic amines on hydrolysis. Demethylated degradation products derived from the phenylurea herbicides behave similarly and would therefore interfere with the analysis of parent phenylurea herbicide residues when the colorimetric procedure is used.

Despite these drawbacks, this procedure is good for the analysis of single urea herbicide residues in waters and soils when the identity of the compound is known. For this reason, automated analytical procedures utilizing colorimetric detection have been developed for the analysis of known phenylurea herbicide residues in soils.[29,30]

B. Gas Chromatography (GC)

Although the gas chromatograph is probably the most commonly used analytical instrument for pesticide residue determination, its use for the direct quantitative estimation of urea herbicides has proved relatively unsuccessful. Since substituted urea compounds are slightly volatile and contain both chlorine and nitrogen atoms they should be amenable to gas chromatographic separation with quantitation using EC or nitrogen-specific detectors. However, it has been shown that phenylurea herbicides undergo thermal decomposition in the injector port, detector, and on the column of GC systems, to phenyl *iso*-cyanates and aliphatic amines.[6,31] The cause of this thermal breakdown has been attributed to the presence of the amide hydrogen atom since substitution of this position with a methyl group appears to stabilize the molecule and allow satisfactory GC analysis.[6,32]

$$\text{aryl} - \underset{\underset{H}{|}}{N} - CO - N \diagdown^{\diagup} \longrightarrow \text{aryl} - \underset{\underset{CH_3}{|}}{N} - CO - N \diagdown^{\diagup}$$

Attempts have been made to reduce the thermal breakdown of the phenylurea herbicides and so allow their direct GC separation and detection. Short chromatographic columns packed with selected supports and operated at relatively low temperatures do not appear to be fully satisfactory. Some researchers have recommended the use of metal columns and specially constructed electron-ECDs,[33-35] while others have maintained that glass columns and unmodified detectors are adequate.[36-38]

Despite the problems of thermal decomposition encountered with the phenylurea herbicides, GC procedures have been reported for the detection of residues extracted from soils and waters.[34,35,37,39] In other laboratories, such analytical procedures have proved unsuccessful. In the authors' laboratory only diuron can be satisfactorily analyzed gas chromatographically.[38]

It therefore seems that the direct GC determination of the phenylurea herbicides is more of an art than a science, and can thus not be recommended as a general and routine multi-residue procedure for the analysis of these compounds. An exception appears to be diuron, whose analysis by such means does appear possible.

To overcome these difficulties with thermal decomposition, analytical procedures based on the GC detection of the aniline moiety, derived by hydrolysis of the parent herbicide, have been devised. To increase sensitivity by ECD, conversion of the substituted amine to brominated derivatives has been suggested.[40,41] However, it has been reported that 3- and di-substituted anilines can form more than one brominated derivative, thus making quantitative analysis difficult.[42] Substituted anilines can also be converted to iodobenzenes by means of a Sandmeyer iodination reaction, and an ana-

Table 8
RELATIVE RETENTION TIMES OF
SUBSTITUTED IODOBENZENES DERIVED
FROM PHENYLUREA HERBICIDES[9]

Herbicide	Substituted iodobenzene	Relative retention
Fenuron, siduron	Iodobenzene	1.0
Fluometuron	3-Trifluoromethyliodobenzene	1.2
Buturon, mono-linuron, monu-ron	4-Chloroidobenzene	2.1
Metobromuron	4-Bromoiodobenzene	3.0
Diuron, linuron, neburon	3,4-Dichloroiodobenzene	4.3
Chlorbromuron	4-Bromo-3-chloroiodobenzene	6.4

Note: Column: glass 1.2 m × 3 mm i.d. Packing: 10% QF-1 on 80 to 100 mesh Gas-Chrom Q.®

lytical procedure utilizing the GC detection of iodobenzenes derived from the phenylurea herbicides is available.[9,42] In this method, the samples are hydrolyzed directly in alkaline solution, the anilines removed by distillation, and diazotized with nitrous acid as described in the previous section. The diazonium salt is then treated with potassium iodide in the presence of catalytic amounts of iodine to carry out conversion to the substituted iodobenzene.

$$\text{aryl} - \text{NH} - \text{CO} - \text{N} \overset{\text{NaOH}}{\longrightarrow} \text{aryl} - \text{NH}_2 \overset{\text{HNO}_2}{\longrightarrow} \text{aryl} - \text{N}_2^+\text{Cl}^- \overset{\text{KI}}{\underset{\text{I}_2}{\longrightarrow}} \text{aryl} - \text{I}$$

The iodinated derivatives, so obtained, are then subjected to GC separation with ECD. The procedure appears to be reproducible and has been used for the analysis of a variety of phenylurea herbicides in different substrates. Detection limits are of the order of 0.01 to 0.05 μg/g.[9,42] The main advantage of this procedure is that multiresidue analysis is possible since the substituted iodobenzenes derived from several phenylurea herbicides can be separated from each other (see Table 8).

The main disadvantages of this approach are that the iodination procedure is time consuming and that the reaction and work-up conditions must be carefully controlled. Other drawbacks are that the same aniline, and therefore the same iodobenzene, can be formed from different phenylureas, and that demethylated degradation products would give rise to the same derivative as the parent compound. However, this procedure is more specific and sensitive than the colorimetric method, and is better suited to analysis of samples containing unknown treatments of phenylurea herbicides.

C. High-Performance Liquid Chromatography (HPLC)

Column chromatography using mobile liquid phases has been used in analytical laboratories for many years as a means of separating compounds from each other and unwanted impurities. Theoretical and technological advances have revolutionized its application in the field of microanalysis, and the technique of HPLC has been used since the middle 1960s as an analytical procedure for the determination of pesticide residues. Modern HPLC uses high-pressure pumps to force the eluting solvent, at a flow rate of 1 to 5 mℓ/min, through small-diameter columns (1 to 3 mm) that are usually less than a meter in length and packed with supports of particle size ≤50 μm.

The most common detectors for monitoring the column effluent are based on UV and visible spectrophotometry. While some detectors operate at a fixed wavelength, others have a variable wavelength adjustment so that the detector may be tuned to the absorption maximum of the compound under test. This latter system permits a greater sensitivity and can reduce the detection of interfering substances. Other detectors based on fluorimetry, polarography, and refractometry are also available.

The main advantage of HPLC over GC is that the nature of the mobile phase can be changed which can result in greater separation of compounds with similar properties. Since these instruments are usually operated at ambient temperature another advantage is that thermal decomposition of the compounds being analyzed does not occur. The major drawback is that detection systems are not as sensitive as the ECD used in GC.

Most phenylurea herbicides have absorption maxima in the region 240 to 250 nm (Table 4) and are thus amenable to detection using an UV detection system attached to the liquid chromatograph. HPLC was first used in 1969 to separate linuron, monuron, diuron, and fenuron[43] and since then has become the method of choice for the determination of phenylurea herbicides in water, soil, and plant samples.[44,45] The residues are first solvent extracted from the substrate, then following solvent evaporation, and, if necessary, some cleanup stage the extracts are subjected directly to liquid chromatographic separation and UV detection.

By judicious use of solvent systems and column packings, analysis of several phenylurea herbicides in the presence of each other is possible. Using this method, demethylated urea degradation products can also be separated from the parent herbicide and determined.

Using HPC, phenylurea herbicides can be detected in waters at the 0.01 μg/g level, while in soils and plants the limits of detection are approximately 0.2 μg/g.[45]

D. Discussion of Analytical Procedures

For the routine monitoring of any class of compounds, such as the urea herbicides, in water, sediment, and plant or animal tissue the ideal requirement is for a common extraction procedure and an analytical system which allows for the separation and determination of all the compounds in the presence of each other, so that a polluting compound can be quickly identified. With over 15 urea compounds in regular agricultural use it will take several years of work to achieve this difficult goal.

It has already been mentioned that multi-residue procedures for the analysis of phenylurea herbicides in aquatic environments are limited and at best confined to a maximum of eight residues, consisting of the most commonly encountered herbicides.

Of the three main analytical procedures suitable for the determination of phenylurea herbicides, the colorimetric method lacks specificity but can be used as a quick check to monitor for the presence of phenylurea residues. Thus a water sample can be collected, solvent extracted, the solvent evaporated, and the residue hydrolyzed with hot acid. If an azo-derivative can be formed, the presence of phenylurea compounds in the water sample may be inferred. By sampling 500-mℓ aliquots of the waters, this colorimetric monitoring test should indicate the possible presence of phenylurea herbicides at the 0.01 μg/g level. HPLC procedures can then be used to identify and quantify the polluting residues. It must be remembered that the herbicides barban, chlorpropham, propanil, and propham can also be hydrolyzed to aromatic amines capable of forming azo-derivatives.

Diuron is the only urea herbicide currently recommended for use in aquatic environments, and so an analytical procedure capable of determining diuron at the 0.005 μg/g level, or lower, is necessary. Neither the colorimetric nor the liquid chromatographic procedures are capable of this low level of detection. It is therefore fortunate that diuron seems to be one of the few phenylurea herbicides that analysts have been able to analyze successfully using GC means, and the detection limits are of the order of 0.001 to 0.002 μg/g.[34,38]

The analysis of urea herbicides in sediments has received little attention and only one method has been reported.[37] Similarly, the extraction and analysis of urea compounds from aquatic vegetation and fish has not been described, though some of the procedures mentioned earlier could, no doubt, be adapted to such analyses.

In view of this general lack of analytical procedures for the phenylurea herbicides, the authors will, in the appropriate section, provide details of a colorimetric screening procedure for assaying the possible presence of phenylurea herbicides in water samples. A GC procedure for the analysis of diuron in water and soil samples will also be described. Finally, HPLC methodology will be presented for the analysis of commonly used phenylurea herbicides in water samples and in the presence of each other.

IV. ANALYTICAL METHODOLOGY

A. Extraction of Phenylurea Herbicides from Waters, Soils, and Plant Tissues

Solvents which have been used to extract commonly used phenylurea residues from water, soil, and plant tissues are listed in Tables 9, 10, and 11. The herbicide recoveries, together with the analytical procedure used, are also listed.

Dichloromethane appears to be the solvent selected by most analysts for the extraction of the phenylurea herbicides from water samples. After extraction the organic solvent is dried over a suitable drying agent, such as sodium sulfate, and the dried solution evaporated to dryness, prior to analysis.

For the extraction of phenylurea compounds from soils, methanol seems to be suitable, and their recoveries from fortified samples of differing soil types appears to be in excess of 70% (Table 10). A study to compare the relative amounts of linuron extracted from treated field soils using different extraction solvents has indicated that methanol is superior to ethanol, methylene chloride, and acetone.[39] The study also showed that simple shaking of the soil with methanol recovered more linuron than did Soxhlet extraction with the same solvent.[39] The only report to describe the extraction of a phenylurea herbicide (diuron) from sediments has shown that methanol is also suitable for such purposes.[37]

It therefore seems that methanol is an adequate extraction solvent for the removal of phenylurea residues from treated soils, and possibly sediments, and that the soils should be shaken with the solvent on a simple mechanical shaker. Following shaking and filtration or centrifugation the methanol is evaporated and the residue dissolved in a suitable solvent for analytical determination.

The recovery of phenylurea herbicides from plant tissue has been achieved using a variety of solvents (Table 11). However, hot ethanol should be avoided[46] since monuron, diuron, fenuron, and chloroxuron, which contain 1,1-dimethyl substituents, are converted to the corresponding carbamates:

$$aryl-NH-CO-N(CH_3)_2 \longrightarrow aryl-NH-COOC_2H_5$$

B. GC Conditions for the Detection of Diuron

The evaporated extracts, following solvent extraction of the water and soil samples are taken up in *iso*-octane and injected directly into the GC without further cleanup. The GC conditions that have been reported for the determination of diuron residues from a variety of substrates are listed in Table 12. The analysis of diuron metabolites 1-(3,4-dichlorophenyl)-3-methylurea and 1-(3,4-dichlorophenyl) urea by GC means has been mentioned only by Bowmer and Adeney,[37] who observed that both these com-

Table 9

EXTRACTION AND DETERMINATION OF PHENYLUREA HERBICIDES FROM WATERS

Phenylurea herbicide	Extraction solvent	Analytical procedure	Fortification level (μg/ml)	Recovery range (%)	Ref.
Diuron	Dichloromethane	EC-GC	0.001—1.0	89—99	34
Diuron	Dichloromethane	EC-GC	0.1—2.5	98—104	37
Diuron	Dichloromethane	EC-GC	0.01—1.0	98—117	38
Diuron	Dichloromethane	EC-GC	—	81	15
Linuron	Dichloromethane	EC-GC	—	81	15
Diuron			0.02—0.1	81—105	
Fenuron			0.04—0.15	88—116	
Linuron	Chloroform	Colorimetry	0.02—0.2	76—109	26
Monuron			0.06—0.16	89—111	
Neburon			0.05—0.13	82—108	
Chlorbromuron				98—100	
Chloroxuron				95—101	
Chlortoluron		HPLC with UV		97—101	
Diuron	Dichloromethane	detector at 240 nm	0.01	98—102	45
Linuron				96—101	
Metobromuron				97—101	
Monolinuron				98—100	
Monuron				98—101	

Table 10

EXTRACTION AND DETERMINATION OF PHENYLUREA HERBICIDES FROM SOILS

Phenylurea herbicide	Extraction solvent	Analytical procedure	Fortification level (μg/g)	Recovery range (%)	Ref.
Monuron	Acetonitrile + 10% acetic acid/shake	Colorimetry	0.25—2.5	53—91	23
Linuron	Petroleum ether/Soxhlet	Colorimetry	1.6	97—102	29
Linuron	Dichloromethane/shake	Colorimetry	2.0	88—106	47
Linuron	Methanol/shake	Colorimetry	1.0	>90	48
Chlorbromuron				80—84	
Diuron				80—86	
Fluometuron	Methanol/shake	EC-GC	0.1—1.0	90—100	35
Linuron				76—104	
Metobromuron				73—80	
Neburon				90—104	
Diuron	Methanol/shake	EC-GC	2.0—50	73—105	37
Diuron	Methanol/shake	Flame-GC	0.02—0.1	80—90	49
Linuron	Methanol/shake	EC-GC	0.1	80—97	39
Chlorbromuron			1.0	91	
Chlortoluron			1.0	84—99	
Diuron			0.2—1.0	80—108	
Fenuron	Acetone/Soxhlet	EC-GC of	1.0	104	41
Linuron		brominated	0.2—1.0	85—100	
Metobromuron		anilines	0.2	80	
Monolinuron			1.0	85—100	
Monuron			0.2	80	
Chlortoluron	Methanol/shake	HPLC with UV detector at 240 nm	0.25—2.5	71—95	50

Table 10 (continued)
EXTRACTION AND DETERMINATION OF PHENYLUREA HERBICIDES
FROM SOILS

Phenylurea herbicide	Extraction solvent	Analytical procedure	Fortification level (μg/g)	Recovery range (%)	Ref.
Chlorbromuron				98—101	
Chloroxuron				97—103	
Chlortoluron				100—103	
Diuron	Methanol/shake	HPLC with UV detector at 240 nm	2.0	98—101	45
Linuron				94—101	
Metobromuron				97—101	
Monolinuron				101—103	
Monuron				99—102	

Table 11
EXTRACTION AND DETERMINATION OF PHENYLUREA HERBICIDES
FROM PLANT TISSUE

Phenylurea herbicide	Plant tissue	Extraction solvent	Analytical procedure	Fortification level (μg/g)	Recovery range (%)	Ref.
Linuron	Corn, soybean, crabgrass, wheat	Acetone/ blend	Colorimetry	0.03—0.48	80—138	51
Diuron	Wheat	Acetone/ Soxhlet	EC-GC	0.5—2.0	92—99	52
Chlorbromuron					85—93	
Chlortoluron					86—94	
Chloroxuron			HPLC with		87—94	
Diuron	Wheat	Methanol/ shake	UV detector at 240 nm	0.5	87—93	45
Linuron					86—96	
Metobromuron					84—93	
Monolinuron					87—96	
Monuron					89—96	
Chlorbromuron						53
Chloroxuron						
Diuron	Cabbage, corn, potatoes, turnips, wheat	Acetone/ blend	HPLC with UV detector at 254 nm	0.01—1.0	>80	
Fenuron						
Fluometuron						
Linuron						
Monuron						

pounds had the same retention times as diuron on the SE-30 + QF-1 column used in their study. Thus, if diuron is suspected in soil and water samples, its identity must be confirmed using TLC procedures (see later). Based on a 100-mℓ water sample, diuron residues at the 0.001 μg/g level should be detectable, while residues of 0.05 μg/g should be analyzable from a 20-g soil sample.

C. HPLC Conditions for Urea Herbicides

After extraction of the water samples, the solvent is evaporated to dryness and the residue taken up in a small volume of methanol. This solution may be injected directly into the liquid chromatograph without further cleanup. Details of column dimensions, column packings, mobile phases, analytical wavelengths, etc. that have been reported for the analysis of commonly used phenylurea herbicides are summarized in Table 13. Only four column and solvent systems have been described for the separation and analysis of mixtures of phenylurea residues (Table 14). Of these only Spherisorb ODS[45]

Table 12

GC CONDITIONS FOR ANALYSIS OF DIURON

Column	Packing	Temps (°C) inj	col	det	Carrier gas (ml/min)	Detector	Approx ret. time (min)	GC	Ref.
1.5 m × 3.5 mm stainless steel	5 % E-301 on 60 to 80 mesh Gas Chrom Q®	265 265	150 155	200 200	N₂ (50)	Modified EC	0.95 0.90	Varian Aerograph, 1520	33—35
1.8 m × 4 mm glass	1.5 % SE-30 + 1.5 % QF-1 on 80 to 100 mesh Chromosorb W, AW, DMCS	270	150	230	A/CH₄ (50)	EC	1.33	Hewlett-Packard, Series 7620A	37
0.91 m × 3.5 mm glass	5 % Dexsil®-300 on 80 to 100 mesh Chromosorb W, HP	200	160	285	A/CH₄ (25)	EC	1.2	Hewlett-Packard, Series 7610A	38
2.0 m × 4 mm	5 % Dexsil®-300 on 80 to 100 mesh Chromosorb W, HP	220	180	350	A/CH₄ (40)	EC	2.5	Tracor, Series 650	38
1.8 m × 6 mm stainless steel	5 % SE-30 on 60 to 80 mesh Gas Chrom Q®	250	180	250	A/CH₄ (50)	EC	—	Hewlett-Packard, Series 5700A	15
1.5 m × 4 mm glass	20 % Apiezon-L on 80 to 100 mesh Chromosorb W, HP	215	215	290	N₂ (80)	Flame ionization	3.5	Pye, Series 104	49
1.8 m × 4 mm glass	1.5 % SE-30 + 1.5 % QF-1 on Anakrom ABS	225	150	200	N₂ (55)	EC	1.5	Hewlett-Packard, Series 7713	52

Table 13

LIQUID CHROMATOGRAPHIC SYSTEMS FOR SOME COMMON UREA HERBICIDES[55]

Phenylurea herbicide	Column packing (5 μm)	Dimensions (mm)	Mobile phase	Substrate	UV wavelength (nm)
Chlorbromuron	LiChrosorb Si 60	150 × 4.6	0.1% Me OH in dichloromethane	Std	245
	C₁₈ Silica	250 × 4.6	60% MeOH in H₂O	Std	240
	Spherisorb ODS	300 × 4.6	60% MeOH in 0.6% NH₃/H₂O	River water	240
	LiChrosorb Si 60	250 × 2.8	10% 2-Propanol in *iso*-octane	Wheat	254
Chloroxuron	LiChrosorb Si 60	150 × 4.6	1% MeOH in dichloromethane	Std	245
	LiChrosorb Si 60C₁₈	250 × 4.6	70% MeOH in H₂O	Std	240
Chloroxuron	Spherisorb ODS	300 × 4.6	60% MeOH in 0.6% NH₃/H₂O	River water	240
	LiChrosorb Si 60	250 × 2.8	20% 2-Propanol in *iso*-octane	Cabbage	254
Chlortoluron	LiChrosorb Si 60	150 × 4.6	1% MeOH in dichloromethane	Std	245
	Spherisorb ODS	300 × 4.6	60% MeOH in 0.6% NH₃/H₂O	River water	240
	PE C₁₈ Sil-X-11	500 × 1.7	20% MeOH in H₂O	Soil, water	215
	Merkosorb Si 60 (10 μm)	200 × 4.0	15% 2-Propanol in hexane	Soil	240
Diuron	LiChrosorb Si 60	150 × 4.6	0.5% MeOH in dichloromethane	Std	245
	C₁₈ Silica	250 × 4.6	65% MeOH in H₂O	Std	240
	Spherisorb ODS	300 × 4.6	60% MeOH in 0.6% NH₃/H₂O	River water	240
	LiChrosorb Si 60	250 × 2.8	10% 2-Propanol in *iso*-octane	Corn	254
	LiChrosorb Si 60	250 × 2.0	20% Hexane in 79% dichloro- methane + 1% ethenol	Std	247
Fenuron	Corasil C₁₈ (37—50 μm)	609 × 2.3	50% MeOH in H₂O	Std	254
	LiChrosorb Si 60	250 × 2.8	20% 2-Propanol in *iso*-octane	Std	254
	LiChrosorb Si 60	250 × 2.0	20% Hexane in 79% dichloro- methane + 1% ethanol	Std	247
Fluometuron	LiChrosorb Si 60	250 × 2.8	20% 2-Propanol in *iso*-octane	Potato	254
Linuron	LiChrosorb Si 60	150 × 4.6	Dichloromethane	Std	245
	C₁₈ Silica	250 × 4.6	65% MeOH in H₂O	Std	240
	Spherisorb ODS	300 × 4.6	60% MeOH in 0.6% NH₃/H₂O	River water	240
	PE C₁₈ Sil-X-11	500 × 1.7	20% MeOH in H₂O	Soil, water	250
	LiChrosorb Si 60	250 × 2.8	10% 2-Propanol in *iso*-octane	Potato, turnip	254

Table 13 (continued)

LIQUID CHROMATOGRAPHIC SYSTEMS FOR SOME COMMON UREA HERBICIDES[55]

Compound	Column	Dimensions	Mobile phase	Sample	λ
Metobromuron	LiChrosorb Si 60	150 × 4.6	Dichloromethane	Std	245
	LiChrosorb Si 60C$_{18}$	250 × 4.6	60% MeOH in H$_2$O	Std	240
	Spherisorb ODS	300 × 4.6	60% MeOH in 0.6% NH$_3$/H$_2$O	River water	240
	LiChrosorb Si 60	250 × 2.0	20% Hexane in 79% dichloromethane + 1% ethanol	Std	247
Monolinuron	LiChrosorb Si 60	150 × 4.6	Dichloromethane	Std	245
	LiChrosorb Si 60 C$_{18}$	250 × 4.6	60% MeOH in H$_2$O	Std	240
	Spherisorb ODS	300 × 4.6	60% MeOH in 0.6% NH$_3$/H$_2$O	River water	240
Monuron	LiChrosorb Si 60	150 × 4.6	1% MeOH in dichloromethane	Std	245
	C$_{18}$ Silica	250 × 4.6	60% MeOH in H$_2$O	Std	240
	Spherisorb ODS	300 × 4.6	60% MeOH in 0.6% NH$_3$/H$_2$O	Wheat, soil, River water	240
	PE C$_{18}$ Sil-X-11	500 × 1.7	20% MeOH in H$_2$O	Soil, water	250
	LiChrosorb Si 60	250 × 2.8	10% 2-Propanol in *iso*-octane	Corn	254
	LiChrosorb Si 60	250 × 2.0	20% Hexane in 79% dichloromethane + 1% ethanol	Std	247

Table 14

HPLC SEPARATION OF PHENYLUREA HERBICIDES[45,53,54]

	Approximate elution time (min)			
Phenylurea herbicide	Spherisorb ODS, 60% methanol in water + 0.6% ammonia at 0.6 ml/min	LiChrosorb Si 60®, 20% 2-propanol in *iso*-octane at 0.5 ml/min	LiChrosorb Si 60®, dichloromethane at 1.2 ml/min	C$_{18}$ bonded silica, 60% methanol in water, at 0.5 ml/min
Chlorbromuron	22	6	5	38
Chloroxuron	28	13	45	52
Diuron	18	11	22	26
Fenuron	—	11	—	—
Fluometuron	—	9	—	—
Linuron	20	6	5	33
Metobromuron	12	—	7	21
Monolinuron	11	—	7	18
Monuron	11	13	37	14

and LiChrosorb Si 60® (using *iso*-propanol in *iso*-octane)[53] have been used for residue analysis. A C$_{18}$-bonded silica column and LiChrosorb Si 60® have been used to separate formulations of phenylurea compounds,[54] a situation where sample concentration is high. Clearly more work is necessary in the area of column packings and mobile phases to enable better separation of phenylurea mixtures for practical residue analysis. Currently the Spherisorb ODS column with an eluting solvent comprised of methanol containing ammonium hydroxide appears to be the most satisfactory system for the separation of phenylurea mixtures at low concentration levels. There is no information regarding the analysis of phenylurea demethylated metabolites using this chromatographic system, though with previous studies with chlortoluron, separation of the parent urea from its metabolites has been achieved.[44,56] By extracting 1 l of water, phenylurea herbicide residues at the 0.01 μg/g level should be detectable. In soils a limit of 0.2 μg/g based on a 50-g sample is possible.

D. Standards

Pure analytical grade reference standards of the various phenylurea herbicides must be used. All solutions when not in use should be stored in the dark in a refrigerator to prevent possible photochemical decomposition and concentration of the standards by evaporation of the solvent.

Standards can be prepared in glass distilled methanol, and a convenient solution for subsequent dilution would contain 1.0 mg phenylurea per milliliter methanol. By diluting 1.0 ml of this solution to 100 ml with methanol, a solution containing 10.0 μg urea per milliliter is obtained, which is satisfactory for use as a standard with the colorimetric procedure. A tenfold dilution of this 10.0 μg/ml solution (with spectrograde methanol) yields a urea concentration of 1.0 μg/ml or 1.0 ng/μl, a strength suitable for use with HPLC. A standard for use with the GC procedure would require a diuron concentration of 0.05 ng/μl, and can be made by diluting 5 ml of the above liquid chromatographic standard to 100 ml with glass distilled *iso*-octane. Standards with phenylurea concentrations intermediate between those mentioned may be made by the analyst as required.

E. Sample Collection

For a description of sampling equipment and collection techniques see the chapter on the triazine herbicides.

F. Sample Preparation

Phenylurea herbicides are relatively stable and do not undergo rapid chemical or biological degradation. Thus water and sediment samples collected from field sites can be shipped in glass containers without special treatment. On arrival at the laboratory, water samples may be filtered through glass wool to remove solid material and then stored in a refrigerator at 4°C to await analysis. If the analysis cannot be carried out within 2 weeks, the sample should be frozen at −20°C.

Sediment samples should be frozen at −20°C on arrival at the laboratory, and prior to analysis should be thawed, filtered to remove excess water, and air-dried. The samples can then be ground and mixed.[37]

G. Confirmatory Techniques

Residue analysis requires confirmation of the identity of the residue by an analytical procedure different from the one used to initially identify the herbicide. Recognized methodologies include TLC, GC, and HPLC using different parameters from those originally used in the analysis and identification through known chemical derivatives.

Thin-layer techniques have been reported for the separation and identification of phenylurea herbicides.[57,58] However, it must be remembered that at least 2 μg of the phenylurea is necessary for satisfactory detection by observation under UV irradiation. Hence, confirmation of such residues in waters and soils containing trace amounts of these chemicals may not be possible.

Some researchers have converted phenylurea herbicides to derivatives such as alkylated ureas and have identified these compounds with the corresponding products prepared from known phenylurea standards using TLC, GC, or HPLC procedures.[5,6,32,53,59]

Currently the most practical confirmation for phenylurea residues, recovered from water and soil samples, is to compare their retention times on at least two GC or HPLC systems using different columns and/or mobile phases. Those listed in Tables 12 and 13 would appear satisfactory. The identity of the unknown residue may be assumed if its retention time is identical with that of the authentic standard, under all conditions.

Since it has been shown that the demethylated degradation products derived from diuron have the same retention time as the parent herbicide on at least one GC column,[37] suspected diuron residues must be confirmed additonally by independent non-gas chromatographic tests. Providing the herbicide concentration in water is in excess of 0.01 μg/g liquid chromatography should provide adequate confirmation for diuron residues.

Silica gel impregnated glass fiber plates have also been described[37] as a means of separating demethylated metabolites from diuron. Following development of treated plates with a mixture of methanol:chloroform:benzene (1:9:25 by volume), the plates were examined under UV radiation to locate the positions of the various compounds. The R_f values of diuron, 1-(3,4-dichlorophenyl)-3-methylurea, and 1-(3,4-dichlorophenyl) urea were reported as 0.84, 0.62, and 0.39, respectively.[37]

V. DETAILED ANALYTICAL PROCEDURES

A. Colorimetric Screening for Qualitative Phenylurea Determination in Water

This method is based on the procedure described by Katz.[26] As mentioned earlier other herbicides such as barban, chlorpropham, propanil, and propham are hydrolyzed with acid to substituted anilines and could so be erroneously assumed as phenylurea compounds. Demethylated phenylurea herbicide metabolites would also interfere.

Reagents — *n*-Butanol, distilled in glass; dichloromethane, distilled in glass; hydrochloric acid, reagent grade, 6 *N*; sodium chloride, analytical grade; sodium nitrite, analytical grade, 2% aqueous solution, prepared daily; ammonium sulfamate, analytical grade, 10% aqueous solution, prepared weekly; *N*-(1-naphthyl)ethylenediamine dihydrochloride, analytical grade, 1% aqueous solution, prepared daily; phenylurea herbicide standard, diuron in methanol, 10 μg/mℓ.

Procedure — Place 500 mℓ well-shaken water sample in a 1-ℓ separatory funnel and extract with 3 × 100 mℓ portions of dichloromethane, shaking vigorously for at least 1 min each time. The combined organic extracts are evaporated in a 250-mℓ round bottomed flask to dryness, using a rotary evaporator at 35°C. Add 15 mℓ 6 *N* hydrochloric acid to the flask, connect to a water-cooled condenser, and heat under reflux conditions for 6 hr. Cool and rinse the condenser with 25 mℓ distilled water, adding the washings to the flask. Add 50 mℓ water to the flask, then add 1.0 mℓ sodium nitrite reagent. Mix well and let stand for 5 min. Destroy excess nitrous acid by adding 1.0 mℓ ammonium sulfamate solution, and shake several times over a 10-min period. Add 2.0 mℓ of the coupling reagent solution; mix well and let stand for 20 min. If the presence of fenuron is suspected, the color will take nearly 3 hr to develop. Azo-derivatives may be extracted by adding 25 g sodium chloride and 10 mℓ *n*-butanol and shaking vigorously for 1 min, when any red complex is quantitatively extracted into the butanol. Reagent blanks must be conducted and 500-mℓ samples of distilled water containing known amounts of a phenylurea herbicide, such as diuron (10 μg and 5 μg), must also be extracted and treated exactly as described. Comparison of the color extracted into the butanol solution with that derived from the diuron standards should allow the analyst to deduce whether phenylurea residues are present in the water specimens. Five micrograms of phenylurea, recovered from a 500-mℓ water sample, should be detectable using this procedure. Thus a detection limit of 0.01 μg/mℓ should be possible.

B. Analysis of Diuron in Water by GC

This method is based on the procedures described by McKone and Hance,[34] Bowmer and Adeney,[37] and Grover and Kerr.[38] Recoveries should be in excess of 90% (see Table 9).

Reagents — Dichloromethane, distilled in glass; *iso*-octane, distilled in glass; sodium chloride, analytical grade, saturated aqueous solution; diuron standard, 0.05 ng/$\mu\ell$ in *iso*-octane.

GC Conditions — The gas chromatograph, column packings, and operating conditions may be selected from those listed in Table 12.

Procedure — Place 100 mℓ of the well-shaken water in a 250 mℓ separatory funnel and extract with 2 × 50 mℓ portions of dichloromethane, shaking vigorously for at least 1 min each time. The combined dichloromethane extracts are placed in a 250-mℓ round-bottomed flask and evaporated to dryness using a rotary evaporator at 35°C. The residue is dissolved in 5.0 mℓ *iso*-octane, 5 mℓ of saturated sodium chloride solution is added, and the flask, and contents, shaken vigorously for 1 min. The organic layer is decanted into a 10-mℓ glass-stoppered flask and aliquots of this solution are taken for GC examination. The presence of diuron is inferred by comparing the retention times of any substances recovered from the water samples with that of authentic diuron. The herbicide concentration in the specimens are obtained by comparing sample peak heights with a standard curve constructed by plotting peak height (in the linear response region) against nanogram diuron injected. The *iso*-octane solutions derived from water samples are diluted if necessary so that the diuron concentration is within the calibration range of the standard curve. Since 0.1 μg diuron in a 100-mℓ water sample should be detectable, a detection limit of 0.001 μg/mℓ is possible.

C. Analysis of Diuron in Sediment by GC

This method is based on the procedure described by McKone[35] and Bowmer and Adeney.[37] Recoveries of diuron residues should be in excess of 80% (see Table 10).

Reagents — *iso*-octane, distilled in glass; methanol, distilled in glass; sodium chloride, analytical grade, saturated aqueous solution; diuron standard, 0.05 ng/$\mu\ell$ in *iso*-octane.k

GC Conditions — The GC, column packings, and operating conditions may be selected from those listed in Table 12.

Procedure — Subsamples of the air-dried soil or sediment (20 g) are placed in a 125-mℓ flask and shaken with 50 mℓ methanol on a wrist-action shaker for 1 hr. After shaking, the slurry is centrifuged at 4000 rpm for 5 min and a 25-mℓ aliquot of supernatant liquor, equivalent to 10 g soil, is placed in a 250-mℓ round-bottomed flask and evaporated to dryness using a rotary evaporator at 35°C. The residue is dissolved in 5.0 mℓ *iso*-octane, 5 mℓ of saturated sodium chloride is added, and the flask, and contents, shaken vigorously for 1 min. The organic layer is decanted into a 10-mℓ glass-stoppered flask and aliquots of this solution are taken for GC examination (see above).

D. Analysis of Diuron in Aquatic Vegetation and Biota by GC

No methods published.

E. HPLC Procedure for Phenylureas in Water

This procedure can be used either for the analysis of single known phenylurea herbicides or for analysis of unknown mixtures. The procedure is based on that described by Farrington, Hopkins, and Ruzicka.[46] Recoveries should be in excess of 95% (see Table 9).

Reagents — Dichloromethane, distilled in glass; methanol, spectrograde; urea standards, 1.0 ng/$\mu\ell$ in spectrograde methanol.

LC Conditions — LC— any suitable commercially available instrument, equipped with an UV detector set at 240 nm. Column-A stainless steel column (300 × 4.6 mm i.d.) packed with 5-μm Spherisorb ODS. Mobile phase-A solution of 60% spectrograde methanol in distilled water containing 0.6% v/v ammonia (sp gr 0.88). Flow rate of mobile phase is 0.6 mℓ/min.

Procedure — Place 500 mℓ of the well-shaken water sample in a 1-ℓ separatory funnel and extract with 2 × 100 mℓ volumes of dichloromethane, shaking vigorously for 1 min each time. The combined organic extracts are evaporated to dryness, in a 250-mℓ round-bottomed flask, using a rotary evaporator at 35°C. The residue is dissolved in 5.0 mℓ spectrograde methanol and 5 $\mu\ell$ aliquots analyzed. The identity of the phenylurea herbicides is tentatively assigned by comparing their retention times with those of known standards, and it should be possible to distinguish between residues of chlorbromuron, chloroxuron, diuron, linuron, metobromuron, monolinuron, and monuron (see Table 14). The quantitative phenylurea content of the samples is calculated by comparing the sample peak heights with those obtained from injections of standard solutions. Detection limits of 0.01 μg/mℓ should be possible.

F. HPLC Procedures for Phenylureas in Sediments, Aquatic Plants, and Biota

No methods published.

REFERENCES

1. *Herbicide Handbook,*4th ed., Weed Science Society of America, Champaign, Ill., 1979.
2. Fryer, J. D. and Makepeace, R. J., Eds., *Weed Control Handbook,* Vol. 1, 6th ed., Blackwell, Oxford, 1977.
3. Geissbühler, H., Martin, H., and Voss, G., The substituted ureas, in *Herbicides, Chemistry, Degradation, and Mode of Action,* Vol. 1, 2nd ed., Kearney, P. C. and Kaufman, D. D., Eds., Marcel Dekker, New York, 1975, chap. 3.
4. Pribyl, J. and Herzel, F., Flüssigchromatographische Parameter herbidizer Wirkstoffgruppen. III. Harnstoffherbizide-ergänzung, *J. Chromatogr.,* 166, 272, 1978.
5. Greenhalgh, R. and Kovacicova, J., A chemical confirmatory test for organophosphorus and carbamate insecticides and triazine and urea herbicides with reactive NH moieties, *J. Agric. Food Chem.,* 23, 325, 1975.
6. Büchert, A. and Løkke, H., Gas chromatographic-mass spectrometric identification of phenylurea herbicides after N-methylation, *J. Chromatogr.,* 115, 682, 1975.
7. Hill, G. D., McGahen, J. W., Baker, H. M., Finnerty, D. W., and Bingeman, C. W., The fate of substituted urea herbicides in agricultural soils, *Agron. J.,* 47, 93, 1955.
8. Crosby, D. G., Herbicide photodecomposition, in *Herbicides, Chemistry, Degradation, and Mode of Action,* Vol. 2, 2nd ed., Kearney, P. C. and Kaufman, D. D., Eds., Marcel Dekker, New York, 1975, chap. 18.
9. Voss, G., Gross, D., Becker, A., and Guth, J. A., Fluometuron, metobromuron, chlorbromuron, and chloroxuron, in *Analytical Methods for Pesticides and Plant Growth Regulators,* Vol. 7, Sherma, J. and Zweig, G., Eds., Academic Press, New York, 1973, 569.
10. Holden, A. V., Effects of pesticides on fish, in *Environmental Pollution by Pesticides,* Edwards, C. A., Ed., Plenum Press, London, 1973, 213.
11. Walker, C. R., The toxicological effects of herbicides and weed control on fish and other organisms in the aquatic ecosystem, in *Proc. Eur. Weed Res. Coun. 3rd Int. Symp. Aquatic Weeds,* Oxford, 1971, 119.
12. Tooby, T. E., The toxicity of aquatic herbicides to freshwater organisms- a brief review, in *Proc. Eur. Weed Res. Coun. 3rd Int. Symp. Aquatic Weeds,* Oxford, 1971, 129.
13. Seiler, J. P., Herbicidal phenylureas as possible mutagens. I. Mutagenicity tests with some urea herbicides, *Mutation Res.,* 58, 353, 1978.
14. International Agency for Research on Cancer, Monuron, in *Cancer Monograph on the Evaluation of Carcinogenic Risk of Chemicals to Man,* Vol. 12, World Health Organization, Geneva, 1976, 167.
15. Willis, G. H., Rogers, R. L., and Southwick, L. M., Losses of diuron, linuron, fenac, and trifluralin in surface drainage water, *J. Environ. Qual.,* 4, 399, 1975.
16. Green, R. E., Goswami, K. P., Mukhtar, M., and Young, H. Y., Herbicides from cropped watersheds in stream and estuarine sediments in Hawaii, *J. Environ. Qual.,* 6, 145, 1977.
17. Baldwin, F. L., Santelmann, P. W., and Davidson, J. M., Movement of fluometuron across and through the soil, *J. Environ. Qual.,* 4, 191, 1975.
18. Bovey, R. W., Burnett, E., Meyer, R. E., Richardson, C., and Loh, A., Persistence of tebuthiuron in surface runoff water, soil, and vegetation in the Texas Blacklands prairie, *J. Environ. Qual.,* 7, 233, 1978.
19. LaFleur, K. S., Wojeck, G. A., and McCaskill, W. R., Movement of toxaphene and fluometuron through Dunbar soil to underlying ground water, *J. Environ. Qual.,* 2, 515, 1973.
20. Grover, R. and Kerr, L. A., unpublished data, 1979.
21. Grover, R., Smith, A. E., and Korven, H. C., A comparison of chemical and cultural control of weeds in irrigation ditchbanks, *Can. J. Plant Sci.,* 60, 185, 1980.
22. Bratton, A. C. and Marshall, E. K., A new coupling component for sulfanilamide determination, *J. Biol. Chem.,* 128, 537, 1939.
23. Lowen, W. K. and Baker, H. M., Determination of macro and micro quantities of 3-(p-chlorophenyl)-1,1-dimethylurea, *Analyt. Chem.,* 24, 1475, 1952.
24. Dalton, R. L. and Pease, H. L., Determination of residues of diuron, monuron, fenuron, and neburon, *J. Assoc. Off. Agric. Chem.,* 45, 377, 1962.
25. Lowen, W. K., Bleidner, W. E., Kirkland, J. J., and Pease, H. L., Monuron, diuron, and neburon, in *Analytical Methods for Pesticides, Plant Growth Regulators, and Food Additives,* Vol. 4, Zweig, G., Ed., Academic Press, New York, 1964, 157.
26. Katz, S. E., Determination of the substituted urea herbicides linuron, monuron, diuron, neburon, and fenuron in surface waters, *J. Assoc. Off. Agric. Chem.,* 49, 452, 1966.

27. **Bleidner, W. E.**, Application of chromatography in determination of micro quantities of 3-(*p*-chlorophenyl)-1,1-dimethylurea, *J. Agric. Food Chem.*, 2, 682, 1954.

28. **Pease, H. L.**, Separation and colorimetric determination of monuron and diuron residues, *J. Agric. Food Chem.*, 10, 279, 1962.

29. **Friestad, H. O.**, Determination of linuron in soil by application of an automated diazotization and coupling procedure, *Bull. Environ. Contam. Toxicol.*, 2, 236, 1967.

30. **Guth, J. A. and Voss, G.**, Automated colorimetric procedure for the determination of total and unchanged urea herbicide residues in soil, *Weed Res.*, 11, 111, 1971.

31. **Spengler, D. and Hamroll, B.**, Trennung und Bestimmung von Carbamat- und Harnstoff-herbiziden durch Reaktions-Gaschromatographie, *J. Chromatogr.*, 49, 205, 1970.

32. **Saunders, D. G. and Vanatta, L. E.**, Derivatization and gas chromatographic measurement of some thermally unstable ureas, *Anal. Chem.*, 46, 1319, 1974.

33. **McKone, C. E. and Hance, R. J.**, The gas chromatography of some substituted urea herbicides, *J. Chromatogr.*, 36, 234, 1968.

34. **McKone, C. E. and Hance, R. J.**, A method for estimating diuron (N'-(3,4-dichlorophenyl)-*N,N*-dimethylurea) in surface water by electron-capture gas chromatography, *Bull. Environ. Contam. Toxicol.*, 4, 31, 1969.

35. **McKone, C. E.**, The determination of some substituted urea herbicide residues in soil by electron-capture gas chromatography, *J. Chromatogr.*, 44, 60, 1969.

36. **Cochrane, W. P. and Wilson, B. P.**, Electrolytic conductivity detection of some nitrogen-containing herbicides, *J. Chromatogr.*, 63, 364, 1971.

37. **Bowmer, K. H. and Adeney, J. A.**, Residues of diuron and phytotoxic degradation products in aquatic situations. I. Analytical methods for soil and water, *Pestic. Sci.*, 9, 342, 1978.

38. **Grover, R. and Kerr, L. A.**, unpublished data, 1978.

39. **Khan, S. U., Greenhalgh, R., and Cochrane, W. P.**, Determination of linuron residues in soil, *Bull. Environ. Contam. Toxicol.*, 13, 602, 1975.

40. **Gutenmann, W. H. and Lisk, D. J.**, Electron affinity residue determination of CIPC, monuron, diuron, and linuron by direct hydrolysis and bromination, *J. Agric. Food Chem.*, 12, 46, 1964.

41. **Caverly, D. J. and Denney, R. C.**, Determination of substituted ureas and some related herbicides in soils by gas chromatography, *Analyst*, 103, 368, 1978.

42. **Baunok, I. and Geissbuehler, H.**, Specific determination of urea herbicide residues by EC chromatography after hydrolysis and iodine derivative formation, *Bull. Environ. Contam. Toxicol.*, 3, 7, 1968.

43. **Kirkland, J. J.**, High-speed liquid chromatography with controlled surface porosity supports, *J. Chromatogr. Sci.*, 7, 7, 1969.

44. **Byast, T. H.**, Reversed-phase high-performance liquid chromatography of some common herbicides, *J. Chromatogr.*, 134, 216, 1977.

45. **Farrington, D. S., Hopkins, R. G., and Ruzicka, J. H. A.**, Determination of residues of substituted phenylurea herbicides in grain, soil, and river water by use of liquid chromatography, *Analyst*, 102, 377, 1977.

46. **Lee, S. S. and Fang, S. C.**, Conversion of monuron to ethyl-*N*-*p*-chlorophenyl carbamate by soxhlet extraction of plant tissue with ethanol, *J. Assoc. Off. Anal. Chem.*, 54, 1361, 1971.

47. **Friestad, H. O.**, Selective colorimetric determination of linuron in soils in the presence of metabolites, *J. Assoc. Off. Anal. Chem.*, 61, 1486, 1978.

48. **Smith, A. E. and Emmond, G. S.**, Persistence of linuron in Saskatchewan soils, *Can. J. Plant Sci.*, 55, 145, 1975.

49. **Khan, S. U., Marriage, P. B., and Saidak, W. J.**, Persistence and movement of diuron and 3,4-dichloroaniline in an orchard soil, *Weed Sci.*, 24, 583, 1976.

50. **Smith, A. E. and Lord, K. A.**, Method for determining trace quantities of the herbicide chlortoluron in soils by liquid chromatography, *J. Chromatogr.*, 107, 407, 1975.

51. **Katz, S. E.**, Determination of linuron and its known and/or suspected metabolites in crop materials, *J. Assoc. Off. Anal. Chem.*, 50, 911, 1967.

52. **Parouchais, C.**, Determination of diuron in wheat, *J. Assoc. Off. Anal. Chem.*, 56, 831, 1973.

53. **Lawrence, J. F.**, High-pressure liquid chromatography analysis of urea herbicides in foods, *J. Assoc. Off. Anal. Chem.*, 59, 1066, 1976.

54. **Sidwell, J. A. and Ruzicka, J. H. A.**, The determination of substituted phenylurea herbicides and their impurities in technical and formulated products by use of liquid chromatography, *Analyst*, 101, 111, 1976.

55. **Lawrence, J. F. and Turton, D.**, High-performance liquid chromatographic data for 166 pesticides, *J. Chromatogr.*, 159, 207, 1978.

56. **Smith, A. E. and Briggs, G. G.,** The fate of the herbicide chlortoluron and its possible degradation products in soils, *Weed Res.,* 18, 1, 1978.
57. **Hance, R. J.,** The chromatographic identification of substituted urea herbicides, *J. Chromatogr.,* 44, 419, 1969.
58. **Abbot, D. C., Blake, K. W., Tarrant, K. R., and Thomson, J.,** Thin-layer chromatographic separation, identification and estimation of some carbamate and allied pesticides in soil and water, *J. Chromatogr.,* 30, 136, 1967.
59. **Lawrence, J. F. and Laver, G. W.,** Analysis of some carbamate and urea herbicides in foods by gas-liquid chromatography after alkylation, *J. Agric. Food Chem.,* 23, 1106, 1975.

Chapter 3

THE TRIAZINE HERBICIDES

Allan E. Smith, Derek C. G. Muir, and Raj Grover

TABLE OF CONTENTS

I.   Introduction 214
        A.   History 214
        B.   Use Patterns 216
        C.   Formulations and Application Rates 216
        D.   Synthesis 216
        E.   Physical Properties 219
        F.   Chemical Properties 219
        G.   Toxicological Properties 220

II.   Environmental Aspects 220

III.   Review of Analytical Procedures 222
        A.   Spectrophotometry 222
        B.   High-Performance Liquid Chromatography (HPLC) 223
        C.   Gas Chromatography (GC) 223
        D.   Discussion of Analytical Procedures 224

IV.   Analytical Methodology 224
        A.   Extraction of Triazine Herbicides from Water Samples 224
        B.   Extraction of Triazine Herbicides from Soil and Sediment Samples .. 224
        C.   Extraction of Triazine Herbicides from Plant Tissue 226
        D.   Extraction of Triazine Herbicides from Animal and Fish Tissue 226
        E.   Cleanup Procedures for Sediment, Plant, and Animal Extracts 226
        F.   GC Columns 228
        G.   GC Detectors 228
        H.   Confirmatory Techniques 228

V.   Sample Collection, Preservation, and Preparation 229
        A.   Introduction 229
        B.   Sampling Apparatus 229
        C.   Sampling Procedures 230
                1.   Water 230
                2.   Sediments 230
                3.   Plant and Animal Samples 230
        D.   Sample Preparation for Extraction 230
                1.   Water 231
                2.   Sediments 231
                3.   Plant Material 231
                4.   Fish Tissue 231

VI.   Detailed Analytical Procedures 231
        A.   Analysis of Atrazine, Simazine, and Terbutryn in Water 231
                1.   Reagents 231

	2.	GC Conditions	231
	3.	GC Determination	232
	4.	Procedure	233
B.		Analysis of Atrazine, Simazine, and Terbutryn in Sediments	233
	1.	Reagents	233
	2.	GC Conditions	233
	3.	Extraction Procedure	233
	4.	Liquid-Liquid Partition Cleanup	233
	5.	Column Cleanup	234
	6.	GC Analysis	234
C.		Analysis of Atrazine, Simazine, and Terbutryn in Plant Tissue	234
	1.	Reagents	234
	2.	Apparatus	234
	3.	GC Conditions	235
	4.	Extraction Procedure	235
	5.	Liquid-Liquid Partition Cleanup	235
	6.	Column Cleanup	235
	7.	GC Analysis and Calculations	235
D.		Analysis of Atrazine, Simazine, and Terbutryn in Fish Tissue	235

References 236

I. INTRODUCTION

A. History

With few exceptions, the triazine herbicides are based on a *s*-triazine structure containing two substituted amino groups, while the third ring carbon atom possesses a chloro-, methoxy-, or azido-grouping.

$X = Cl, OCH_3, SCH_3,$ or N_3

Currently there are about 25 triazine derivatives that have been developed as commercial herbicides or are in experimental use (Table 1).

The Geigy laboratories in Basel, Switzerland synthesized triazine compounds in 1952, and Gast, Knuesli, and Gysin first described their selective herbicidal properties.[1,2] Soon thereafter, worldwide field trials established the commerical importance of the now well-known herbicides, atrazine and simazine. Since then numerous attempts have been made to prepare several other substituted triazines and test their herbicidal properties. Substitution of the chloro-, methoxy-, methylthio-, or azido-groups, with other substituents has led to compounds with reduced or no biological activity. However, changes in the alkylamino side chains to alkoxyalkylamino and, more recently, to cyclo- and cyano-alkyl derivatives, have provided compounds such as cyprazine and cyanazine, which are both biologically active and show variable selectivity.

In general, the triazine herbicides have little or no effect on germinating weed seeds but affect metabolic processes within the plant following root uptake. Control of sus-

Table 1
THE TRIAZINE HERBICIDES

Common name	Chemical name	Formula	Mol. wt.	Structure
				(triazine ring, positions 2, 4, 6)
ametryn*	2-(ethylamino)-4-(isopropylamino)-6-(methylthio)-s-triazine	$C_9H_{17}N_5S$	227	CH_3CH_2NH- $(CH_3)_2CHNH-$ CH_3S-
atraton	2-(ethylamino)-4-(isopropylamino)-6-(methoxy)-s-triazine	$C_9H_{17}N_5O$	211	CH_3CH_2NH- $(CH_3)_2CHNH-$ CH_3O-
atrazine*	2-chloro-4-(ethylamino)-6-(isopropylamino)-s-triazine	$C_8H_{14}ClN_5$	216	$Cl-$ CH_3CH_2NH- $(CH_3)_2CHNH-$
aziprotryn	2-azido-4-(isopropylamino)-6-(methylmercapto)-s-triazine	$C_7H_{11}N_7S$	225	N_3- $(CH_3)_2CHNH-$ CH_3S-
chlorazine	2-chloro-4,6-(diethylamino)-s-triazine	$C_{11}H_{20}ClN_5$	258	$Cl-$ $(C_2H_5)_2N-$ $(C_2H_5)_2N-$
cyanazine*	2-chloro-4-(ethylamino)-6-(1-cyano-1-methylethyl-amino-s-triazine	$C_9H_{13}ClN_6$	241	$Cl-$ CH_3CH_2NH- $N\equiv C-C\begin{smallmatrix}CH_3\\ .NH-\\ CH_3\end{smallmatrix}$
cyprazine*	2-chloro-4-(cyclopropylamino)-6-(isopropylamino)-s-triazine	$C_9H_{14}ClN_5$	228	$\triangleright-NH-$ $(CH_3)_2CHNH-$
dimethametryn	2-(methylthio)-4-(ethylamino)-6-(1,2-dimethylpropyl-amino-s-triazine	$C_{11}H_{21}N_5S$	255	CH_3S- CH_3CH_2NH- $(CH_3)_2CH-CHNH-\,CH_3$
dipropetryn*	2-(ethylthio)-4,6-bis(isopropylamino)-s-triazine	$C_{11}H_{21}N_5S$	255	C_2H_5S- $(CH_3)_2CHNH-$ $(CH_3)_2CHNH-$
ipazine	2-chloro-4-(diethylamino)-6-(isopropylamino)-s-triazine	$C_{10}H_{18}ClN_5$	244	$Cl-$ $(C_2H_5)_2N-$ $(CH_3)_2CHNH-$
methoprotryne	2-(isopropylamino)-4-[(methoxypropyl)amino]-6-(methylthio)-s-triazine	$C_{11}H_{21}N_5OS$	261	$(CH_3)_2CHNH-$ $CH_3O(CH_2)_3NH-$ CH_3S-
procyazine	2-chloro-4-(cyclopropylamino)-6-(1-cyano-1-methyl-ethylamino)s-triazine	$C_{10}H_{12}ClN_6$	252	$\triangleright-NH-$ $N\equiv C-C\begin{smallmatrix}CH_3\\ .NH-\\ CH_3\end{smallmatrix}$
prometon*	2,4-bis(isopropylamino)-6-(methoxy)-s-triazine	$C_{10}H_{19}N_5O$	225	$(CH_3)_2CHNH-$ $(CH_3)_2CHNH-$ CH_3O-
prometryn*	2,4-bis(isopropylamino)-6-(methylthio)-s-triazine	$C_{10}H_{19}N_5S$	241	$(CH_3)_2CHNH-$ $(CH_3)_2CHNH-$ CH_3S-
propazine*	2-chloro-4,6-bis(isopropylamino)-s-triazine	$C_9H_{16}ClN_5$	230	$Cl-$ $(CH_3)_2CHNH-$ $(CH_3)_2CHNH-$
secbumeton*	2-sec-butylamino-4-(ethylamino)-6-(methoxy)-s-triazine	$C_{10}H_{19}N_5O$	225	$\begin{smallmatrix}CH_3\\ CHNH\\ C_2H_5\end{smallmatrix}$ CH_3CH_2NH- CH_3O-
simetone	2-methoxy-4,6-bis(ethylamino)-s-triazine	$C_8H_{15}N_5O$	233	CH_3O- CH_3CH_2NH- CH_3CH_2NH-
simetryn*	2,4-bis(ethylamino)-6-(methylthio)-s-triazine	$C_8H_{15}N_5S$	249	CH_3CH_2NH- CH_3CH_2NH- CH_3S-
simazine*	2-chloro-4,6-bis(ethylamino)-s-triazine	$C_7H_{12}ClN_5$	202	$Cl-$ CH_3CH_2NH- CH_3CH_2NH-
terbumeton	2-tert-butylamino-4-(ethylamino)-6-(methoxy)-s-triazine	$C_{10}H_{19}N_5O$	225	$(CH_3)_3CNH-$ CH_3CH_2NH CH_3O-
terbuthylazine*	2-tert-butylamino-4-chloro-6-(ethylamino)-s-triazine	$C_9H_{15}ClN_4$	215	$(CH_3)_3CNH-$ $Cl-$ CH_3CH_2NH-
terbutryn*	2-tert-butylamino-4-(ethylamino)-6-(methylthio)-s-triazine	$C_{10}H_{19}N_5S$	241	$(CH_3)_3CNH-$ CH_3CH_2NH- CH_3S-
trietazine	2-chloro-4-(diethylamino)-6-(ethylamino)-s-triazine	$C_9H_{16}ClN_5$	230	$Cl-$ $(C_2H_5)_2N-$ CH_3CH_2NH-
hexazinone*	3-cyclohexyl-6-(dimethylamino)-1-(methyl)-s-triazine-2,4-(1H,3H)-dione	$C_{12}H_{20}N_4O_2$	252	(structure)
metribuzin*	4-amino-6-[1,1-dimethyl(ethyl)]-3-(methylthio)-1,2,4-triazin-5-(4H)-one	$C_8H_{14}N_4OS$	214	(structure)

* WSSA common names

ceptible plants is due to inhibition of the photosynthetic Hill reaction, resulting in chlorosis and eventual death.[3]

B. Use Patterns

The most commonly used triazine herbicides are undoubtedly atrazine and simazine, which are extensively used throughout the world. The use patterns of the most commonly encountered triazine herbicides are summarized in Table 2, together with their application rates.[4,5]

Simazine and terbutryn are the only triazine herbicides used for the selective control of algae and submerged weeds in aquatic situations such as ponds, swimming pools, recirculating water cooling towers and fountains. In the U.S., terbutryn has been tested for the control of weeds in ponds and lakes and its registration for aquatic use is being actively pursued by the manufacturer, but its major use at present is for selective control of weeds in winter cereals, such as wheat, barley, and sorghum.

C. Formulations and Application Rates

These are summarized in Table 2. The triazine herbicides are generally formulated as wettable powders or granules. However, liquid and emulsifiable concentrates of several derivatives are also available.[4,5] For selective control of weeds, the triazine herbicides are applied as pre-emergence treatments to bare soil. Several triazines also show post-emergent activity to varying degrees, and can therefore be applied as post-emergent treatments to young emerged weeds in certain crops. For these formulations, water is usually the carrier.

The rates of application of triazine herbicides as selective treatments in crops vary from 0.25 to 4 kg/ha, depending on the herbicide, crop, and soil type. For general vegetation control in noncrop land (including irrigation ditches) the rates vary from 10 to 50 kg/ha, being dependent upon the type of weed species present and the duration of control desired. Simazine is applied for aquatic weed control to give concentrations of 0.5 to 2 ppm on a water volume basis. The recommended concentration for terbutryn applications to water is 0.1 ppm applied as a surface spray or as a granular formulation.

D. Synthesis

Cyanuric chloride, the basic starting material for synthesis of most of the triazine herbicides, is formed by trimerization of cyanogen chloride, which in turn is synthesized from hydrocyanic acid and chlorine. Cyanuric chloride may be considered as a triple imide chloride, the chlorine atoms of which are highly reactive and readily ex-

Table 2
USE PATTERNS OF TRIAZINE HERBICIDES[4,5]

Herbicide	Application method	Formulations	Carrier	Rate (kg/ha)	Type of weed control	Crops or other uses
Ametryn	Pre- or post-emergence	WP	Water	2—8	Annual broadleaf and grass weeds	Pineapple, sugarcane, bananas, potatoes
Atrazine	Post-directed	WP	Water		Desiccant	Corn, potato vine
	Post-emergence	WP/L	Water/oil	2—4	Annual broadleaf and grass weeds	Corn, sorghum, rangeland, sugarcane, macadamia orchards, pineapple, turf grass, sod, conifer reforestation, Christmas tree plantations, raspberry
	Pre-plant, pre-emergence	WP/L/G	water			
				<1	Nonselective control of weeds	Chemical summerfallow and noncrop land, industrial and rights-of-way areas
				10—20	Vegetation control, soil sterilant	
Aziprotryn	Pre- or post-emergence	WP	Water	1—4	Annual broadleaf and grass weeds	Brassica crops, onions, leeks
Cyanazine	Pre- or post-emergence	WP/L/G	Water	1—4	Annual grasses and broadleaf weeds	Corn, grain sorghum, cotton, peanuts, wheatfallow, soybeans, peas
Cyprazine	Post-emergence	EC	Water	<1	Annual seedling weeds and grasses	Corn, sorghum, sugarcane
Desmetryn	Post-emergence	WP	Water	0.25—0.5	Fat hen and other annual broadleaf weeds	Brassica crops
Dipropetryn	Pre-emergence	WP	Water	1—2	Annual broadleaf and grass weeds	Cotton
Metribuzin	Pre-plant, pre-emergence, post-emergence	WP	Water	0.25—4	Annual grasses and broadleaf weeds	Potatoes, sugarcane, established alfalfa, established asparagus, tomatoes, soybeans, citrus, turf grasses, established cereals, established peas, some range and pasture grasses
Prometon	Pre- or post-emergence	WP/L/EC	Water/oil	10—60	Nonselective control of weeds	Noncrop land

Table 2 (continued)
USE PATTERNS OF TRIAZINE HERBICIDES[4,5]

Herbicide	Application method	Formulations	Carrier	Rate (kg/ha)	Type of weed control	Crops or other uses
Prometryn	Pre- or post-emergence	WP	Water	1—3	Annual broadleaf weeds and grasses	Cotton, celery, peas, potatoes
Propazine	Pre-emergence	WP/L	Water	1—3	Annual broadleaf weeds and grasses	Sorghum
Secbumeton	Post-emergence	EC	Water	1—8	Annual seedling grasses and broadleaf weeds	Established alfalfa, sugarcane
Simazine	Pre-emergence	WP/G	Water	2—4	Annual broadleaf and grass weeds	Corn, citrus, deciduous fruits and nuts, olives, pineapple, established alfalfa, perennial grasses grown for seed or pasture, turf grasses grown for sod, ornamental nursery plantings, Christmas tree and shelterbelt plantations, sugarcane, asparagus, beans and artichokes.
	Pre-emergence	WP/G	Water	10—20	General vegetation control	Noncrop land, industrial and rights-of-way areas
	Spray/spread direct to water	WP/L/G	Water	2—8	Algae and submerged aquatic weeds	Ponds, swimming pools, large aquaria, ornamental fish ponds, fountains, recirculating water cooling towers
Terbuthylazine	Pre-emergence	WP	Water	2—4	Annual broadleaf and grass weeds	Sorghum, corn, vineyards, citrus and pome fruits
Terbutryn	Pre- or post-emergence	WP	Water	1—2	Annual broadleaf and grass weeds	Winter wheat, winter barley, sorghum, peas, potatoes

Note: EC, emulsifiable concentrate; WP, wettable powder; L, liquid; G, granular.

Table 3
SOME PHYSICAL AND TOXICOLOGICAL PROPERTIES OF COMMERCIALLY USED TRIAZINE DERIVATIVES[3,4]

Common name	Melting point (°C)	Solubility in water (°C) (ppm)	Vapor pressure at 20°C (mmHg)	pK$_a$ at 21°C	LD$_{50}$ (rats) (mg/kg)
Ametryn	84—85	185 (20)	8.4×10^{-7}	4.1	1110
Atrazine	173—175	33 (27)	3.0×10^{-7}	1.7	3080
Aziprotryn	95	50 (20—25)	2.0×10^{-6}	—	5833
Cyanazine	166—167	171 (25)	1.6×10^{-9}	1.0	334
Cyprazine	167—168	6.9 (25)	3.0×10^{-7}	—	1200
Desmetryn	84—86	580 (20—25)	1.0×10^{-6}	4.0	595
Dipropetryn	104—106	16 (20—25)	7.4×10^{-7}	4.3	5000
Metribuzin	126—127	1220 (20)	$<1.0 \times 10^{-5}$	—	1090—1206
Prometon	91—92	750 (20)	2.3×10^{-6}	4.3	2980
Prometryn	118—120	48 (20)	1.0×10^{-6}	4.1	3750
Propazine	212—214	8.6 (20)	2.9×10^{-8}	1.7	>5000
Secbumeton	86—88	620 (20)	7.3×10^{-6}	4.4	1000
Simazine	225—227	3.5 (20)	6.1×10^{-9}	1.7	>5000
Terbuthylazine	177—179	8.5 (20—25)	1.1×10^{-6}	2.0	3690
Terbutryn	104	25 (20)	9.6×10^{-6}	4.3	2500

changed for nucleophilic groups such as amines, alkoxides, mercaptides, and azides. The details of the reaction media, the temperature, and other necessary conditions for these stepwise transformations are described elsewhere.[3]

E. Physical Properties

Some of the physical properties of the triazine herbicides, such as melting points, water solubilities, vapor pressures, and pKa values are listed in Table 3.[3-5] In general, the triazine herbicides are white crystalline solids with low vapor pressures and high melting points. The solubilities in water are relatively low, the order for the dialkyamino-s-derivatives being 2-chloro- < 2-methylthio- < 2-methoxy-analogs. A pronounced increase in the solubilities of the s-triazines occurs at pH levels near their respective pKa values, where strong protonation occurs.

The dialkylamino-s-triazines are weak bases in aqueous solution, with the nonamino substituents determining the degree of basicity. The methoxy-s-triazines are the most basic and the chloro-s-triazines the least. The relative complexity of the N-alkyl substituents also affects the degree of basicity of these compounds, but to a lesser extent.

F. Chemical Properties

The triazine herbicides can undergo several chemical reactions, especially those involving the substituents in the 2,4, and 6 carbon positions. Several of these reactions have been instrumental in the development of derivatives with enhanced or modified biological effectiveness.

The acid or alkaline hydrolysis of the s-triazines leads to the replacement of the chloro-, methoxy-, or methylthio- groups with that of the hydroxyl. This reaction can occur also catalytically and as well is the basis for the detoxification of chlorotriazines in corn.

N-dealkylation of s-triazine herbicides, an important degradation mechanism in biological systems, has been demonstrated to occur chemically, using Fenton's reagent (hydrogen peroxide and ferrous sulfate).[6]

Direct chemical cleavage of the triazine ring of the s-triazine herbicides occurs only under drastic conditions, such as heating with concentrated sulfuric acid, melting with alkali, or heating at elevated temperatures.[3] The *s*-triazine herbicides are susceptible to photolysis, both under UV wavelengths and in sunlight.[7,8] The nonamino (C-2) position is the main reaction site and the hydroxytriazines are the major photoproducts. Methylthio-*s*-triazines (but not the chloro- or methoxy-derivatives) also yield desmethylthio analogs.[8] The alkyl side chains can also be removed by photolysis.[8] However, there is little evidence that the bioactivity and degradation of the *s*-triazines is affected by sunlight, in water or on soil surfaces, under field conditions.[7]

G. Toxicological Properties

The LD_{50} values of the triazine herbicides, using rats, range from 334 to 5000 ppm, indicating low to moderate mammalian toxicity (Table 3).[4] The LC_{50} values for selected triazine herbicides to fish and several other aquatic organisms are summarized in Table 4.[4,9] There is little or no information on the long-term toxic effects of triazine herbicides to organisms, or on mammalian mutagenicity.

II. ENVIRONMENTAL ASPECTS

With the exception of terbutryn and simazine, the triazine herbicides are not recommended for aquatic weed control. However, the symmetrical triazines are among the most widely used herbicides in agriculture today and are applied both to the growing crops and directly to the soil. Thus, the fate of these herbicides in the soil environment is of importance since such factors as soil persistence, leaching, and movement of residues by run-off can be major influences governing the unwanted contamination of streams, rivers, ponds, and lakes. Triazine residues resulting from droplet drift from terrestrial applications and accidental spills may also cause pollution of aquatic systems. Vapor movement is not a problem since the triazine herbicides are relatively nonvolatile.

Triazines, especially atrazine and simazine, are among the most persistent herbicides in use,[3,10] and residues can remain in the soil at phytotoxic levels for up to 1 year after application.[11] Leaching of *s*-triazines under normal field conditions is not a major path for loss of the compounds,[12] though low ppb levels have been detected in ground water in sandy soils.[13] Movement of atrazine, simazine, and other triazine herbicides in run-off waters from treated fields following rainfall events can be an important pathway of loss of the compounds. Run-off studies of triazine herbicides have been summarized in a recent review.[14] Atrazine residues have been reported in river water, especially during the spray application period, in corn-growing areas of the U.S. and Canada.[15-17] Applications of atrazine and simazine to (dry) irrigation ditches can also result in residues of the compounds in irrigation water, especially during the initial filling of the ditches.[18,19]

The triazines are degraded by chemical and biological processes.[3,20] The major breakdown products of chloro-, methoxy-, and methylthio-triazines in soils and waters are the respective hydroxytriazines, which are formed primarily by chemically induced hydrolytic reactions, though soil microorganisms can also mediate this conversion.[3,21] Hydroxytriazines are strongly adsorbed to soils and thus are difficult to extract from such samples. The other major degradation pathway involves biological cleavage of the alkyl groups on the amino side chains. The structures of the two major breakdown products derived from atrazine are shown on page 222.

Table 4
ACUTE TOXICITY TO FISH AND OTHER AQUATIC INVERTEBRATES FOR SELECTED TRIAZINE HERBICIDES.[4,9]

LC$_{50}$(ppm)[a]

Herbicide	Rainbow trout	Bluegill sunfish	Fathead minnow	Brown shrimp	Pink shrimp	Oyster	Goldfish	Crab	Water flea	Catfish	Largemouth bass	Carp
Ametryn	8.8	4.1	5.7	>1(48)	—	>1.0	14.0	—	—	—	0.71	—
Atrazine	4.5	24	15	>1(48)	—	>1.0	60	>1000	6.9(48)	—	—	—
Cyprazine	6.2	—	—	—	—	—	—	—	—	—	—	—
Dipropetryn	2.3	3.7	—	—	—	—	—	—	—	—	—	—
Metribuzin	>100	>100	—	—	—	—	—	—	—	—	—	—
Prometon	20	>32	—	—	>1(48)	>1.0	8.6	—	—	—	—	—
Prometryn	2.5	10	—	—	>1(48)	>1.0	3.5	—	—	—	—	—
Propazine	17.5	>100	—	—	—	>1.0	>32	—	—	—	—	—
Simazine	2.8	16	6.4	—	113	>1.0	>32	>1000	—	—	—	—
Terbutryn	3	4	2.9	—	—	—	—	—	—	3	—	4

[a] All LC$_{50}$ values for a 96-hr exposure, unless otherwise stated.

Both hydroxytriazines and dealkylated metabolites have been observed in field soils and contaminated waters.[11,13,17]

The persistence of triazines in the aquatic environment has received some attention and studies have shown that simazine and terbutryn residues can remain in pond water and sediment for several months.[22,23] Although simazine has been reported in fish raised in treated ponds, there is no evidence that triazine herbicides are concentrated along major links in the food chain.[22]

III. REVIEW OF ANALYTICAL PROCEDURES

There are three main procedures for the analysis of triazine residues in waters, soils, and plant and animal tissues. These methods are based on spectrophotometry, high-performance liquid chromatography (HPLC), and gas chromatography (GC). In this section the principles of each procedure will be outlined together with the advantages and disadvantages inherent in the three methods.

A. Spectrophotometry

Chloro-*s*-triazines, such as atrazine and simazine, will react with pyridine to give quaternary pyridinium halides, which can then undergo addition of a hydroxyl group to form a carbinol base.[24] In the presence of alkali the heterocyclic ring of this system is cleaved to yield an anil of glutaconic dialdehyde, which exists in equilibrium with the enol form. This mixture absorbs at about 440 nm, so that spectrophotometric determination of solution absorbance at this wavelength may be used to calculate the amount of chlorotriazine (Tr-Cl) originally present. The overall reaction may be summarized as:

$$Tr-NH-CH=CH-CH=CHCHO \rightleftharpoons Tr-N=CH-CH=CH-CH=CHOH$$

A general analytical procedure based on this reaction has been developed for the quantitative determination of chloro-*s*-triazines extracted from soil samples.[24,25] This procedure, however, is unsatisfactory since the color generated is not stable and fades rapidly. Also, extracts derived from untreated soil samples can give rise to highly colored blanks. Thus this method is of very limited use.

A more useful and versatile spectrophotometric procedure that is capable of determining chloro-, methoxy-, and methylthio-triazines is based on the principle that *s*-triazines are converted to their respective hydroxytriazines in 50% aqueous sulfuric

Table 5

LIQUID CHROMATOGRAPHIC SYSTEMS FOR ANALYSIS OF SOME COMMON TRIAZINE HERBICIDES AND THEIR METABOLITES

Triazine	Column packing	Mobile phase	UV detection wavelength (nm)	Ref.
Terbutryn	Permaphase ETH	20% Methanol in water	254	27
Atrazine and other herbicides	PE C18 Sil-X-11	20% Methanol in water	220—245	28
14 Triazine herbicides	µBondapak C18	70% Methanol in water	220—240	29
12 Triazine herbicides	Lichrosorb Si60 (5µ)	2% 2-Propanol in trimethyl pentane	254	26
Atrazine, cyanazine, simazine	Porasil ODS	Chloroform:water (40:60)	230	30
Hydroxytriazines	Lichrosorb Si60 (5µ)	Dichloromethane:methanol (95:5) with 0.1 M propionic acid	254	31
Hydroxytriazines	Lichrosorb Si60 (10µ)	Chloroform:methanol:water (700:300:60) + 0.1% phosphoric acid	240	32
Hydroxyterbutryn	Lichrosorb Si100 (5µ)	Dichloromethane:methanol (95:5) 0.05 M propionic acid	254	23
Hydroxyterbutryn	Lichrosorb KAT (10µ)	Acetonitrile:water (90:10) 1.0—0.2 m M ammonium acetate	254	23

acid.[24,25] The cations of the hydroxytriazines show absorption maxima in the region of 240 nm. Thus water, soil, and plant samples may be solvent extracted and, if necessary, a column cleanup stage carried out. The evaporated extracts are then hydrolyzed with acid to the hydroxy derivatives and the absorbance at 240 nm measured. Absorbance measurements are also made at 225 and 255 nm so that a correction may be made for background absorbance due to co-extracted interferences.[24,25] This procedure is rapid and is applicable for the analysis of many triazine herbicides. However, the method is nonspecific since the hydroxytriazines all absorb at 240 nm, and so this procedure can only be applied to samples containing single, and known, triazine residues. The method is quite sensitive and will detect residues in a variety of substrates at the 0.05 µg/g level.

B. High-Performance Liquid Chromatography (HPLC)

The principles of liquid chromatography were discussed in the previous chapter on the substituted urea herbicides. Since the triazines have absorption maxima in the region of 220 nm they can be analyzed using liquid chromatographic instruments equipped with UV detection systems. It has been noted that the sensitivity to triazine herbicides is 50-fold greater when using a variable wavelength detector tuned to 220 nm than that obtained by using a fixed wavelength of 254 nm.[26] To date several stationary and mobile phases have been described for the detection of chloro-, methoxy-, and methylthio-substituted triazines, as well as hydroxytriazine metabolites, and these are summarized in Table 5. Liquid chromatographic analytical procedures have proved most successful for the determination of triazine residues in water where detection limits of 1 µg/ℓ have been reported;[28] however its use for the determination of hydroxytriazines recovered from crops and aquatic sediments is limited since time-consuming cleanup procedures are required.[23,32]

The applicability of liquid chromatography to the separation and determination of more than one triazine residue in an aquatic sample has still to await evaluation.

C. Gas Chromatography (GC)

With the number of specific detection systems available, GC analysis is the most commonly used procedure for the determination of all triazine residues in water, soil, and plant and animal tissue.[33,34] The samples are first solvent extracted and after a suitable cleanup stage the extracts are examined by GC using nitrogen-, chlorine-, or sulfur-specific detectors. These detectors have the additional advantage over the relatively nonspecific ECD in that sample cleanup prior to determination need not be as rigorous. Using GC, detection of triazine residues at the 0.05 μg/g level, and lower, can be achieved. Such procedures are also suitable for the analysis of mixtures of triazine herbicides and a multi-residue method has been described for the separation and determination of twelve commonly encountered triazines extracted from a variety of substrates.[35]

Dealkylated metabolites derived from atrazine, simazine, and terbutryn can also be determined by GC without any modification; however, the hydroxy metabolites require conversion to their methoxy or silyl derivatives prior to analysis.[36-39]

D. Discussion of Analytical Procedures

Since the spectrophotometric procedures are nonspecific and the liquid chromatographic methods are not as sensitive as those based on GC, and have yet to be adapted to multi-residue analysis, the favored methods for analysis of the triazine herbicides are those involving GC separation in conjunction with nitrogen-, sulfur-, or chlorine-specific detectors. These are simple procedures which allow for multi-residue analysis as well as the determination of metabolites.

Special emphasis will be placed in the appropriate section on procedures for the extraction and determination of atrazine, simazine, and terbutryn in water, sediment, plant, and fish samples, since these compounds are almost likely to be used in aquatic situations.

IV. ANALYTICAL METHODOLOGY

A. Extraction of Triazine Herbicides from Water Samples

Procedures for the extraction of triazine herbicides from water samples have been reviewed up until 1972.[33,39,40] Table 6 summarizes some of the more recent reports for the extraction of atrazine, simazine, terbutryn, and related *s*-triazines from natural waters. Dichloromethane has been the most widely used solvent for water extraction. Several analysts have recommended that the pH of water samples be adjusted to 8 or 9 prior to extraction since methylthio-*s*-triazines, with pKa values in the range of 4 to 5 do not partition efficiently into the organic phase at acidic pH values.[27,28,41]

Recently the use of macroreticular resins has been investigated for the recovery of pesticides and other organic pollutants from water samples. These samples are percolated through the resin, onto which they become adsorbed, and from which they can be subsequently desorbed following elution with a suitable solvent. Advantages to using such resin systems are that extraction of large sample volumes is possible which permits a lower level of pesticide detection than can be obtained by using regular solvent extraction procedures. The macroreticular resins XAD-2 and XAD-4 have been used for the recovery of atrazine from river water samples.[30,42,43] Few recovery studies have been performed with water samples containing *s*-triazine herbicides at the subppb level, even though atrazine is often found at these levels in contaminated waters.[16,17]

B. Extraction of Triazine Herbicides from Soil and Sediment Samples

Methods for the extraction of *s*-triazine residues from soils and sediments have been reviewed up to 1972.[33,39] Table 7 summarizes more recent studies, as well as reports in which sediments from filtered run-off water, rivers, and ponds have been analyzed for *s*-triazine herbicide residues.

Table 6
EXTRACTION OF TRIAZINE HERBICIDES FROM NATURAL WATERS

Triazine	Type of water	Solvent and water/solvent ratio	Fortification level ($\mu g/\ell$)	Ref.
Atrazine, ametryn, terbutryn	Pond or canal	Dichloromethane (10:1)	1—100	44
Atrazine	River	Dichloromethane (5:1)	10	48
Terbutryn	Pond, irrigation and canal	Dichloromethane (10:1)	1—100	27
Atrazine, cyanazine, metribuzin	Tile-drain	Ethyl acetate (6:1)	1—100	13
Simazine	Irrigation	Dichloromethane (2:1)	500	45
Simazine	Irrigation	Dichloromethane (4:1)	50—1000	18
Atrazine, simazine	NS[a]	Dichloromethane (10:1)	66.5	46
Atrazine, ametryn	Filtered runoff	Ethyl acetate (10:1)	NS[a]	47
Terbutryn	Pond, pH 7—9	Dichloromethane (5:1)	0.5—50	23

[a] Not stated.

Table 7
EXTRACTION AND CLEANUP PROCEDURES FOR TRIAZINE HERBICIDES FROM SOILS AND SEDIMENTS

Triazine	Soil or sediment	Solvent system	Cleanup technique	Ref.
Simazine, cyanazine, degradation products (^{14}C)	Soil (not air dry)	Methanol:water (4:1) (shaking)	None	50
Atrazine cyanazine, and metabolites	Soil (not air dry)	Acetonitrile:water (2:1) (shaking)	Chloroform partition	41
Atrazine	Soil (air dry)	Acetonitrile:water (9:1) Acetonitrile:water (2:1) (shaking)	Dichloromethane partition, alumina	48
Atrazine	Soil (sieved, not dried)	Hot acetonitrile:water (9:1) (blending)	Hexane:ether (2:1) partition	52
Atrazine (^{14}C)	Soil (dry) (20% water added)	Methanol (reflux)	Dichloromethane partition, alumina	35
Atrazine, cyanazine	Soil (dry) (10% water added)	Methanol (ultrasonic)	Chloroform partition	53
Atrazine and metabolites (^{14}C)	Soil (not dried)	Methanol (Soxhlet)	None	54
Simazine	Sediment	Column-eluted with diethyl ether	Acetonitrile partition	55
Atrazine, ametryn	Sediments (drained and dried)	Methanol + ethyl acetate (reflux)	Alumina	47
Atrazine	Silt (water saturated)	Acetone + ethyl acetate (shaking)	Celite, Norit® SG 1 and magnesium oxide	56 57
Terbutryn (^{14}C) and metabolites	Pond sediment	Acetonitrile:water (4:1) reflux)	Alumina	23

There has been little work undertaken to elucidate the factors affecting the recovery of s-triazine herbicides from soils and sediments to determine whether method of sample fortification, moisture content of sample, soil type, and extraction technique affect triazine recovery. The hydrolysis of s-triazine herbicides to the hydroxytriazines in moist soils also makes fortification recovery studies difficult.

Analysis of treated field soils or soils fortified with ^{14}C-labeled chloro-s-triazines has indicated that hot methanol or acetonitrile-water mixtures are the most efficient solvent systems for the extraction of triazine residues.[25,35] Shaking or tumbling of the soil with the methanol or aqueous acetonitrile have also been shown to be effective extraction techniques.[41,48,49]

C. Extraction of Triazine Herbicides from Plant Tissue

Extraction and cleanup procedures for s-triazine herbicides from plant material prior to 1972 have been reviewed.[33] Table 8 summarizes some of the more recent approaches that have been used for the extraction of triazine residues from plant material. Few triazine recovery experiments from aquatic plant substrates have been reported, though atrazine[58] and terbutryn[23] have been recovered from such samples using solvent systems similar to those used with crops. There have been some experiments to compare the relative efficiencies of triazine recovery from plant tissue using different extraction techniques and solvent systems and researchers have demonstrated that methanol, or aqueous methanol, are more efficient than chloroform or ethyl acetate for the recovery of ^{14}C-labeled triazine residues from plant material.[35,49,50] Chloro-s-triazines are rapidly hydrolyzed in plants such as corn and sorghum to the corresponding hydroxytriazines which are difficult to solvent extract.[59,32]

D. Extraction of Triazine Herbicides from Animal and Fish Tissue

The herbicides simazine and terbutryn would be expected in fish tissue following treatment of water with these compounds,[22,64,65] and in such cases the residues reside chiefly in the liver and intestines. Table 9 lists some of the extraction and cleanup procedures that have been used for animal and fish samples containing triazine herbicide residues.

As in the case of moist soils, sediments, and certain plants, the triazine herbicides are converted to N-dealkyl- and hydroxytriazine derivatives in fish tissue.[64] A variety of solvents (Table 9) have been tried for the recovery of triazine residues from fish, but no comparative extraction data are available. However, methanol or acetonitrile would seem adequate for the extraction of triazines from fish samples since they are used for triazine recovery from soil and plant samples.

E. Cleanup Procedures for Sediment, Plant, and Animal Extracts

For GC determination using the nitrogen-specific conductivity or AFDs, water extracts containing triazine residues do not normally require cleanup.[25,35,41,48]

Cleanup procedures that have been used for triazine-containing extracts recovered from soil, plant, and animal tissue samples are listed in Tables 7, 8, and 9. Many authors have reported that for extracts derived from soil and crop samples only a simple solvent partitioning from the extraction solvent is necessary, prior to GC analysis using a conductivity detector.[41,48,60] Several analysts have partitioned acetonitrile, or aqueous acetonitrile extracts, containing s-triazine herbicides, with hexane to remove co-extractive materials.[23,48,65] This procedure is particularly effective for cleanup of fish tissue extracts.[23] Crop and animal tissue extracts have also been cleaned up by partitioning between hexane and acidic solutions.[35,69]

The column cleanup procedure originally described by Knuesli et al.[24] using basic alumina (activity V) has been widely used for crop and soil extracts containing s-tria-

Table 8

EXTRACTION AND CLEANUP PROCEDURES OF TRIAZINE HERBICIDES FROM PLANT TISSUE

Herbicide	Substrate	Solvent system and technique	Cleanup	Ref.
Cyanazine, simazine, atrazine, and degradation products (^{14}C)	Corn	Methanol:water (4:1) (shaking)	None	50
Cyanazine and degradation products (^{14}C)	Maize leaves, grain	Methanol:water (4:1) (tumbling)	Ethyl acetate partition	49
Atrazine simazine, terbutryn, etc.	(1) Fruits, vegetables (2) Hay, grain	Methanol, homogenizing then shaking	(1) Alumina (2) Hydrochloric acid/hexane cleanup, alumina	35
Atrazine, metribuzin, simazine, etc.	Root crops	Methanol; methanol-water; or Acetonitrile: water mixtures (blending)	Dichloromethane partition, alumina	60
Atrazine and conjugated metabolites	Sorghum roots	Methanol:water (4:1) (blending)	Chloroform partition	61
Cyanazine	Algae	Acetone (blending), then 0.025 N hydrochloric acid reflux	TLC	62
Atrazine	Planktonic algae	Benzene:hexane (80:20) (blending)	NoOne	63
Atrazine	Zooplankton	Ethyl acetate:hexane (1:1) (blending)	Celite®, Norit®, SG 1, magnesium oxide	57
Terbutryn	Cattails	Acetonitrile:water (95:5) (blending)	Dichloromethane partition, alumina	23

Table 9

EXTRACTION AND CLEANUP PROCEDURE FOR TRIAZINE HERBICIDES FROM ANIMAL AND FISH SAMPLES

Triazine	Substrate	Solvent system and technique	Cleanup	Ref.
Cyanazine (^{14}C) and metabolites	Crabs, snails, fish	Acetone (blending) then 0.025 N hydrochloric acid reflux	None	62
Simazine	Fish	Column-diethyl ether elution	Hexane: acetonitrile partition	55
Simazine	Fish	Chloroform (blending)	Hexane: acetonitrile partition	65
Simazine	Animal tissues	Chloroform (blending)	Alumina	66
Atrazine and metabolites	Chicken eggs, tissues, excretion	Methanol (shaking)	Alumina	67 68
Atrazine	Fish	Ethyl acetate: hexane (1:1) (blending)	Celite®, Norit®, SG1 Magnesium oxide	57
Terbutryn and metabolites	Fish	Acetonitrile:water (95:5) then acetonitrile: 0.01 N hydrochloric acid (95:5)	Alumina	23
Atrazine, desethyl atrazine	Fish, fish eggs	MeOH (sodium sulfate added) (blending)	Hexane, hydrochloric acid	69

zine residues.[9,23,25,35,48,49] Cleanup on acid washed alumina (activity I) has also been used,[24] especially where recovery of hydroxytriazine degradation products is required.[23,67,68,70] In general, acidic alumina has been reported to be less efficient for removing co-extractive materials from extracts containing chloro-*s*-triazines than basic alumina.[71]

Florisil® columns have also been reported for the cleanup of crop and soil extracts containing cyanazine and metribuzin.[49,72] Root crop extracts, containing simazine and atrazine have been cleaned up using low temperature precipitation of co-extracted lipids.[73] Gel permeation chromatography has also been advocated for cleanup of crop extracts containing chloro-*s*-triazines and hydroxytriazines.[32,74]

F. GC Columns

Several GC column liquid phases have been reported for the separation of the triazine herbicides; these include XE-60, SE-30, DC-200, Carbowax® 20M, and OV-17. The retention times for some commonly used triazine herbicides and their *N*-dealkyl derivatives on Carbowax® 20M, OV-17 and XE-60 are summarized in Table 10.[75,76]

G. GC Detectors

Since the triazine herbicides contain four or five nitrogen atoms, all triazines may be detected using a GC system equipped with a Coulson or Hall electrolytic conductivity detector, operated in the nitrogen mode.[48,73] Similarly, alkali-flame ionization detectors(AFID), with their selective response to nitrogen-containing compounds, have proved suitable for the detection of triazine residues.[35,77] Either of these two detection systems can therefore be used for the estimation of individual, or mixtures of, triazine residues.

The element-specific detectors, such as the microcoulometric or electrolytic conductivity detectors operable in the chloride mode are recommended for determination, or confirmation, of chloro-*s*-triazine residues, while flame photometric detectors (FPD) operated in a sulfur mode are similarly used for methylthio-*s*-triazines.[35]

The ECD is not of great use for detecting triazine residues, since a more rigorous cleanup of extracts is required than that necessary when using nitrogen-, chloride-, and sulfur-specific detection systems.

H. Confirmatory Techniques

Residue analysis requires confirmation of the identity of the residue(s) using an analytical procedure different from that used to identify the herbicide(s) initially. Confirmatory procedures include PC and TLC; GC and LC using different parameters (such as column systems) from those originally used in the analysis; and identification through known chemical derivatives.

PC and TLC techniques have been reported for the separation and identification of the triazine herbicides, and the subject has been extensively reviewed.[34,78] However, at least 1 to 2 µg of the triazines are usually necessary for satisfactory detection, so that confirmation of such residues in water and sediment samples containing trace amounts of these chemicals may not be possible.

Triazine herbicides have also been converted to a variety of derivatives suitable for GC confirmation analysis, and such techniques have been reviewed.[79-81]

Currently, the most practical confirmation for triazine herbicides recovered from water, sediment, plant, and animal samples is to compare the retention times using at least two GC systems with different columns. The identity of the unknowns may be assumed if their retention times are identical with those of the authentic standards under all conditions. The chloride- and sulfur-specific detectors are also helpful for

Table 10
GC CHARACTERISTICS OF SOME COMMONLY USED
S-TRIAZINES AND THEIR *N-DEALKYLATED DERIVATIVES*

Triazine	Stationary phase	Column temp (°C)	Carrier flow ml/min	Approx ret. time (min)
Propazine	5% Carbowax® 20M[a]	215	75	4.8
Atrazine				6.2
Simazine				7.8
Prometryn	5% Carbowax® 20M[a]	215	50	6.2
Ametryn				7.8
Simetryn				9.8
Prometryn	10% Carbowax® 20M[a]	240	112	7.0
Ametryn				8.8
Simetryn				10.9
Propazine	5% XE-60[a]	190	88	6.8
Atrizine				7.9
Simazine				8.6
Atrazine				3.8
N-Deethylatrazine				8.6
Simazine	3% Carbowax® 20M[b]	220	40	5.2
N-Deethylsimazine				11.4
Terbutryn				5.0
N-Deethylterbutryn				10.4
Atrazine				5.8
N-Deethylatrazine				4.0
Simazine				6.2
N-Deethylsimazine	3% OV-17[b]	200	4.5	
Terbutryn				9.0
N-Deethylterbutryn				6.5

[a] Column conditions: 5 ft × 0.25 in., aluminum, stationary phases all on Anakrom ABS. Data from Reference 75.
[b] Stationary phases on Chromosorb W-HP, 80/100 mesh in 6 ft × 0.25 in., glass column. Data from Reference 76.

identifying chloro- and methylthio-triazines. GC-MS systems can also be used as a final confirmation.

V. SAMPLE COLLECTION, PRESERVATION, AND PREPARATION

A. Introduction

The sampling procedures described here are those recommended in standard texts and in reviews.[82-85] Sampling is an integral part of analysis, especially for trace residue work, and while analysts may not be involved in this aspect of the work, they should nevertheless be aware of the procedures and problems associated with it. Eberhard et al.[85] have discussed approaches to the design of sampling programs to meet specific objectives.

The procedures outlined here assume that the s-triazine herbicide has been applied for weed control in a pond, a section of a lake or irrigation canal, or that a survey of a watershed that has been treated with s-triazine herbicide is to be carried out.

B. Sampling Apparatus

Glass jars — Amber glass of 1 *l* capacity, with Teflon® or aluminum foil-lined

screw caps. The jars should be rinsed with distilled water, methanol, and dichloromethane before use.

Sediment and biota containers — Glass sample jars of 500-m*l* capacity, with Te-
flon® or aluminum foil-lined screw caps.

Water samplers — Van Dorn, or equivalent for lakes and ponds. A long steel pole
to which a bottle can be attached may be suitable for small streams. Weighted bottle
type depth integrating samplers are recommended for rivers.[84] A variety of automated
samplers are also available.[82,83]

Sediment samplers — Corer type[86] or Ekman or Shipek dredges would be adequate.

C. Sampling Procedures
1. Water
Factors such as time of year, location of sampling site, number of samples per site,
number of sites, stability of the residue, and statistical considerations must all be considered before field sampling begins.[84] Lakes and ponds should be sampled in a systematic fashion such as at sites one-third and two-thirds the length of the pond, while
rivers should be sampled at a bridge or easily accessible site at several points, such as
one-third and two-thirds the width of the river. For statistical purposes a minimum of
three samples per site, analyzed separately, would allow comparisons between sites to
be made. As a general rule it is better to "over-sample" and select the samples to be
analyzed at a later date.[85]

In lakes or deep ponds where thermal stratification occurs during the summer and
winter months, a number of samples should be taken at different depths with a Van
Dorn type sampler and pooled in the sample bottle. For "dip" samples the container
is lowered into the water rinsed with the water to be sampled and then filled just below
the surface and capped.[39,84] The sample should be held in a refrigerated container and
transferred to the laboratory for storage at 2 to 5°C. If the samples are not analyzed
within 24 hr, 10 m*l* dichloromethane is added and the treated specimens stored at 2
to 5°C until analysis. As a precaution, a storage stability study should be carried out
if the samples are to be stored for a long time.

2. Sediments
General procedures are available in standard texts.[82,83] Sediment may also be recovered as suspended sediment from filtered run-off water or river water by filtration
or centrifugation. In shallow waters a coring device should be used since sample depth
can be controlled by selecting the appropriate portion of the core (usually 0 to 3 cm
depth).[86] Dredge devices (Ekman, Shipek) are useful in deep water or where bottom
sediments are too soft for a corer.

Since metribuzin has been shown to degrade nonbiologically in frozen soils,[87] sediment samples should be stored at −40°C until analysis. As a precaution, especially if
low-temperature freezers are not available, a storage stability study should be carried
out by fortifying untreated check samples with an aqueous solution of the triazine
herbicides at 50 to 500 μg/kg levels.

3. Plant and Animal Samples
General sampling procedures are available for sampling fish and other aquatic animals and aquatic plants.[82] Samples are stored at −40°C until analysis. A storage stability study on the ground or chopped sample may be undertaken by fortifying untreated check samples, as with sediments.

D. Sample Preparation for Extraction
General procedures for crops and animal products (including fish) are available.[88]

1. Water

The sample is brought to rt and shaken well before transferring to a measuring cylinder. Sample volume is recorded, and the sample transferred quantitatively to a 2-ℓ separatory funnel.

2. Sediments

From the few reports available, it appears that sediment samples may be drained of excess water and analyzed in the manner recommended for soil samples.[82] Samples should be analyzed without further drying to minimize losses of s-triazine residues by hydrolysis to the hydroxytriazine derivatives.[51] To obtain the moisture content of the sediment being analyzed, a portion (10 g) of the sample is heated in an oven at 100°C for 24 hr, and then re-weighed. Unused material is re-frozen.

3. Plant Material

The entire sample (500 to 1000 g) is cut into small pieces using a food cutter, chopper, or knife. A portion (10 g) is analyzed for moisture content, as described above. The unused sample is frozen at −40°C.

4. Fish Tissue

The fish is rinsed in distilled water and the excess moisture removed using a paper towel. The fish, or several if small specimens, are ground in a small meat grinder several times to obtain a homogeneous sample. Any unused material is refrozen.

VI. DETAILED ANALYTICAL PROCEDURES

A. Analysis of Atrazine, Simazine, and Terbutryn in Water

This method is based on the procedures described by Muir,[23] Ramsteiner et al.,[35] Sirons,[41] and Purkayastha and Cochrane.[48] Recoveries of these herbicides from water samples should be in excess of 90% (see Table 11). This procedure will also recover the N-dealkylated triazine metabolites, but is not satisfactory for the determination of the hydroxytriazine degradation products.

1. Reagents

Solvents — All solvents should be distilled in glass or be of pesticide residue grade, or their equivalent. Water must be distilled in glass and residue free.

Sodium sulfate — Granular, anhydrous, that has been heated overnight at 400°C.

Alumina — Woelm Basic grade, adjusted to activity V and stored in a sealed flask.

Analytical standards — Obtained from a National Standards repository and stored in a freezer until needed. Solutions of each triazine standard are prepared at a concentration of 10 mg/ 100 mℓ ethyl acetate or acetone. Methanol should not be used for methylthio-s-triazines, since such herbicides undergo conversion to the methoxy-s-triazines in this solvent. The stock solutions, in amber glass flasks are stored in a freezer at approximately −20°C.

For GC analysis, **working standards** are prepared in acetone, ethyl acetate, or methanol (except for methylthio-s-triazines) containing 0.1, 0.5, 1.0, 2.0, and 5.0 μg/mℓ (ng/$\mu\ell$).

2. GC Conditions

Instruments — Any machine capable of taking a Hall conductivity detector or Coulson conductivity detector operated in the nitrogen mode; or an AFID. The conditions described are for Tracor GC Models 222 or 560. Retention times are displayed in Table 10.

Table 11
TYPICAL RECOVERIES FROM SAMPLES FORTIFIED
WITH ATRAZINE, SIMAZINE AND TERBUTRYN

Triazine	Substrate	Fortification level ($\mu g/g$)	Recovery (%)	GLC[a] detector	Ref.
Atrazine	Water	0.01	102	Coulson	48
		0.05	92.5	AFD	77
	Corn	0.05	76	Coulson	48
			101	AFD	77
	Soil	0.02	95	Coulson	48
		1.10	87.5	AFD	77
Simazine	Potatoes	0.10	92	MCD	35
	Bananas	0.02	95	Coulson	35
	Soil	0.3	92	AFD	35
		0.05	96	Coulson	35
	Fish	0.1	98	EC	55
Terbutryn	Cattails	0.9	92	AFD	23
	Sediment	0.8	97	AFD	23
		0.08	106	AFD	23
	Water	0.0005	108	AFD	23
		0.05	101	AFD	23
		0.5	78	Coulson	35
	Fish	0.05	89	AFD	23

[a]   Coulson conductivity detector, Nitrogen mode; AFD; MCD (Cl mode)

Columns — 3% OV-17 and/or 3% Carbowax® 20M on 80 to 100 mesh Chromosorb W, HP, and packed in 1.8 × 4 mm i.d. glass tubing. With the OV-17 column, inlet, oven, and detector temperatures are maintained at 220, 200, and 250°C, while the corresponding temperatures for the Carbowax® 20M column are 230, 220, and 250°C, respectively. Carrier gas flow rates of 40 mℓ/min are used. Procedures for packing and conditioning of GLC columns are available.[82,88]

Hall conductivity detector — Furnace at 850°C solvent systems of 2-propanol: water (1:1) at a flow rate of 0.5 to 1.0 mℓ/min. The hydrogen flow rate is 30 mℓ/min.

Under these conditions, and with the columns described above, an injection of 5 ng triazine should give approximately a half-scale deflection at an attenuation of × 4 on a 1-mV recorder.

Alkali-flame detector — Operated with a hydrogen flow of 3 mℓ/min, and an air flow of 120 mℓ/min. A low attenuation of 1 × 2, or 1 × 4 × 10^{-12} amps/mV is satisfactory and conserves the bead life. An injection of 5 ng should give a half-scale deflection at an attenuation of 1 × 4, with a 1-mV recorder.

3. GC Determination

Standardize the instrument using the 3% Carbowax® 20M column so that atrazine, simazine, or terbutryn will give a half-scale deflection with approximately 5 ng of standard injected with either the Hall conductivity or AFD. Check that the baseline is stable and that the peak heights of the standard are reproducible (±5%). Determine detector linearity by injecting a series of standards that give from 10% to greater than 100% of full scale (bring on scale with the attenuator or use an integrator) response. Use similar injection volumes to reduce errors associated with small volume injections. Obtain slope, intercept, and correlation coefficient (r) and plot the results on graph paper. Both detectors should be linear (r > 0.95) over a 100- to 1000-fold range, i.e., from about 0.5 to 500 ng.

Inject 5 $\mu\ell$ sample volumes. If the triazine herbicide peak height is off scale dilute and reinject: alternatively, run at higher attenuation. Inject working standards frequently (e.g., every second sample) since neither detector is noted for its stability.

4. Procedure

Check water sample pH and adjust to pH 7 to 9 if necessary. Shake for 1 min in a 2ℓ separatory funnel 3 times with dichloromethane (150, 75, and 75 mℓ). Allow phases to clear, and drain the organic phase through a small column (2 cm i.d. with Teflon® stopcock) of sodium sulfate (2 cm height), into a 500-mℓ round-bottomed flask. Leave any emulsion in the separatory funnel. If a layer of emulsion is still present on the last extraction step allow it to separate for 15 to 30 min or add 50 mℓ of a saturated sodium chloride solution, swirl, and allow to separate. Combine the extracts and evaporate almost to dryness (0.5 to 1 mℓ) on a rotary evaporator at 35°C. Transfer extract to a test tube with ethyl acetate using a disposable glass pipette. Evaporate the ethyl acetate to 1.0 mℓ under a stream of dry air for direct GC analysis.

The presence of atrazine, simazine, terbutryn, or their metabolites, is inferred by comparing the retention times of peaks on the chromatograms with those of authentic standards. The herbicide concentrations in the specimens are obtained by comparing sample peak heights with those from the standard curve.

A detection limit of 0.2 μg/ℓ water sample should be possible.

B. Analysis of Atrazine, Simazine, and Terbutryn in Sediments

This procedure is based on those of Muir,[23] Ramsteiner et al.,[35] Sirons,[41] and Purkayastha and Cochrane.[48] Recoveries of the triazine herbicides should be greater than 95% (cf. Table 11). This procedure will also recover the dealkylated triazine metabolites, but is not satisfactory for the determination of the hydroxytriazine products.

1. Reagents

Same as those described in the section on water analysis.

2. GC Conditions

Same as those described in the section on water analysis.

3. Extraction Procedure

Weigh 100 g wet sediment (usually about 60% water) or proportionally less dry material into a 500-mℓ round-bottomed flask. Add 200 mℓ acetonitrile: water (4:1) and attach to a reflux condenser. Apply sufficient heat for the mixture to reflux gently for 16 hr (overnight). Allow to cool. Alternatively, the mixture may be shaken on a wrist-action shaker for 1 hr, allowed to stand overnight, and then reshaken 20 min.

Filter on a Buchner with a glass fiber filter paper. Wash the extraction flask and residuum with a total of 100 acetonitrile. Measure the volume of filtrate and transfer the desired aliquot (½ to ⅔) to a round-bottomed flask for evaporation of acetonitrile on a rotary evaporator at 40°C. Store the remaining extract in a small (125-mℓ) storage bottle in case a check on the results of GC analysis is required.

As an alternative to evaporation on a rotary evaporator the extract may be transferred to a beaker and evaporated on a steam bath under a stream of air. Acetonitrile-water mixtures can foam initially on evaporation under vacuum, and adjustment of the level of vacuum with a needle valve may be necessary. Reduce the extract to 20 to 30 mℓ and subject to liquid-liquid partition cleanup.

4. Liquid-Liquid Partition Cleanup

Transfer the aqueous phase to a separatory funnel (250 mℓ) with water, to give a total volume of approximately 50 mℓ. Check solution pH and adjust to pH 7 to 9 if necessary. Rinse evaporation flask or beaker with 25 mℓ dichloromethane and transfer washings to the separatory funnel. Add an additional 25 mℓ dichloromethane and shake 2 min with sufficient force to disperse the organic solvent. Allow phases to clear and drain the dichloromethane through a small column of sodium sulfate into a 250-mℓ round-bottomed flask as described for water evaporation. Extract the aqueous phase with another 50 mℓ dichloromethane. Handle emulsions as described previously. Evaporate the combined organic extracts just to dryness and proceed to column chromatography.

Sediment extracts may not require a column cleanup stage and this can be determined by GC analysis of untreated sample blanks. If column cleanup is unnecessary, dissolve the above residues in ethyl acetate (5 mℓ) and evaporate to 1 mℓ for direct GC analysis.

5. Column Cleanup

Prepare a small chromatographic column (20 mm × 5 mm) adding 2 g alumina Activity V (to give a column of 7 cm height). Top with 0.5 cm sodium sulfate and prewash with hexane:ethyl acetate (2:1, 3 mℓ), then hexane (3 mℓ). Dissolve the sediment residue from the liquid-liquid partitioning stage in 2 mℓ toluene and transfer to the column. Wash the flask with hexane (2 × 1 mℓ) and add to the column. When the last of the hexane enters the sodium sulfate layer add 5 mℓ hexane:ethyl acetate (2:1) and immediately collect the eluate in a test tube. Evaporate to 1 mℓ for GC analysis.

6. GC Analysis

Suitable aliquots of the above extracts are injected into the GC and the presence of atrazine, simazine, terbutryn, or their metabolites is inferred by comparing the retention times of peaks on the chromatograms with those of authentic standards. Herbicide concentration in samples are determined by comparing their peak heights with those of the requisite standard curve, to obtain nanograms injected.

To calculate the concentration of triazine in the original sediment the following equation is used:

$$\text{Triazine } (\mu g/g) = \frac{\text{ng injected}}{\text{sample wt (g)}} \times \frac{\text{sample extract vol (mℓ)}}{\text{injection vol (mℓ)}} \times \frac{\text{aliquot}}{\text{correction}} \times \frac{\text{moisture}}{\text{correction}}$$

where moisture correction = $1/(\% \text{ moisture} \div 100)$ and aliquot correction = 1/(fraction of total extract volume used). A detection limit of 0.005 to 0.01 μg/g (dry weight) should be possible.

C. Analysis of Atrazine, Simazine, and Terbutryn in Plant Tissue

This procedure is based on those of Muir,[23] Ramsteiner et al.,[35] Sirons,[41] and Purkayastha and Cochrane.[48] Recoveries of the triazine herbicides should be in excess of 80% (see Table 11).

1. Reagents

Same as those described in the section on water analysis.

2. Apparatus

A high-speed blender with glass or stainless steel cups and screw cap, or a Polytron homogenizer, or a Virtis blender with 50-mℓ flask.

3. GC Conditions

Same as those described in the section on water analysis.

4. Extraction Procedure

Use acetonitrile:water (4:1) for all plant material, but add 10% water by weight for dry (<10% water) material. Weigh a representative 100 g wet (60 to 90% water) sample (proportionately less for dry material) into a Waring blender cup. Add 200 mℓ extraction solvent and blend at high speed for 5 min. Transfer the macerate to a centrifuge tube (300 mℓ) with the aid of 50 mℓ acetonitrile, cap and shake 2 hr on a wrist-action shaker. Centrifuge (2500 rpm, 10 min) and transfer supernatant to a graduated cylinder. Record the volume, and calculate the fraction (i.e., supernatant volume/extraction volume + moisture content of sample [mℓ]) of extract recovered. Proceed with acetonitrile evaporation as described for sediment, then to liquid-liquid partition and column cleanup stage.

5. Liquid-Liquid Partition Cleanup

Same as that described in section on sediment analysis.

6. Column Cleanup

Same as that described in section on sediment analysis.

7. GC Analysis and Calculations

Same as those described in section on sediment analysis. A detection limit of 0.005 to 0.01 μg/g (dry weight) should be possible.

D. Analysis of Atrazine, Simazine, and Terbutryn in Fish Tissue

This procedure is based on that of Muir[23] for terbutryn; a recovery of 90% should be possible.

Reagents — Same as those described in the section on water analysis.

Apparatus — Same as that described in the section on plant tissue analysis.

GC conditions — Same as those described in the section on water analysis.

Extraction procedure — Weigh 5 g of ground tissue into a 50-mℓ test tube or Virtis blender flask add 30 mℓ acetonitrile:water (95:5) and blend with a Polytron or Virtis blender (2 min) at high speed. Centrifuge the mixture (2500 rpm, 5 min) in a 50-mℓ tube (glass or stainless steel) and transfer the supernatant to a 250-mℓ separatory funnel. Repeat the extraction with an additional 30 mℓ acetonitrile:water (95:5). Combine supernatants adjust to pH 3-4 with hydrochloric acid and extract by shaking with 60 mℓ hexane. Allow phases to separate and drain the aqueous phase into a 250-mℓ round-bottomed flask to evaporate the acetonitrile as described for sediment extracts. Neutralize the aqueous phase (pH 7 to 9) before liquid-liquid partition step and column cleanup stage.

Liquid-liquid partition cleanup — Same as that described in section on sediment analysis.

Column cleanup — Fish extracts following liquid-liquid partitioning cleanup may not require column cleanup.

GC analysis and calculations — Same as those described in the section on water analysis. Report concentrations on a μg/g (whole fish) basis. A detection limit of terbutryn in fish of 0.04 μg/g is possible.

REFERENCES

1. Gast, A., Knuesli, E., and Gysin, H., Ueber Pflanzenwachstumsregulatoren. I. Mitteilung: Chlorazin, eine phytotoxisch wirksame Substanz, *Experientia*, 11, 107, 1955.
2. Gast, A., Knuesli, E., and Gysin, H., Ueber Pflanzenwachstumsregulatoren. II. Mitteilung: Ueber weitere phytotoxische Triazine, *Experientia*, 12, 146, 1956.
3. Esser, H. O., Dupuis, G., Ebert, E., Marco, G. J., and Vogel, C., *s*-Triazines, in *Herbicides, Chemistry, Degradation, and Mode of Action*, Vol. 1, 2nd ed., Kearney, P. C. and Kaufman, D. D., Eds., Marcel Dekker, New York, 1975, chap. 2.
4. *Herbicide Handbook*, 4th ed., Weed Science Society of America, Champaign, Ill., 1979.
5. Fryor, J. D. and Makepeace, R. J.,Eds., *Weed Control Handbook*, Vol. 1, 6th ed., Blackwell, Oxford, 1977.
6. Plimmer, J. R., Kearney, P. C., and Klingebiel, U. I., *s*-Triazine herbicide dealkylation by free-radical generating systems, *J. Agric. Food Chem.*, 19, 572, 1971.
7. Crosby, D. G., Herbicide photodecomposition, in *Herbicides, Chemistry, Degradation, and Mode of Action*, Vol. 2, 2nd ed., Kearney, P. C. and Kaufman, D. D., Eds., Marcel Dekker, New York, 1975, chap. 18.
8. Burkhardt, N. and Guth, J. A., Photodegradation of atrazine, atraton, and ametryne in aqueous solution with acetone as a photosensitizer, *Pestic. Sci.*, 7, 65, 1976.
9. Tweedy, B. G. and Kahrs, R. A., *s*-Triazines, in *Analytical Methods for Pesticides and Plant Growth Regulators*, Vol. 10, Zweig, G. and Sherma, J., Eds., Academic Press, New York, 1978, 493.
10. Sheets, T. J., Persistence of triazine herbicides in soils, *Res. Rev.*, 32, 287, 1970.
11. Sirons, G. J., Frank, R., and Sawyer, T., Residues of atrazine, cyanazine, and their phytotoxic metabolites in a clay loam soil, *J. Agric. Food Chem.*, 21, 1016, 1973.
12. Helling, C. S., Movement of *s*-triazine herbicides in soils, *Res. Rev.*, 32, 175, 1970.
13. Muir, D. C. and Baker, B. E., Detection of triazine herbicides and their degradation products in tile-drain water from fields under intensive corn (maize) production, *J. Agric. Food Chem.*, 24, 122, 1976.
14. Wauchope, R. D., The pesticide content of surface water draining from agricultural fields — a review, *J. Environ. Qual.*, 7, 459, 1978.
15. Richard, J. J., Junk, G. A., Avery, M. J., Nehring, N. L., Fritz, J. S., and Svec, H. J., Analysis of various Iowa waters for selected pesticides: atrazine, DDE and dieldrin — 1974, *Pestic. Monit. J.*, 9, 117, 1975.
16. Muir, D. C. G., Yoo, J. Y., and Baker, B. E., Residues of atrazine and *N*-deethylated atrazine in water from five agricultural watersheds in Quebec, *Arch. Environ. Contam. Toxicol.*, 7, 221, 1978.
17. Frank, R. and Sirons, G. J., Atrazine: its use in corn production and its loss to stream waters in southern Ontario, 1975—1977, *Sci. Total Environ.*, 12, 223, 1979.
18. Anderson, L. W. J., Pringle, J. C., Raines, R. W., and Sisneros, D. A., Simazine residue levels in irrigation water after ditchbank application for weed control, *J. Environ. Qual.*, 7, 574, 1978.
19. Grover, R., Smith, A. E., and Korven, H. C., A comparison of chemical and cultural control of weeds in irrigation ditchbanks, *Can. J. Plant Sci.*, 60, 185, 1980.
20. Jordan, L. S., Farmer, W. J., Goodwin, J. R., and Day, B. E., Nonbiological detoxication of the *s*-triazine herbicides, *Res. Rev.*, 32, 267, 1970.
21. Kaufman, D. D. and Kearney, P. C., Microbial degradation of *s*-triazine herbicides, *Res. Rev.*, 32, 235, 1970.
22. Mauck, W. L., Mayer, F. L., and Holz, D. D., Simazine residue dynamics in small ponds, *Bull. Environ. Contam. Toxicol.*, 16, 1, 1976.
23. Muir, D. C. G., Determination of terbutryn and its degradation products in water, sediments, aquatic plants, and fish, *J. Agric. Food Chem*, 28, 714, 1980.
24. Knuesli, E., Burchfield, H. P., and Storrs, E. E., Simazine, in *Analytical Methods for Pesticides, Plant Growth Regulators, and Food Additives*, Vol. 4, Zweig, G., Ed., Academic Press, New York, 1964, 213.
25. Mattson, A. M., Kahrs, R. A., and Murphy, R. T., Quantitative determination of triazine herbicides in soils by chemical analysis, *Res. Rev.*, 32, 371, 1970.
26. Lawrence, J. F. and Turton, D., High-performance liquid chromatographic data for 166 pesticides, *J. Chromatogr.*, 159, 207, 1978.
27. Byast, T. H. and Cotterill, E. G., Two methods for estimating terbutryn residues in water using HPLC and GLC with a conductivity detector, *J. Chromatogr.*, 104, 211, 1975.
28. Byast, T. H., Reversed-phase high-performance liquid chromatography of some common herbicides, *J. Chromatogr.*, 134, 216, 1977.

29. **Jork, H. and Roth, B.**, Vergleichende chromatographische untersuchungen bei *s*-Triazinen, *J. Chromatogr.*, 144, 39, 1977.

30. **Pashal, D., Bicknell, R., and Siebenmann, K.**, Determination of atrazine in runoff water by high-performance liquid chromatography, *J. Environ. Sci. Health*, B(13), 105, 1978.

31. **Lawrence, J. F. and Leduc, R.**, High-performance liquid chromatography of some acidic and basic organic compounds on silica gel with mobile phases containing organic acids, *Anal. Chem.*, 50, 1161, 1978.

32. **Ramsteiner, K. A. and Hormann, W. D.**, High-pressure liquid chromatographic determination of hydroxy-*s*-triazine residues in plant material, *J. Agric. Food Chem.*, 27, 934, 1979.

33. **Cochrane, W. P. and Purkayastha, R.**, Analysis of herbicide residues by gas chromatography, *Toxicol. Environ. Chem. Rev.*, 1, 137, 1973.

34. **Fishbein, L.**, Chromatography of triazines, *Chromatogr. Rev.*, 12, 167, 1970.

35. **Ramsteiner, K. A., Hormann, W. D., and Eberle, D. O.**, Multiresidue method for the determination of triazine herbicides in field-grown agricultural crops, water, and soils, *J. Assoc. Off. Anal. Chem.*, 57, 192, 1974.

36. **Muir, D. C. G. and Baker, B. E.**, A method for the routine semiquantitative determination of hydroxy-*s*-triazines in soils, *J. Agric. Food Chem.*, 26, 420, 1978.

37. **Khan, S. U., Greenhalgh, R., and Cochrane, W. P.**, Chemical derivatization of hydroxyatrazine for gas chromatographic analysis, *J. Agric. Food Chem.*, 23, 430, 1975.

38. **Stoks, P. G. and Schwartz, A. W.**, Determination of *s*-triazine derivatives at the nanogram level by gas-liquid chromatography, *J. Chromatogr.*, 168, 455, 1979.

39. **Chesters, G., Pionke, H. B., and Daniel, T. C.**, Extraction and analytical techniques for pesticides in soil, sediment, and water, in *Pesticides in Soil and Water*, Guenzi, W. D., Ed., Soil Science Society America, Madison, 1974, 451.

40. **Faust, S. D. and Suffet, I. H.**, Analysis of pesticides and herbicides in the water environment, in *Water and Water Pollution Handbook*, Vol. 3, Ciaccio, L. L., Ed., Marcel Dekker, New York, 1972, 1249.

41. **Sirons, G. J.**, unpublished data, 1979.

42. **Junk, G. A., Richard, J. J., Grieser, M. D., Witiak, D., Witiak, J. L., Arguello, M. D., Vick, R., Svec, H. J., Fritz, J. S., and Calder, G. V.**, The use of macroreticular resins in the analysis of water for trace organic contaminants, *J. Chromatogr.*, 99, 745, 1974.

43. **Rees, G. A. V. and Au, L.**, Use of XAD-2 macroreticular resin for the recovery of ambient trace levels of pesticides and industrial organic pollutants from water, *Bull. Environ. Contam. Toxicol.*, 22, 561, 1979.

44. **McKone, C. E., Byast, T. H., and Hance, R. J.**, A comparison of some methods for the determination of triazine herbicides in water, *Analyst*, 97, 653, 1972.

45. **Smith, A. E., Grover, R., Emmond, G. S., and Korven, H. C.**, Persistence and movement of atrazine, bromacil, monuron, and simazine in intermittently-filled irrigation ditches, *Can. J. Plant Sci.*, 55, 809, 1975.

46. **Thompson, J. F., Reid, S. J., and Kantor, E. J.**, A multiclass, multiresidue analytical method for pesticides in water, *Arch. Environ. Contam. Toxicol.*, 6, 143, 1977.

47. **Green, R. E., Goswami, K. P., Mukhtar, M., and Young, H. Y.**, Herbicides from cropped watersheds in stream and estuarine sediments in Hawaii, *J. Environ. Qual.*, 6, 145, 1977.

48. **Purkayastha, R. and Cochrane, W. P.**, Comparison of electron capture and electrolytic conductivity detectors for the residue analysis of *s*-triazine herbicides, *J. Agric. Food Chem.*, 21, 93, 1973.

49. **Beynon, K. I.**, The analysis of crops and soils for the triazine herbicide cyanazine and some of its degradation products. I. Development of method, *Pestic. Sci.*, 3, 389, 1972.

50. **Beynon, K. I., Stoydin, G., and Wright, A. N.**, A comparison of the breakdown of the triazine herbicides cyanazine, atrazine and simazine in soils and maize, *Pestic. Biochem. Physiol.*, 2, 153, 1972.

51. **Ott, D. E., Formica, G., Liebig, G. F., Eberle, D. O., and Gunther, F. A.**, Mechanized extraction and cleanup of atrazine residues in soil prior to gas chromatographic analysis, *J. Assoc. Off. Anal. Chem.*, 54, 1388, 1971.

52. **Hormann, W. D., Formica, G., Ramsteiner, K. and Eberle, D. O.**, Automated method for extraction, cleanup and gas chromatographic determination of triazine herbicides in soil. *J. Assoc. Off. Anal. Chem.*, 55, 1031, 1972.

53. **Hill, B. D. and Stobbe, E. H.**, The use of ultrasonic extraction in the determination of some *s*-triazine herbicides in soils, *J. Agric. Food Chem.*, 22, 1143, 1974.

54. **Best, J. A. and Weber, J. B.**, Disappearance of *s*-triazines as affected by soil pH using a balance sheet approach, *Weed Sci.*, 22, 364, 1974.

55. **Hesselberg, R. J. and Johnson, J. L.**, Column extraction of pesticides from fish, fish food and mud, *Bull. Environ. Contam. Toxicol.*, 7, 115, 1972.

56. **Kadoum, A. M. and Mock, D. E.**, Herbicide and insecticide residues in tailwater pits: water and pit bottom soil from irrigated corn and sorghum fields, *J. Agric. Food Chem.,* 26, 45, 1978.

57. **Klaassen, H. E. and Kadoum, A. M.**, Distribution and retention of atrazine and carbofuran in farm pond ecosystems, *Arch. Environ. Contam. Toxicol.,* 8, 345, 1979.

58. **Pillay, C. G. P., Weete, J. D., and Davis, D. E.**, Metabolism of atrazine by *Spartina alterniflora.* I. Chloroform-soluble metabolites, *J. Agric. Food Chem.,* 25, 852, 1977.

59. **Frear, D. S., Hodgson, R. H., Shimabukuro, R. H., and Still, G. G.**, Behaviour of herbicides in plants, *Adv. Agron.,* 24, 327, 1971.

60. **Lawrence, J. F.**, Comparison of extraction methods for triazine herbicides in root crops using electrolytic conductivity detection, *J. Agric. Food Chem.,* 22, 137, 1974.

61. **Shimabukuro, R. H., Walsh, W. C., Lamoureux, G. L., and Stafford, L. E.**, Atrazine metabolism in sorghum: chloroform-soluble intermediates in the *N*-dealkylation and glutathione conjugate pathways, *J. Agric. Food Chem.,* 21, 1031, 1973.

62. **Yu, C-C., Booth, G. M., and Larsen, J. R.**, Fate of triazine herbicide cyanazine in a model ecosystem, *J. Agric. Food Chem.,* 23, 1014, 1975.

63. **Butler, G. L., Deason, T. R., and O'Kelley, J. C.**, Loss of five pesticides from cultures of twenty-one planktonic algae, *Bull. Environ. Cont. Toxicol.,* 13, 149, 1975.

64. **Johannes, H., Heri, W., and Reynaert, J.**, Triazines for the control of algae and submersed vascular plants, Ciba-Geigy, Basel, 1973.

65. **Mauck, W. L.**, A Review of the Literature on the Use of Simazine in Fisheries, Bureau of Sport Fisheries and Wildlife, Washington, 1974.

66. **Duggan, R. E., Barry, H. C., Enos, H. F., Fishbach, H., and Ramsey, L. L.**, Pesticide Analytical Manual, Vol. 2, U.S. Department of Health, Education and Welfare, Washington, D.C., 1975.

67. **Khan, S. U. and Foster, T. S.**, Residues of atrazine (2-chloro-4-ethylamino-6-isopropylamino-*s*-triazine) and its metabolites in chicken tissues, *J. Agric. Food Chem.,* 24, 768, 1976.

68. **Foster, T. S. and Khan, S. U.**, Metabolism of atrazine by the chicken, *J. Agric. Food Chem.,* 24, 566, 1976.

69. **Sirons, G. J.**, personal communication, 1979.

70. **Flint, G. T. and Aue, W. A.**, The gas-liquid chromatographic determination of 2-hydroxy-*s*-triazine, *J. Chromatogr.,* 52, 487, 1970.

71. **McGlamery, M. D., Slife, F. W., and Butler, H.**, Extraction and determination of atrazine from soil, *Weed Sci.,* 15, 35, 1967.

72. **Thornton, J. S. and Stanley, C. W.**, Gas chromatographic determination of sencor and metabolites in crops and soil, *J. Agric. Food Chem.,* 25, 380, 1977.

73. **Lawrence, J. F. and McLeod, H. A.**, A low temperature cleanup procedure for triazine herbicides in root crops, *Bull. Environ. Contam. Toxicol.,* 12, 752, 1974.

74. **Karlhuber, B. A. and Eberle, D. O.**, Advances towards automation of pesticide residue determinations, *Anal. Chem.,* 47, 1094, 1975.

75. **Mattson, A. M., Kahrs, R. A., and Schneller, J.**, Use of microcoulometric gas chromatograph for triazine herbicides, *J. Agric. Food Chem.,* 13, 120, 1965.

76. **Muir, D. C. G.**, unpublished results, 1979.

77. **Khan, S. U. and Purkayastha, R.**, Application of a thermionic detector in the analysis of *s*-triazine herbicides, *J. Agric. Food Chem.,* 23, 311, 1975.

78. **Sherma, J.**, Thin-layer chromatography: recent advances, in *Analytical Methods for Pesticides and Plant Growth Regulators,* Vol. 7, Sherma, J. and Zweig, G., Eds., Academic Press, New York, 1973, chap. 1.

79. **Cochrane, W. P.**, Confirmation of insecticide and herbicide residues by chemical derivatization, *J. Chromatogr. Sci.,* 13, 246, 1975.

80. **Kahn, S. U.**, Chemical derivatization of herbicide residues for gas liquid chromatographic analysis, *Res. Rev.,* 59, 21, 1975.

81. **Cochrane, W. P.**, Application of chemical derivatization techniques for pesticide analysis, *J. Chromatogr. Sci.,* 17, 124, 1979.

82. Manual of Analytical Quality Control for Pesticides in Human and Environmental Media, Environmental Toxicology Division, U.S. Environmental Protection Agency, Triangle Park, N.C., 1979.

83. Standard Methods for Examination of Water and Wastewater, 14th ed., American Public Health Association, New York, 1975.

84. **Felz, H. R. and Culbertson, J. K.**, Sampling procedures and problems in determining pesticide residues in the hydrologic environment, *Pestic. Monitor J.,* 6, 171, 1972.

85. **Eberhardt, L. L., Gilbert, R. O., Hollister, H. L., and Thomas, J. M.**, Sampling for contaminants in ecological systems, *Environ. Sci. Technol.,* 10, 917, 1976.

86. **Holtz, D. D., May, F. L., and Tindle, R. C.**, A core-type sampler for pesticide studies, *Prog. Fish-Culturist,* 34, 117, 1972.

87. **Webster, G. R. B. and Reimer, G. J.** Cold storage degradation of the herbicide metribuzin in field soil samples awaiting analysis, *Pest. Sci.*, 7, 292, 1976.

88. Analytical Methods for Pesticide Residues in Foods, Health Protection Branch, Department of National Health and Welfare, Ottawa, 1973.

INDEX

A

Absorption, 134
Acetylation, 95
Acetylcholine (ACh), 26
ACh, see Acetylcholine
Acid digestion, 73
Acid hydrolysis, 119
Acyl, 126
Acylation, 97, 130
Acyl derivatives of amines, 119
Adsorption, 29, 34, 67, 68, 131
Adsorption chromatography, 77
AFID, see Alkali flame ionization detector
Aglycone, 44, 120
Aldicarb, 24, 34, 37, 41, 43, 44, 48—49, 91, 92,
    118, 145, 153
Alkali flame ionization detector (AFID), 51, 82,
    88, 228
Alkaline hydrolysis, 117
Alkylation, 124, 129, 130
Alumina, 78
Ametryn
    structure of, 215
    use patterns of, 217
Amine hydrolysis of N-methylcarbamates,
    116—119
Amines, 125, 126
    acyl derivatives of, 119
Aminocarb, 44, 68, 73, 114, 117, 142
Aminophenyl N-methylcarbamates, 73—75
Analysis of residue, 49—66
Analytical hydrolysis, 119—121
Anhydride, 102
Anilines, 124, 125, 128
Animal samples, 227, 230
Animal tissue, 71
    extraction of triazine herbicides from, 226
    N-methylcarbamates in, 55—56
Antidote for carbamate poisoning, 27
Aryl N-methylcarbamates, 22—24
Atraton structure, 215
Atrazine, 214
    fish tissue and, 235
    plant tissue and, 234—235
    sediment and, 233—234
    structure of, 215
    use patterns of, 217
    water and, 231—233
Atropine, 27
"Aue" packing, 85
Aziprotryn
    structure of, 215
    use patterns of, 217

B

Barban, 196, 198

C

Base hydrolysis, 120
Benzamide, 119
Benzthiazuron structure, 187
Benzyl, 126
Benzylation, 130
Benzyl chloride, 130
Biodegradation, 206, 220
    of urea herbicides, 192
Bird tissue, 71
    N-methylcarbamates in, 55—56
Breakdown pathways, 44
Bromination, 97, 104, 125
Bufencarb, 38
Buturon
    retention time of, 197
    structure of, 185

C

Capillary GC columns, 89
Carbamate phenols
    derivatization of, 104—116
    direct GLC of, 89—90
    isolation of, 80
Carbamates, 1—182
    antidotes for, 27
    chemical properties of, 16—22
    colorimetric detection of, 143—145
    confirmation of, 151—152
    derivitazation of, 123—130
    enzymetic determination of, 138—140
    fluorescence detection of, 141—143
    GC detection of, 80—93
    half-lives of, 29
    HPLC determination of, 130—136
    metabolites of, 45—49
    N-methylation of, 94—123
    modes of action of, 26—29
    MS determination of, 146—151
    nomenclature of, 6—16
    oxime, 60—62, 75, 91, 92
    parent, 79—80
    parent compounds of, 22—26
    persistence of, 29
    physical properties of, 16—22
    poisoning from, 27
    separation of phenols from parents in, 79—80
    structures of, 7—15
    synthesis of, 6
    synthesis of metabolites of, 45—49
    synthesis of parent compounds for, 22—26
    TLC determination of, 136—138
    use of, 4
Carbamic acid, 5
Carbamylation, 27
Carbanolate, 108, 110
Carbaryl, 21, 22, 29, 34, 36, 38, 42, 44, 71, 85,
    108—110, 118, 126, 134, 137, 138,
    141—145, 151, 153

half-life of, 33
persistence of, 33
Carbofuran, 22, 29, 31—34, 37, 38, 44, 70, 73,
    78, 85, 101, 103, 105, 106, 110, 122, 136,
    142, 144, 145, 150, 153
    derivatives of, 48
Carbofuran-7-phenol, 110
Carbon, 78
Carbonic acid derivatives, 5
CCD, see Coulson conductivity detector
Cellulose column, 195
Chemical derivatization, 94, 95
Chemical hydrolysis, 29
Chemical properties
    of carbamate insecticides, 16—22
    or triazine herbicides, 219—220
Chlorazine structure, 215
Chlorbromuron, 203
    extraction of, 200, 201
    physical properties of, 190
    retention time of, 197
    structure of, 185
    use patterns of, 184
Chloreturon structure, 185
Chloroacetylation, 108
Chloroxuron, 203
    extraction of, 200, 201
    physical properties of, 190
    structure of, 185
    use patterns of, 184
Chlorpropham, 196, 198
Chlortoluron, 203
    extraction of, 200—201
    physical properties of, 190
    structure of, 185
Cholinesterase inhibitors, 138
Chromatography
    adsorption, 77
    gas, see Gas chromatography
    high-performance liquid, see High-performance
        liquid chromatography
    liquid, 130, 131, 203—204
    paper, 228
    reaction gas, 89
    thin-layer, see Thin-layer chromatography
Chromogenic agents, 137
Chromogenic reagents, 143
CIPC, 24, 30, 35, 91, 124—126, 128, 145, 146
    TMS derivatives of, 124
Cleanup
    of N-methylcarbamates, 53—60
    of oxime carbamates, 60—62
    of N-phenylcarbamates, 63
    of sample extracts, 76—80
    of thiocarbamate herbicides, 64—65
Coagulation, 72, 77
Colorimetry, 129, 143—145, 191, 194—196,
    206—207
Columns
    capillary GC, 89
    cellulose, 195
    glass, 196

metal, 196
silica, 205
XAD-2 resin, 69
Concentration of sample extracts, 76
Confirmation, 84, 95, 96, 98, 119, 123, 124, 136,
    138, 151, 152
    of triazine herbicides, 228—229
    of urea herbicides, 206
Conjugated residues, 70
Conjugates, 49
Conjugation, 36, 41, 43—44
Coulson conductivity detector (CCD), 81, 84
Crop extracts, 226
Cyanazine
    structure of, 215
    use patterns of, 217
Cycluron structure, 187
Cyprazine
    structure of, 215
    use patterns of, 217

D

Dansyl, 136
Dansyl chloride, 141, 143
Decomposition, 89
Degradation, 40—49
    biological, see Biodegradation
Derivatives
    acyl, 119
    carbofuran, 48
    carbonic acid, 5
    hydrolysis products of N-methylcarbamates,
        103—119
    hydroxyalkyl, 48
    intact N-methylcarbamates, 95—103
    N-Methylcarbamates, 95—119
    N-perfluoro, 98
    TFA, 97
    N-thiomethyl, 96
    TMS, 96, 124
Derivatization, 123—130
    amine hydrolysis products of N-
        methylcarbamates, 116—119
    carbamate phenols, 104—116
    chemical, 94, 95
    intact compounds, 124
    N-methylcarbamates, 94—123
    post-column, 136
Desmetryn use patterns, 217
Desorption, 34
Detection
    EC, 109
    UV, 223
Detectors
    alkali flame ionization, 51, 82, 88, 228
    Coulson conductivity, 81, 84
    electron capture, see Electron capture detector
    flame ionization, 81, 82, 84, 88, 89, 93, 97,
        109, 121, 125
    flame photometric, 51, 81, 82, 84, 92, 93, 96,
        105, 130, 228

gas chromatographic, 81—82
Hall, 81
microcoulometric, 81, 87
nitrogen-phosphorus (N-P), 81, 82, 128
Determination
 colorimetric, 129
 N-methylcarbamates, 52—60, 82—89
 oxime carbamates, 60—62
 N-phenylcarbamates, 63
 thiocarbamate herbicides, 64—65
Diazotization, 145
Dicofol, 90
Difenoxuron structure, 185
Diflubenzuron structure, 185
Digestion by acid, 73
Dimefuron structure, 185
Dimethametryn structure, 215
Dimethylamine, 125
N,N-Dimethylcarbamates, 24
Dimetilan, 141
Dinitrophenyl (DNP), 114, 116, 118
Dinitrophenylation, 95, 109, 110, 116, 118
Dipropetryn
 structure of, 215
 use patterns of, 217
Direct GC, 146
Direct GLC, 90—94, 129
 of carbamate phenols, 89—90
 of *N*-methylcarbamates, 82—89
Distillation, 75
Dithiocarbamates, 25—26, 76, 93, 129, 145
Diuron, 192, 198—203, 207, 208
 extraction of, 200—201
 physical properties of, 190
 retention time of, 197, 206
 structure of, 185
 use patterns of, 184
DNP, see Dinitrophenyl
DNT, 114, 116
 anilines of, 118

E

EBDC, see Ethylenebisdithiocarbamates
EC, 51
ECD, see Electron capture detector
Ecosystems, 38
Efficiency of extraction, 72
EI, see Electron impact
Electron capture detector (ECD), 4, 49, 81, 82,
  87, 93—98, 101, 104, 105, 109—112, 114,
  125, 127—129, 196, 228
Electron impact (EI) MS, 146
Environmental persistence, 29—40
Enzymatic techniques, 138—140
Enzyme inhibition, 138
EPTC, 32, 75, 125
Ethers, 112
Ethidimuron structure, 187
Ethylenebisdithiocarbamates (EBDC), 25—26,
  43, 76, 93, 129—130, 145

Ethylenediamine, 129
Ethylenethiourea (ETU), 43, 129—130, 145
Ethylenethiuram monosulfide (ETM), 29
ETM, see Ethylenethiuram monosulfide
ETU, see Ethylenethiourea
Extraction, 66—76, 120, 194, 198, 199, 207, 208,
  230—235
 chlorbromuron, 200, 201
 chloroxuron, 200, 201
 chlortoluron, 200—201
 diuron, 200—201
 efficiency of, 72
 fenuron, 200, 201
 fluometuron, 200, 201
 linuron, 200—201
 N-methylcarbamates, 52—60
 metobromuron, 200—201
 monolinuron, 200—201
 monuron, 200—201
 neburon, 200
 oxime carbamates, 60—62
 N-phenylcarbamates, 63
 thiocarbamate herbicides, 64—65
 triazine herbicides, 224—226
Extracts
 cleanup of, 76—80
 concentration of, 76
 crop, 226
 soil, 226

F

FDNB, 116, 126
Fenuron, 191, 198, 203
 extraction of, 200, 201
 physical properties of, 190
 retention time of, 197
 structure of, 186
 use patterns of, 184
FID, see Flame ionization detector
Fish, 36, 71, 199
 atrazine in, 235
 N-methylcarbamates in, 55—56
 samples of, 227, 231
 simazine in, 235
 terbutryn in, 235
 triazine herbicides from 226
Flame ionization detector (FID), 81, 84, 89, 93,
  97, 109, 121, 125
 alkali, 82, 88, 151, 228
Flame photometric detector (FPD), 51, 81, 82,
  84, 92, 93, 96, 105, 130, 228
Florisil, 78
Fluometuron, 192, 203
 extraction of, 200, 201
 physical properties of, 190
 retention time of, 197
 structure of, 186
 use patterns of, 184
Fluorescence, 89, 136, 141—143
Fluorimetry, 198

Fluorogenic labeling, 134, 142
FPD, see Flame photometric detector
Fractionation-cleanup, 114
Fragmentography, 101
Freezing out, 71
Fungicides
    carbamate, see Carbamates
    dithiocarbamate, 25—26, 76, 93, 129, 145
    EBDC, see Ethylenebisdithiocarbamates
Funnels, 69

G

Gas chromatography (GC), 4, 51, 80—94, 101,
        125, 127, 128, 136, 150, 151, 196—197,
        199—202, 206, 207, 223—224, 226, 228,
        231—233
    capillary columns for, 89
    direct, 146
    reaction, 89
Gas liquid chromatography (GLC), 49, 70, 76,
        78—80, 83—90, 93, 96, 98, 109, 112, 116,
        119, 121, 123, 125, 127, 128, 130
    direct, see Direct GLC
GC, see Gas chromatography
Gels, 78, 108
Glass columns, 196
GLC, see Gas liquid chromatography
Glucuronides, 44
Glycoside, 44

H

Half-lives
    of carbamates, 29
    of carbaryl, 33
    of N-methylcarbamates, 31
Hall electrolytic conductivity detector, 81
Halogenation, 97
Herbicides
    carbamate, see Carbamates
    heterocyclic urea, 187
    phenylurea, 185—186
    pre-emergence, 184
    substituted urea, 183—211
    thiocarbamate, see Thiocarbamates
    triazine, see Triazine
    urea, see Urea
Heterocyclic urea herbicide structures, 187
Hexazinone structure, 215
HFBA, 126, 128
HFB derivatives, 150
High-performance liquid chromatography
        (HPLC), 68, 84, 94, 130—136, 141, 143,
        197—198, 201—206, 208, 223
    reverse-phase, 131
HPLC, see High-performance liquid
        chromatography
Hydrolysis, 21, 41—42, 45, 71, 103, 124—129,
        136—138, 145

acid, 119
amine, 116—119
analytical, 119—121
base, 120
chemical, 29
N-methylcarbamates, 103—119, 121
Hydroxy, 48
Hydroxyalkyl derivatives, 48
3-Hydroxy carbofuran, 29, 37, 73, 98, 106, 110,
        153
3-Hydroxy carbofuran-7-phenol, 120
N-Hydroxymethyl, 48
3-Hydroxy-7-phenol, 110

I

Inhibition of enzymes, 138
Insecticides
    carbamate, see Carbamates
    o.p., see Organophosphorus pesticides
    organophosphorus, see Organophosphorus
        pesticides
Intact compound derivatization, 124
Intact N-methylcarbamate derivatives, 95—103
Interferences, 195, 223
Ipazine structure, 215
IPC, 30, 35, 42, 44, 91, 124, 125, 145, 146
    TMS derivatives of, 124
Isolan, 141
Isolation of phenols, 80, 121
Isonoruron structure, 187
Isoproturon
    physical properties of, 190
    structure of, 186

K

Karbutilate
    physical properties of, 190
    structure of, 186
    use patterns of, 184
Keeper, 76
3-Keto carbofuran, 29, 33, 37, 101, 106, 153
3-Keto-7-phenol, 110

L

Labeling, fluorogenic, 134, 142
Landrin, 110, 143
LC, see Liquid chromatography
Linuron, 192, 198, 203
    extraction of, 200—201
    physical properties of, 190
    retention time of, 197
    structure of, 186
Liquid chromatography (LC), 130, 131, 203—204
    high-performance, see High-performance liquid
        chromatography
Liquid-liquid partitioning, 77, 233—235

M

Mass fragmentography, 101
Mass spectrometry (MS), 84, 94, 103, 108, 136,
        146, 150—152
MCD, see Microcoulometric detector
Metabolic pathways, 44
Metabolism, 40—49, 129
    studies in, 66
Metabolites, 153
    of carbamates, 45—49
    synthesis of, 45—49
Metal columns, 196
Methabenzthiazuron
    physical properties of, 190
    structure of, 187
Methiocarb, 43, 85, 89, 142, 143
Methomyl, 41, 49, 91, 92, 118, 122, 129, 134
Methoproptryne structure, 215
Methoxylation, 125
Methylamine, 125, 142
Methylation, 124
    on-column, 96, 121
N-Methylcarbamates, 66—73
    amine hydrolysis products of, 116—119
    animal tissue and, 55—56
    bird tissue and, 55—56
    cleanup of, 52—60
    derivatives of, 95—103
    derivatives of hydrolysis products of, 103—119
    derivatization of, 94—123
    direct GLC determination of, 82—89
    extraction of, 52—60
    fish tissue and, 55—56
    half-lives of, 31
    hydrolysis of, 121
    plant tissue and, 57—60
    sediments and, 53—54
    soils and, 53—54
    water and, 52—53
Metobromuron, 204
    extraction of, 200—201
    physical properties of, 190
    retention time of, 197
    structure of, 186
Metoxuron
    physical properties of, 190
    structure of, 186
Metribuzin
    structure of, 215
    use patterns of, 217
Mexacarbate, 73, 84, 85, 114, 117, 118, 141,
        144, 153
Microbial activity, 192
Microcoulometric detector (MCD), 81, 87
MNT, see 2-Nitro-4-trifluoromethyl
Mobam, 122, 134, 142, 144
Mobility, 34
Model ecosystems, 38
Molinate, 31
Monolinuron, 204
    extraction of, 200—201

physical properties of, 190
retention time of, 197
Monuron, 191, 198, 204
    extraction of, 200—201
    physical properties of, 190
    retention time of, 197
    structure of, 186
    use patterns of, 184
MS, see Mass spectrometry

N

Nabam, 25, 44
Naphthol, 153
1-Naphthol, 36, 108, 112, 137, 141, 142, 144
NBD-Cl, 14
Neburon
    extraction of, 200
    retention time of, 197
    structure of, 186
p-Nitrobenzenediazonium fluoroborate, 137, 144
Nitrogen-phosphorus detector, 81, 82, 128
2-Nitro-4-trifluoromethyl (MNT), 125
2-Nitro-4-trifluoromethyl (MNT) anilines, 118
NMR, see Nuclear magnetic resonance
Nomenclature of carbamates, 6—16
Norea
    physical properties of, 190
    structure of, 187
    use patterns of, 184
N-P detector, see Nitrogen-phosphorus detector
Nuclear magnetic resonance (NMR), 151

O

On-column methylation, 96, 121
On-column reactions, 121—123, 125
o.p.s., see Organophosphorus pesticides
Organophosphorus pesticides, 26, 27, 138
Oxamyl, 41, 91, 129
Oxidation, 36, 41—43
Oxime carbamates, 75, 91, 92
Oxime N-methylcarbamates, 24

P

Packing, 85
Paper chromatography (PC), 228
Parent carbamates, 79—80
Parent compounds, 22—26
Partitioning
    coefficient of, 34, 67
    liquid-liquid, 77, 233—235
Partitionng values, 16, 120, 152
Pebullate, 44
Pentafluorobenzyl (PFB), 114, 116, 127
    ethers of, 112
Pentafluorobenzylation, 115

Pentanochlor, 124
N-Perfluoro derivatives, 98
Persistence
    carbamate insecticides, 29
    carbaryl, 33
    environmental, 29—40
PFB, see Pentafluorobenzyl
PFPA, 128
Phenols, 114, 126
    carbamates, 80, 89—90, 104—118
    direct GLC of, 89—90
    isolation of, 80, 121
    separation of parent carbamates from, 79—80
N-Phenylcarbamates, 24, 75, 91, 126
    cleanup of, 63
    determination of, 63
    extraction of, 63
Phenylurea herbicide structures, 185—186
N-Phenylureas, 124
Photolysis, 30
Physical properties, 190
    of carbamates, 16—22
    of triazine herbicides, 219
Picryl, 115
Pirimicarb, 24, 42, 68
Plants, 36—37, 71—73, 195, 199, 193, 227
    atrazine in, 234—235
    extraction of pesticides from, 201, 226
    N-methylcarbamates in, 57—60
    samples of, 230, 231
    simazine in, 234—235
    terbutryn in, 234—235
Poisoning antidotes for carbamate, 27
Polarography, 198
Post-column derivatization, 136
Pre-emergence herbicides, 184
Pre-emergency treatments, 216
Preservation of samples, 51—66
Priming, 85
Procyazine structure, 215
Promacyl, 110
Promecarb, 85
Prometon
    structure of, 215
    use patterns of, 217
Prometryn
    structure of, 215
    use patterns of, 218
Propanil, 196, 198
Propazine
    structure of, 215
    use patterns of, 218
Propham, 196, 198
Propoxur, 44, 107, 108, 110, 142, 143, 145
Pummerer reaction, 98
P-values, see Partitioning values
Pyridine as catalyst, 102
Pyrolan, 141

Q

Quality of water, 32—33

R

Radiolabeling, 40
Reaction gas chromatography, 89
Reflectance spectroscopy, 137
Refractometry, 198
Relative retention time, see Retention times
Residue analysis, 49—66
Residues
    confirmation of, 151—152
    conjugated, 70
Resin column, 69
Retention times, 111, 197, 206
Reverse-phase HPLC, 131
Ring hydroxy, 48
Routes of pesticidal transport, 192
Run-off water, 220, 224

S

Sample extracts
    cleanup of, 76—80
    concentration of, 76
Samples
    animal, 227, 230
    collection of, 205
    fish, 227, 231
    plant, 230, 231
    preparation of, 206
    preservation of, 51—66
    sediment, 224—226, 230, 231
    soil, 224—226
    water, 224, 230, 231
Sampling, 51
    apparatus for, 129—130
    procedures for, 230
Secbumeton
    structure of, 215
    use patterns of, 218
Sediment, 33—36, 69—71, 225, 230, 231
    atrazine in, 233—234
    N-methylcarbamates in, 53—54
    simazine in, 233—234
    terbutryn in, 233—234
    triazine herbicides in, 224—226
Selectivity, 143
Sensitivity, 121, 126, 137, 138, 141
Separation of parent carbamates from their
        phenols, 79—80
Separatory funnel method, 69
Siduron
    physical properties of, 190
    retention time of, 197
    structure of, 186
Silica column, 205
Silica gel, 78, 108
Silylation, 96
Simazine, 214, 216
    in fish tissue, 235
    in plant tissue, 234—235
    in sediment, 233—234

structure of, 215
use patterns of, 218
Simetone structure, 215
Simetryn structure, 215
Simzaine, 231—233
Soil, 33—36, 69—71, 192, 193, 195, 199, 220, 225, 226
  extraction of pesticides from, 200—201
  extracts of, 226
  N-methylcarbamates in, 53—54
  triazine herbicides in, 224—226
Soil sterilant use patterns, 184
Spectroscopy, 222—223
  IR, 151
  mass, see Mass spectrometry
  reflectance, 137
  UV, 198
  visible, 198
Stability, thermal, 82
Standards, 205
Steam distillation, 75
Sterilants, 184
Structure, 187, 215
  activity relationships to, 27
  of carbamate pesticides, 7—15
  of heterocyclic urea herbicides, 187
  of phenylurea herbicides, 185—186
Substituted urea herbicides, 183—211
Sulfonates, 105
Sulfone, 153
Sulfoxidation, 43
Sulfoxide, 153
Swep, 124
Synthesis
  of carbamates, 6, 22—26, 45—49
  of parent compounds for carbamates, 22—26
  of triazine herbicides, 216—219
Systemic pesticides, 36

Thermal stability, 82
Thiazfluron structure, 187
Thin-layer chromatography (TLC), 40, 51, 79, 84, 130, 131, 136—138, 141, 143,152, 201, 206, 228
Thin-layer chromatography (TLC)-enzyme inhibition, 138
Thiocarbamates, 24—25, 35, 43, 49, 75, 92, 119, 145, 149
  cleanup of, 64—65
  determination of, 64—65
  extraction of, 64—65
Thiofanox, 41, 91
N-Thiomethyl derivatives, 96
Thiophosphoryl, 105
TMAH, see Trimethylanilinium hydroxide
TMS, see Trimethylsilyl
Toxicological properties of triazine herbicides, 220
Trace enrichment, 68
Transesterification, 121
Transport routes, 192
Triazine herbicides, 213—239
  chemical properties of, 219—220
  confirmation of, 228—229
  extraction of, 224—226
  physical properties of, 219
  synthesis of, 216—219
  toxicological properties of, 220
  use patterns of, 216, 217—218
s-Triazines, 142
Trichloroacetic acid (TCA), 191
Trichloroacetylation, 106
Trietazine structure, 215
Trifluoroacetyl (TFA), 102,114, 126, 129, 130
  derivatives of, 97
Trifluoroacetylation, 98, 103, 109
Trimethylanilinium hydroxide (TMAH), 121, 124
Trimethylsilyl (TMS), 129
  derivatives of, 96, 124

T

TCA, see Trichloroacetic acid
Tebuthiuron, 192
  physical properties of, 190
  structure of, 187
Terbucarb, 124
Terbumeton structure, 215
Terbuthylazine
  structure of, 215
  use patterns of, 218
Terbutryn, 216
  in fish tissue, 235
  in plant tissue, 234—235
  in sediment, 233—234
  in water, 231—233
  structure of, 215
  use patterns of, 218
TFA, seeTrifluoroacetyl
Thermal decomposition, 89

U

Ultra-Bond, 128
Urea, 126, 137, 146
Urea herbicides
  biodegradation of, 192
  confirmation of, 206
  heterocyclic, 187
  substituted, 183—211
  use patterns of, 188—189
Use patterns, 184, 188—189, 217—218
UV absorption, 134
UV detection, 223
UV spectrophotometry, 198

V

Visible spectrophotometry, 198

W

Water
   run-off, 220, 224
   samples, 224, 230, 231
Water quality objectives, 32—33
Working standards, 205

X

XAD-2 resin column, 69